The Butterflies
of Pakistan

The Butterflies of Pakistan

T. J. ROBERTS
Sitara-i-Imtiaz
Ph.D. (Cantab.), M.S.A. (Brit. Col.)

With a Foreword by
Syed Babar Ali

OXFORD
UNIVERSITY PRESS

Great Clarendon Street, Oxford OX2 6DP

Oxford University Press is a department of the University of Oxford.
It furthers the University's objective of excellence in research, scholarship,
and education by publishing worldwide in

Oxford New York

Athens Auckland Bangkok Bogotá Buenos Aires Calcutta
Cape Town Chennai Dar es Salaam Delhi Florence Hong Kong Istanbul
Karachi Kuala Lumpur Madrid Melbourne Mexico City Mumbai
Nairobi Paris São Paulo Shanghai Singapore Taipei Tokyo Toronto Warsaw

with associated companies in Berlin Ibadan

Oxford is a registered trade mark of Oxford University Press
in the UK and in certain other countries

© Oxford University Press 2001

The moral rights of the author have been asserted

First published 2001

All rights reserved. No part of this publication may be reproduced, translated,
stored in a retrieval system, or transmitted, in any form or by any means,
without the prior permission in writing of Oxford University Press.
Enquiries concerning reproduction should be sent to
Oxford University Press at the address below.

This book is sold subject to the condition that it shall not, by way
of trade or otherwise, be lent, re-sold, hired out or otherwise circulated
without the publisher's prior consent in any form of binding or cover
other than that in which it is published and without a similar condition
including this condition being imposed on the subsequent purchaser.

ISBN 0 19 577995 9

Printed in Pakistan at
Mas Printers, Karachi.
Published by
Ameena Saiyid, Oxford University Press
5-Bangalore Town, Sharae Faisal
PO Box 13033, Karachi-75350, Pakistan.

Contents

Foreword	vii
Acknowledgements	ix
List of Plates and Figures	xi
Checklist of the Butterflies of Pakistan	xiii
1. Introduction	1
2. Form and Function	3
3. Senses and Sensibility	19
4. Co-evolutionary Competition—Predators and Defence	24
5. The Distribution of Butterflies in Pakistan	32
6. Papilionidae—Swallowtails	34
7. Pieridae—Whites and Yellows	50
8. Nymphalidae	74
9. Satyrinae—Browns, Satyrs, and Walls	98
10. Acraeinae	116
11. Lycaenidae—Blues, Coppers, Hairstreaks, and Silverlines	122
12. Riodininae (Erycnidae)—Judys and Punches	156
13. Hesperiidae—Skippers	158
Appendix 1 Bibliography	177
Appendix 2 Glossary of Technical Terms	181
Appendix 3 Gazetteer	183
Index	189

World Wide Fund
For Nature

Office of the President

Avenue du Mont-Blanc
1196 Gland - Switzerland
Telephone +41-22-364 92 92
Telefax   +41-22-364 54 68

April 21, 1998

Lahore address:
Packages Limited
Shahrah-e-Roomi
Lahore-54760, Pakistan
tel: (92-42) 5811548, 5811976
fax: (92-42) 5811978, 5811195

FOREWORD

Compared with the rest of the sub-continent, Pakistan has a rather arid climate with extensive areas of desert. As a consequence of such georgraphic limitations, her flora and fauna are somewhat restricted.

Despite this, Tom Roberts has amply demonstrated in his earlier books that our country possesses quite a unique and fascinating diversity of native plants and wildlife.

This book, the first attempt to describe all of Pakistan's known butterfly fauna, also shows how our western and northern mountainous regions possess an amazingly rich variety of butterflies which have adapted brilliantly to the harsh conditions imposed by low rainfall and high altitude.

Butterflies, with their seemingly fragile form and dancing colours attract everyone's eye, whether within our town garden or while holidaying in the hills. I hope this book will encourage the curious to learn more about them, and will reveal the complexity of their development and adaptations to cope not only with extreme climates but with a host of enemies also.

With their rapid life cycle and infinite variety, it is only recently being appreciated by scientists how their study can help to reveal the secrets of evolutionary development and the self-regulating mechanisms of genetics. Indeed, such studies are helping us to monitor un-due pressures on our environment and are even contributing to a better understanding of genetics which is so vital to our knowledge about the development of many human health problems.

It is only by learning about our wildlife, including butterflies that we can come to appreciate the part they play in sharing the resources of our planet and what they can contribute to future generations of Pakistanis.

S. Babar Ali

Registered as:
WWF - Fondo Mondiale per la Natura
WWF - Fondo Mundial para la Naturaleza
WWF - Fonds Mondial pour la Nature
WWF - Welt Natur Fonds
WWF - World Wide Fund For Nature
(Known as World Wildlife Fund
in Canada and the USA)

President   : S. Babar Ali
Vice - President : Rodney Wagner

Director General : Dr. Claude Martin

President Emeritus :
HRH The Duke of Edinburgh

Acknowledgements

Because I undertook preparation of this book after retirement from Pakistan, I have relied heavily upon the very comprehensive collection material from the Indian Subcontinent housed in the British Museum of Natural History. I am tremendously indebted to Dr Philip Ackery, Manager of the Butterfly Collection, for his patience and unfailing help in dealing with my many queries over the course of numerous visits to the Museum, as well as much help and advice from his colleague Dr Richard Vane Wright.

I am especially indebted to Dr Harish Goankar, compiler of the *Atlas of Indian Butterflies*, and of international repute, for his help in adding to my provisional Pakistan Checklist many high-elevation and less well known species and their more recent distributional records covering the families Papilionidae and Pieridae. Also for help and advice covering the Parnassians from Dr Christoph Häuser of the Koenig Museum in Bonn, and Dr Osamu Yata of Kyushu University in Japan for help with up-to-date classification of Pieridae. I also received much help over taxonomic problems, and criticism of my introductory chapters, from Dr Paul Whalley, formerly of the Entomological Department of the Natural History Museum. I also owe a special thanks to Dr Meena Haribal for criticizing and reviewing a selection of my species accounts.

I wish to thank Farooq Ahmad Khan, Director of the Zoological Survey of Pakistan, for allowing me to study the Survey's butterfly collection.

I am grateful to friends who have allowed me to reproduce their photographs of living butterflies, including Dr Morag Magrath for part of the dust jacket design, David Corfield, and Khan Mohammed Khan of the Punjab Wildlife Department. Also I am much indebted to Isobel Shaw (author of *Pakistan Trekking Guide*, 1993), who generously gave me a selection of her photographs from the more inaccessible far northern regions (*see*, Plate 8), where so many of Pakistan's rarer and more interesting montane species are found.

Though I have spent many weeks in Mumbai (Bombay) studying the Bombay Natural History Society's unrivalled collection of both mammal skins and bird skins, my narrowly focused interests during those visits stopped me from studying their huge butterfly collection, as witnessed by the plates illustrating Meena Haribal's Book (Haribal, 1992). Nevertheless, I must pay tribute to the Society for sponsoring and financing, in most instances, the first comprehensive, up-to-date books on the fauna of the subcontinent. It was in 1957 that M.A. Wynter-Blyth's *Butterflies of the Indian Region* was published by the Society, and this book immediately enabled me to take up a more serious study of Pakistan's butterfly fauna. As the following species accounts will show, the many volumes of the Society's *Journals*, dating from 1886, have also been a rich source of information, particularly covering the early life history of many butterfly species.

I am indebted to Oxford Cartographers for executing the two maps of Pakistan, based on my rough drawings, and not least for the care and skill with which Oxford University Press, Karachi, have tried to follow my wishes in making up the butterfly plates and producing the final publication.

List of Plates and Figures

PLATES

1.	Map showing districts	Between pp. 8-9
2.	Map showing locations in northern areas	,,
3.	Larvae—Swallowtails (Papilionidae) and Whites (Pieridae)	,,
4.	Larvae—Browns (Satyrinae) and Nymphalidae	,,
5.	Larvae—Blues (Lycaenidae), Punches (Riodinidae), Beaks (Libytheinae) and Skippers (Hesperiidae)	,,
6.	Habitat Scenes, including desert	,,
7.	Habitat Scenes, including foothills	,,
8.	Habitat Scenes, including far northern borders	,,
9.	Living butterflies—Papilionidae and Pieridae	,,
10.	Living butterflies—Nymphalidae, Riodinidae, a Satyrid and a Lycaenid	Between pp. 40-41
11.	Butterfly Predators, and Behaviour	,,
12.	Apollos	,,
13.	Apollos	,,
14.	'Yellow' Swallowtails and Apollos	,,
15.	'Black Swallowtails' and Peacocks	,,
16.	'Small Whites', *Euchloe, Baltia* and *Pontia*	,,
17.	Cabbage Whites and Bath Whites	,,
18.	Blackveins and Jezebels	,,
19.	Pioneer, Gull, Wanderer, Yellow Orange Tip and Emigrants	,,
20.	Grass Yellows, and Clouded Yellows	Between pp. 72-73
21.	Clouded Yellows	,,
22.	Brimstones, Arabs	,,
23.	Nawab, Courtesan, Purple Emperor, White Admirals, Map and Castor	,,
24.	Larger Fritillaries	,,
25.	Red band (*Melitaea*) Fritillaries	,,
26.	Common Leopard, Admirals and Commas	,,
27.	Pansies and Tortoiseshells	,,
28.	Eggflies, Sergeant, Sailers and Baron	,,
29.	Tree Browns, Walls and Bush Brown	,,
30.	Meadow Browns and Arguses	,,
31.	Branded Meadow Browns, White Edged Rockbrown and Common Argus	Between pp. 104-105
32.	Rock Browns and Satyrs	,,

33. Satyrs, *Ypthima* Rings and Evening Brown	,,
34. Danaids, Beaks and Punches and Coster	,,
35. Pierrots and Babul Blues	,,
36. *Everes* Cupids, Hedge Blues and Jewel Blues	,,
37. Arguses, *Vaccinia* Jewels and Green Under-wings	,,
38. Green Underwings and *Polyommatus* Meadow Blues	,,
39. Meadow Blues, Lime Blue and Grass Blues	Between pp. 152–153
40. *Euchrysops* Cupids, Pea Blue, Ceruleans and Line-Blues	,,
41. Coppers, and Sapphires	,,
42. Hairstreaks, Oak Blues, and Peacock Royal	,,
43. Silverlines, Cornelian, Guava Blue and Flashes	,,
44. Awls, and Flats	,,
45. Angles and Skippers	,,
46. Inky Skipper, Scrub Hopper, Palm Bob, Demon, Ace, and Darts	,,
47. Palm Dart, Chequered Darter, Swifts and Hopper	,,

FIGURES

1. Wing venation and generalized areas	xxiv
2. Butterfly eggs	4
3. External features of larva	5
4. Butterfly pupae	8
5. Butterfly pupae	9
6. Forelegs (Prothorax Pair) of butterfly families	15
7. Scent scales (Androconia) and location of Brands or Stigmata	16
8. Antennae and Palpi	17
9. Phragis of fertilized female Parnassians	35

Checklist of the Butterflies of Pakistan

In compiling any butterfly Checklist, many difficulties occur in choosing correct nomenclature, especially with reference to the original species description. Often in the past the same species was separately described as two and even three distinct species in publications which occurred at about the same time or in countries which were not aware of other collectors' findings. An example would be the Golden Copper, *Lycaena solskyi*. Added to this, the original description was often ascribed to an outdated or duplicated genus. As more information became available, it became possible to assign many species to more appropriate genera. In such instances, the original describer has his or her name placed in brackets according to accepted convention. If without brackets, it is to be assumed that original species naming still stands. In the large family of Lycaenidae, there have been many more changes in classification and this author has preferred to refer in some cases only to more recent publications covering the Indian subcontinent, as being information more useful to the researcher of literature covering our particular region.

Those species numbered with letter 'b' as suffix are currently excluded from Pakistan's Checklist.

Family PAPILIONIDAE

Subfamily PARNASSIINAE

1. *Hypermnestra helios* (Nickerl, 1846) (Syn: *Helios balucha*, Moore, 1906) – Desert Apollo
2. *Parnassius actius* (Eversmann, 1843)
3. *Parnassius jacquemontii* Boisduval, 1836 – Keeled Apollo
4. *Parnassius tianschanicus* Oberthür, 1879 – (Syn: *Parnassius discobolus*, sensu: Tytler, 1926) – Larger Keeled Apollo
5. *Parnassius epaphus* Oberthür, 1879 – Common Red Apollo
6. *Parnassius hardwickei* Gray, 1831 – Common Blue Apollo
7. *Parnassius acco* Gray, 1853 – Varnished Apollo
8. *Parnassius delphius* (Eversmann, 1843) – Kafir Banded Apollo
9. *Parnassius cardinalis* Grum-Grshimailo, 1887
10. *Parnassius staudingeri* Bang-Haas, 1882
11. *Parnassius stoliczkanus* Felder and Felder, 1865 (sub-spp. *P.s. atkinsoni*, *P.s. chitralica*, and *P.s. tytlerianus*) – Pir Panjal Banded Apollo
12. *Parnassius simo* Gray, 1853 – Black-edged Apollo
13. *Parnassius boedromius* Püngeler, 1901
14. *Parnassius charltonius* Gray, 1853 – Regal Apollo
15. *Parnassius loxias* Püngeler, 1901
16. *Parnassius inopinatus* Kotzsch, 1940

Subfamily PAPILIONINAE

17. *Graphium cloanthus* (Westwood, 1841) (Syn: *Zetides cloanthus*, Evans, 1923) – Glassy Bluebottle
18. *Papilio polytes* (Syn: *Papilio romulus*, Cramer, 1775), female forms *stichius* and *romulus* – Common Mormon
19. *Pachliopta aristolochiae* (Fabricius, 1775) (Sensu: *Tros aristolochiae*, Evans, 1932) – Common Rose
20. *Atrophaneura polyeuctes* (Doubleday, 1842) (Syn: *Tros philoxenus*, Wynter-Blyth, 1957) – Common Windmill
21. *Papilio alexanor* (Esper, 1799) – Southern Swallowtail or Balochi Swallowtail
22. *Papilio machaon* Linnaeus, 1758 – Common Yellow Swallowtail
23. *Papilio ladakensis* Moore, 1884 – Ladak Swallowtail
24. *Papilio* (Syn: *Princeps*) *demoleus* (Linnaeus, 1758) – Lime Butterfly
25. *Papilio polyctor* Boisduval, 1836 – Common Peacock
26. *Papilio arcturus* Westwood, 1842 – Blue Peacock
27. *Chilasa clytia*, Hampson, 1889, form *dissimilis* – Common Mime

Family PIERIDAE

Subfamily PIERINAE 'Whites'

28. *Leptosia nina* (Fabricius, 1793) (Syn: *Leptosia xiphia*, Moore, 1881) – Psyche
29. *Baltia shawi* Moore, 1878 – Shaw's Dwarf
30. *Baltia butleri* (Moore, 1882) – Butler's Dwarf
31. *Euchloe charlonia* (Donzel, 1842) – Lemon White or Greenish Black Tip
32. *Euchloe lucilla* Butler 1886 (Syn: *Synchloe charlonia lucilla*, Butler, 1886) – Pale Lemon White
33. *Euchloe belemia* (Esper, 1800) – Striped White or Green Striped White
34. *Euchloe ausonia* (Hübner, 1799) (Syn: *Synchloe daphalis*, Moore, 1865) – Pearl White or Mountain Dappled White
35. *Pontia chloridice* (Hübner, 1799) (Syn: *Parapieris chloridice*, Moore, 1904) – Lesser or Small Bath White
36. *Pontia callidice* (Hübner, 1799) (Syn: *Parapieris callidice*, de Niceville, 1897) – Lofty Bath white or Peak White
37. *Pontia daplidice* (Linnaeus, 1758) (Sensu: *Pieris daplidice*, Moore, 1904 and Evans, 1932) – Bath White
38. *Pontia glauconome* Klug, 1829 – Desert Bath White
39. *Pieris kreuperi* Staudinger, (1860) (Sensu: *Artogeia kreuperi*, Whalley, 1981) – Green Banded White or Kreuper's Small White
40. *Pieris devta* (Röber, 1907) – Chitral Banded White
41. *Pieris napi* (Linnaeus, 1758) (sub-sp. *P. napi montana*) – Green-Veined White
42. *Pieris ajaka* (Moore, 1865) – The Murree Green Veined White
43. *Pieris deota* (de Niceville, 1883) – Alpine or Kashmir Large White
44. *Pieris canidia* (Sparrman, 1768) – Indian Cabbage White
45. *Pieris rapae* (Linnaeus, 1758) – Small Cabbage White
46. *Pieris brassicae* (Linnaeus, 1758) – Large Cabbage White
47. *Aporia leucodice* (Eversmann) (Syn: *Aporia soracte*) – Himalayan Blackvein
48. *Aporia nabellica* (Boisduval, 1836) – Dusky Blackvein

49. *Delias belladona* (Fabricius, 1793) – Hill Jezebel
50. *Delias sanaca* (Moore, 1857) – Pale Jezebel
51. *Delias eucharis* (Drury, 1773) – Common Jezebel
52. *Belenois aurota* (Moore, 1881), (Sensu: *Anapheis mesentina*, Bingham, 1907) – Pioneer
53. *Cepora nerissa* (Fabricius, 1793) (Syn: *Huphina nerissa*, Doherty, 1886) – Common Gull
54. *Parenonia anais* (Bouge, 1837) (Syn: *Valeria valeria*, sensu: Talbot, 1939, sensu: *Parenonia hyppia* in Puri, 1931) – Common Wanderer
55. *Appias libythea* (Fabricius, 1793) – Striped Albatross

Subfamily COLIADINAE 'Yellows'

56. *Catopsila crocale* (Cramer, 1775) – Common Emigrant   }
                                                               }   Now considered conspecific
57. *Catopsila pomona* (Fabricius, 1775) – Lemon Emigrant   }
58. *Catopsila pyranthe* (Linnaeus, 1758) (Syn: *Catopsila florella*, Fabricius, 1775), African Emigrant – Mottled Emigrant
59. *Ixias pyrene* (Linnaeus, 1764) – Yellow Orange Tip
59b. *Ixias marianne* Butler, 1871 – White Orange Tip
60. *Eurema hecabe* (Linnaeus, 1758) (Sensu: *Terias hecabe* of Wynter-Blyth, 1957) – Common Grass Yellow
61. *Eurema brigitta* (Cramer, 1780) (Sensu: *Terias libythea* in Wynter-Blyth, 1957) – Small Grass Yellow
62. *Eurema laeta* (Boisduval, 1836) (Sensu: *Terias venata* [Moore, 1857], wet season form) – Spotless Grass Yellow
63. *Eurema blanda* (Boisduval, 1836) (Syn: *Terias silhetana*, Wallace, 1867) – Three Spot Grass Yellow
64. *Gonepteryx rhamni* (Linnaeus, 1758) – Common Brimstone
65. *Gonepteryx mahaguru* (Gistel, 1857) (Syn: *Gonepteryx zaneka* [Moore]/*G. aspasia* [Verity]) – Lesser Brimstone
66. *Gonepteryx farinosa* (Zeller, 1837) – Chitral Brimstone
67. *Colias fieldi* Ménétriés, 1855 (Syn: *Colias croceus* [Fourcroy], *Colias electo* [Linnaeus]) – Dark Clouded Yellow
68. *Colias erate* Butler, 1880, sub-sp. *pallida* and *lativata* (Syn: *Colias hyale*, Bingham, 1907) – Pale Clouded Yellow
69. *Colias marcopolo* Grum-Grshimailo, 1888 – Marco Polo's Clouded Yellow
70. *Colias alpheraky* Staudinger, 1882 (sub-sp. *Colias a. chitralensis*, Verity) – Greenish Clouded Yellow
71. *Colias wiskotti* Staudinger and Bang-Haas, 1882, form *leuca* – Broad-bordered Clouded Yellow
72. *Colias stoliczkana* Moore, 1878 – Orange Clouded Yellow
73. *Colias cocandica* Erschoff, 1874 – Pamir Clouded Yellow
74. *Colias ladakensis* C. and R. Felder, 1865 – Ladak Clouded Yellow
75. *Colias eogene* C. and R. Felder, 1865 – Fiery Clouded Yellow
76. *Colias leechi* Grum-Grshimailo, 1893 – Glaucous Clouded Yellow
77. *Colotis amata* Bingham, 1907 (Syn: *Colotis calais* in Talbot, 1939) – Small Salmon Arab
78. *Colotis protractus* (Bingham, 1907) (Syn: *Colotis phisadia* of Talbot, 1939) – Blue Spot Arab
79. *Colotis vestalis* (Butler, 1876) – White Arab
80. *Colotis fausta* (Olivier, 1801) – Large Salmon Arab
81. *Colotis etrida* (Boisduval, 1836) – Little Orange Tip
82. *Colotis danae* (Fabricius, 1775) – Crimson Tip

Family NYMPHALIDAE

Subfamily CHARAXINAE

83. *Eriboea athamus* (Drury), 1770 (Syn: *Polyura athamus*) – Common Nawab

Subfamily APATURINAE (or Tribe APATURINI)

84. *Apatura ambica* Kollar – Indian Purple Emperor
85. *Sephisa dichroa* (Kollar) – Western Courtier
86. *Limenitis trivena* Moore – Sub-spp. *trivena*, *hydaspes* and *gilgitica* – Indian White Admiral

Subfamily CALINAGINAE

87. *Calinaga buddha* Moore – Freak

Subfamily NYMPHALINAE

88. *Ariadne merione* (Cramer) (Syn: *Ergolis merione*) – Common Castor
89. *Phalanta phalantha* (Drury) (Syn: *Atella phalantha*) – Common Leopard
90. *Argynnis lathonia* (Linnaeus) 1758 (Sensu: *Issoria lathonia*, Sakai, 1981) – Queen of Spain Fritillary
91. *Argyreus hyperbius* (Johanssen) 1764 (Syn: *Argynnis hyperbius*) – Indian Fritillary
92. *Argynnis pandora* (Wiener Verzeichniss, 1776) (Syn: *Argynnis maia* in Evans, 1932) – Western Silverstripe or Cardinal of Europe
93. *Argynnis childreni* Gray, 1831 (Syn: *Childrena childreni*) – Large Silverstripe
94. *Fabriciana kamala* (Moore) 1857 (Syn: *Argynnis kamala*) – Common Silverstripe
95. *Argynnis jerdoni* Lang, 1868 – Jerdon's Silverspot
96. *Argynnis aglaia* Linnaeus, 1758 (Syn: *Mesoacidalia aglaja*) – Dark Green Silverspot or Fritillary
96b. *Fabriciana argospilata* – Afghan Fritillary
97. *Argynnis adippe* (Syn: *Fabriciana cydippe* and *Fabriciana niobe*, Sakai, 1981) – High Brown Silverspot or Fritillary, Niobe Fritillary of Russia
98. *Clossiana hegemone* Staudinger (Syn: *Argynnis hegemone* in Evans, 1932) – Whitespot Fritillary
99. *Bolaria pales* Fruhstorfer (Syn: *Argynnis pales* in Evans, 1932) – Straightwing Silverspot or Shepherd's Fritillary
100. *Melitaea shandura* Evans, 1932 – Shandur Fritillary
101. *Melitaea persea* Kollar, 1850 – Desert Fritillary
102. *Melitaea lutko* (Syn: *Melitaea schoenis* and *M. robertsi lutko* in Evans, 1932) – Balochi Fritillary
103. *Melitaea trivia* (Butler, 1880) (sub-sp. *Melitea trivia robertsi*) – Lesser Spotted Fritillary
104. *Melitaea didyma* (Esper, 1777) (Syn: *Melitaea persea saxatalis*) – Redband Fritillary of Evans, 1932, Spotted Fritillary of Europe
105. *Melitaea minerva* (Staudinger, 1895) – Pamir Fritillary, (Syn: *Melitea balba* of Tytler, 1926)
106. *Melitaea arcesia* (Moore, 1857) – Blackvein Fritillary
107. *Melitaea ala* (Syn: *Melitea pseudoala*)
108. *Junonia hierta* (Fabricius) (Syn: *Precis* spp. of Evans, 1932 and Wynter-Blyth, 1957) – Yellow Pansy

109. *Junonia orithya* (Linnaeus) – Blue Pansy
110. *Junonia almana* (Linnaeus) – Peacock Pansy
111. *Junonia lemonias* (Linnaeus) – Lemon Pansy
112. *Junonia iphita* (Cramer) – Chocolate Pansy
113. *Vanessa indica* (Herbst) 1794 – Indian Red Admiral
114. *Vanessa atlanta* Linnaeus, 1758 – European Red Admiral
115. *Cynthia cardui* (Linnaeus) (Syn: *Vanessa cardui*) – Painted Lady
116. *Vanessa canace* Linnaeus, 1767 (Syn: *Kaniska canace*) – Blue Admiral
117. *Polygonia egea* (Cramer) (Syn: *Vanessa egea*) – Eastern or Southern Comma
118. *Polygonia c-album* Linnaeus, 1758 – The European Comma
119. *Nymphalis vau-album* Wiener Verzeichniss, 1776 – False Comma or Comma Tortoiseshell
120. *Vanessa xanthomelas* (Denis and Scheiffermuller) (Syn: *Nymphalis xanthomelas*) – Large Tortoiseshell or Yellow Leg Tortoiseshell
121. *Vanessa polychloros* (Linnaeus) – Black Leg Tortoiseshell
122. *Vanessa cashmiriensis* (Kollar) (Syn: *Aglais cachmiriensis*) – Indian Tortoiseshell
123. *Aglais urticae* Linnaeus (Syn: *Vanessa urticae*) – Mountain or European Tortoiseshell
123b. *Araschnia prorsoides* (Blanchard) – Mongol
124. *Hypolimnas misippus* (Linnaeus) 1758 – Danaid Eggfly
125. *Hypolimnas bolina* (Linnaeus) 1758 – Great Eggfly
126. *Cyrestis thyodamus* Boisduval – Common Map
127. *Pantoporia opalina* (Kollar) (Syn: *Parathyma opalina*) – Himalayan or Hill Sergeant
128. *Neptis hylas* Moore, 1872 – Common Sailer
129. *Neptis mahendra* Moore, 1872 – Himalayan Sailer
130. *Neptis zaida* Doubleday, Hewitson – Pale Green Sailer
131. *Neptis soma butleri* Moore – Sullied Sailer
131b. *Neptis yerbury* Butler – Yerbury's Sailer
132. *Euthalia aconthea* (Moore) (Syn: *Euthalia garuda*) – Common Baron

* See Addendum

Subfamily SATYRINAE

133. *Lethe verma* (Kollar, 1844) – Straight-Banded Tree Brown
134. *Lethe rohria* (Fabricius, 1787) – Common Tree Brown
135. *Lethe confusa* Aurivillius, 1898 – Banded Tree Brown
136. *Pararge schakra* (Kollar, 1844) (Syn: *Lasiommata schakra*, Moore, 1892) – Common Wall
137. *Pararge maerula* (C. and R. Felder, 1867) (Syn: *Lasiommata maerula*, Moore, *P. schakra maerula*) – Scarce Wall
138. *Pararge menava* Evans, 1932 (Syn: *Lasiommata menava*, Moore, 1865) – Dark Wall
139. *Pararge eversmanni* Eversmann, 1847 (Sensu: *Kirinia eversmanni* in Sakai, 1981, *Amecera cashmirensis*, Moore, 1892) – Yellow Wall
140. *Maniola pulchra* (Syn: *Hyponephele pulchra*) (C. and R. Felder, 1867) var. *neoza* Lang – Dusky Meadow Brown
141. *Maniola pulchella* (Syn: *Hyponephele pulchella*) (C. and R. Felder, 1867) – Tawny Meadow Brown
142. *Maniola narica* (Syn: *Hyponephele narica*) (Hübner, 1818) – Tawny Branded Meadow Brown
143. *Maniola lupinus* (Costa, 1835) (Sensu: *Hyponephele lupinus* in Sakai, 1981, *Epinephele interposita*, Marshall and de Niceville, 1883) – Branded Meadow Brown

144. *Maniola davendra latistigma* Fruhstorfer, 1911 (Sensu: *Hyponephele davendra* in Sakai, 1981) – White-Ringed Meadow Brown
145. *Maniola tenuistigma* Moore, 1892 – Lesser White-Ringed Meadow Brown
146. *Maniola wagneri* (Herrich-Schäffer, 1852) – Oval Spot Meadow Brown (*Coenonympha wagneri* in Sakai, 1981)
147. *Maniola hilaris* (Syn: *Hyponephele hilaris*) Staudinger, 1886 – Pamir Meadow Brown
148. *Coenonympha myops* Staudinger, 1881 (Syn: *Lyela myops*) – Balochi Heath
149. *Hipparchia parisatis* (Kollar, 1849) (Sensu: *Eumenis parisatis* in Wynter-Blyth, 1957) – White Edged Rockbrown
150. *Eumenis mniszechii gilgitica* Tytler, 1926 (Syn: *Hipparchia mniszechii* [Herrich-Schäffer, 1852], and *Eumenis baldiva* Moore, 1892) – Tawny Rockbrown
151. *Eumenis thelephassa* Hübner, 1819–27 (Sensu: *Pseudochazara lehana* in Sakai, 1981, or *Hipparchia thelephassa* in Talbot, 1947) – Baluchi Rockbrown
152. *Hipparchia persephone* (Syn: *Chazara persephone* in Sakai, 1981, and *Eumenis persephone* in Wynter-Blyth, 1957) – Dark Rockbrown
153. *Hipparchia heydenreichi* Lederer, 1853 (Syn: *Chazara heydenreichi*, Sensu: *Philarcta shandura* in Leslie and Evans, 1903, or *Eumenis heydenreichi shandura* in Evans, 1932) – Shandur Rockbrown
154. *Karanasa pimpla* C. and R. Felder, 1867 (Syn: *Hipparchia actaea* [Esper, 1780] and *Karanasa actaea pimpla* Evans, 1932) – Dark or Black Satyr
155. *Karanasa digna* Bingham, 1905 (Syn: *Hipparchia digna* Marshall, in Talbot, 1947 and *Kanetisa digna* in Sakai, 1981) – Chitrali Satyr
156. *Karanasa huebneri* Moore, 1892 var: *cadesia* Leslie and Evans, 1903 and *pupilata* Evans, 1932 – Tawny Satyr
157. *Hipparchia moorei* (Evans, 1912) (Syn: *Karanasa moorei*) – Turkestan Satyr or Shandur Satyr in Evans, 1932
158. *Hipparchia boloricus* (Grum-Grshimailo, 1888) (Sensu: *Karanasa regeli* Tytler, 1926) – Turkestan Satyr
159. *Aulocera padma* (Kollar, 1844) (Syn: *Satyrus padma*) – Great Satyr
160. *Aulocera swaha* (Kollar, 1844) – Common Satyr
161. *Aulocera saraswati* (Kollar, 1844) – Striated Satyr
162. *Erebia nirmala* var: *dashka* Moore, 1865, and *materta* (Fruhstorfer, 1916) (Syn: *Callerebia nirmala* Butler, from Murree) – Common Argus
163. *Paralasa* (Syn: *Erebia*) *annada* Moore, 1857 – Ringed Argus
163b. *Paralasa scanda* (*Erebia scanda* Kollar, 1844) – Pallid Argus
164. *Erebia kalinda* Moore, 1865 (Syn: *Paralasa shakti* in Sakai, 1981) – Scarce Mountain Argus
165. *Erebia shallada* (Marshall and de Niceville, 1880) – Mountain Argus
166. *Erebia mani* (Marshall and de Niceville, 1880) – Yellow Argus (of Evans, 1932)
167. *Ypthima inica* Hewitson, 1865 – Lesser Three Ring
168. *Ypthima nareda* (Kollar, 1844) – Large Three Ring
169. *Ypthima asterope* (Klug, 1832) – Common Three Ring
170. *Ypthima sakra* Moore, 1857 – Himalayan Five Ring
171. *Ypthima bolanica* Marshall, 1882 – Desert Four Ring
172. *Ypthima avanta* Moore, 1882 (Syn: *Ypthima ordinata*, and *Ypthima lisandra*) – Jewel Four Ring
173. *Melanitis leda* (Linnaeus), 1758 – Common Evening Brown
174. *Mycalesis perseus* (Fabricius) 1775 – Common Bush Brown

Subfamily ACRAEINAE

175. *Acraea violae* (Fabricius) (Syn: *Telchinia violae* Horsfield, 1829) – Tawny Coster

Subfamily DANAINAE
Danaus, Kluk, 1802 and Danais, Latreille, 1804

176. *Tirumala limniace* Moore, 1882 (Syn: *Danaus limniace* [Cramer]) – Blue Tiger
177. *Danaus genutia* de Niceville, 1882 (Syn: *Danais plexippus* [Moore, 1865]) – Common Tiger
178. *Danaus chrysippus* (Syn: *Danais chrysippus* Wynter-Blyth, 1957) – Plain Tiger
179. *Euploea core* (Horsfield and Moore, 1857) – Common Crow

Subfamily LIBYTHEINAE

180. *Libythia lepita* Moore, 1857 – Common Beak
181. *Libythia celtis* Fuessly – European Beak or Nettle Tree Butterfly

Family LYCAENIDAE

Subfamily POLYOMMATINAE

182. *Castalius rosimon* (Fabricius) – Common Pierrot
183. *Tarucus balkanica* Bethune-Baker (Syn: *Tarucus nigra*) – Black Spotted Pierrot or Little Tiger Blue of Europe
184. *Tarucus nara* Kollar (Syn: *Tarucus alteratus* and *Tarucus extricatus*) – Rounded Pierrot
185. *Tarucus theophrastus* (Evans, 1932) (Syn: *Tarucus indica*) – Pointed Pierrot or Common Tiger Blue of Europe
186. *Tarucus rosacea* Austant (Syn: *Tarucus mediterraneae*) – Balochistan or Mediterranean Pierrot
187. *Tarucus callinara* Butler – Spotted Pierrot
188. *Tarucus venosus* Moore – Himalayan Pierrot
189. *Syntarucus plinius* Fabricius – Zebra Blue
190. *Azanus jesous* Lederer (Syn: *Syntarucus jesous*) – African Babul Blue
191. *Azanus ubaldus* (Cramer) – Bright Babul Blue
192. *Azanus uranus* Butler – Dull Babul Blue
193. *Everes buddhista* (Syn: *Everes shandura* in Evans, 1932, also *Cupido buddhista* in D'Abrera, 1993) – Shandur Cupid
194. *Everes dipora contracta* (Syn: *Argiades diporides* in Evans, 1932 and *Everes diporides* in Wynter-Blyth, 1957) – Chapman's Cupid
195. *Everes argiades indica* (in Evans, 1932) or *Everes hugelii dipora* – Tailed Cupid or Short-tailed Cupid in Europe
196. *Arletta vardhana* Moore (Syn: *Lycaenopsis vardhana*) – Dusky Hedge Blue
197. *Celastrina puspa gisca* Fruhstorfer – Common Hedge Blue
198. *Celastrina ladonides* Hemming – Silvery Hedge Blue
199. *Celastrina argiolus* Westwood – Hill Hedge Blue or Holly Blue of Europe

200. *Philotes vicrama* Hemming (Syn: *Polyommatus vicrama, and Pseudophilotes vicrama* in Russia) – Chequered Blue
201. *Plebejus christophi* Butler (Syn: *Lycaeides christophi*, or *Polyommatus christophi*) – Small Jewel Blue
202. *Plebejus sephyrus* (Evans) (Syn: *Polyommatus pylaon* or *Plebejus pylaon*) – Baluchi Jewel Blue or Zephyr Blue
203. *Aricia agestis* Moore – (*Polyommatus astrarche*) – Orange Bordered Argus or Brown Argus of Europe
204. *Aricia chiron* Swinhoe (*Aricia eumedon* in D'Abrera, 1993) – Streaked Argus
205. *Aricia astorica jermyni* Tytler – Astor Argus
206. *Turanana cytis laspura* Evans – Spotted Argus
207. *Vaccinia kwaja* Evans (Sensu: *Vaccinia hyrcana* in Evans, 1932) – Dark Jewel Blue
208. *Vaccinia iris* (Evans) (Syn: *Polyommatus iris*) – Jewel Argus
209. *Agriades pheretiades* Evans (Syn: *Polyommatus orbitulus walli* in Evans, 1932) – Greenish Mountain Blue
210. *Agriades jaloka* (Syn: *Lycaene orbitulus jaloka* in Tytler, 1926, and *Polyommatus orbitulus jaloka* in Evans, 1932) – Alpine Argus
211. *Albulina pheretes* (sub-sp. *asiatica*, Moore) – Mountain Blue
212. *Albulina omphisa* (Moore) – Dusky Green Underwing
213. *Albulina metallica* (de Niceville) (Syn: *Albulina chitralensis*, Tytler) – Small Green Underwing
214. *Albulina galathea* (Blanchard) (Syn: *Polyommatus galathea*) – Large Green Underwing
215. *Glaucopsyche alexis* Staudinger (Syn: *Polyommatus cyllaris*) – Western Green Underwing or Green-underside Blue of Europe
216. *Iolana gigantea* Tytler – Gilgit Meadow Blue
217. *Polyommatus seiversi* Evans – Pale Jewel Blue
218. *Polyommatus loewii* Evans – Large Jewel Blue (Syn: *Polyommatus baroghila*, Tytler, 1926)
219. *Polyommatus sarta* Swinhoe – Brilliant Meadow Blue
220. *Polyommatus gracilis* Moore (Sensu: *Polyommatus devanica* in Evans, 1932) – Dusky Meadow Blue
221. *Polyommatus bogra* Evans (Sensu: *Polyommatus actis bogra* in Evans, 1932) – Balochi Meadow Blue
222. *Polyommatus icarus* Butler – Violet Meadow Blue or Common Blue of Europe
223. *Polyommatus eros* Swinhoe – Common Meadow Blue or Eros Blue of Europe
224. *Polyommatus florenciae* Tytler, 1926
225. *Chilades laius* (Cramer) – Lime Blue
226. *Zizeeria trochilus* Kollar (Syn: *Chilades trochilus* in Puri, 1931, *Feyeria trochilus* in Cantlie, 1962, and *C. trochylus* in Whalley, 1996) – Grass Jewel
227. *Zizeeria maha* Kollar – Pale Grass Blue
228. *Zizeeria knysa* Moore (Syn: *Zizeeria lysimon* in Evans, 1932) – Dark Grass Blue or African Grass Blue
229. *Zizeeria galba* Evans (Syn: *Freyeria galba* in Cantlie, 1962) – Persian Grass Blue
230. *Zizina otis* Moore – Lesser Grass Blue
231. *Zizula hylax* Fabricius – Tiny Grass Blue
232. *Euchrysops cnejus* (Fabricius) (Syn: *Catochrysops cnejus*) – Gram Blue
233. *Euchrysops parrhasius* (Fabricius) (Syn: *Euchrysops contracta* [Butler]) – Small Cupid
234. *Euchrysops pandava* (Horsfield) (Syn: *Euchrysops parrhasius minuta*) – Plains Cupid
235. *Catochrysops strabo* (Fabricius) – Forget-me-not
236. *Lampides boeticus* (Linnaeus) – Pea Blue or Long-tailed Blue of Europe
237. *Jamides bochus* Cramer – Dark Cerulean
238. *Jamides celeno* Cramer – Common Cerulean
239. *Nacaduba nora* Felder (Syn: *Prosotas nora*) – Common Lineblue
240. *Nacaduba dubiosa* Felder (Syn: *Prosotas dubiosa*) – Tailless Lineblue

Subfamily LYCAENINAE

241. *Lycaena pavana* (Horsfield) – White-Bordered Copper
242. *Lycaena phlaeas* (Linnaeus) – Common Copper or Small Copper in Europe
243. *Lyceaena kasyapa* (Moore) – Green Copper
244. *Lyceaena solskyi* Moore (Syn: *Lyceaena thetis* in Evans, 1932) – Golden Copper
245. *Lyceaena phoenicurus* Lederer – Balochi Copper
246. *Lyceaena caspius* de Niceville (Syn: *Lyceaena hyrcana*, and *Lyceaena athamantis balucha*) – Purple Copper
247. *Thestor callimachus* Eversmann (Syn: *Tomares callimachus* in Evans, 1932) – Red Copper
248. *Heliophorus sena* (Kollar, 1844) – Sorrel Sapphire
249. *Heliophorus tamu* Kollar, 1865 (Syn: *Ilerda viridipunctata* in Marshall and de Niceville, 1890) – Powdery Green Sapphire
250. *Heliophorus bakeri* Evans, 1932 – Western Blue Sapphire

Subfamily THECLINAE

251. *Neolycaena connae* (Evans) – Baloch Hairstreak
252. *Callophrys rubi* Linnaeus – Green Hairstreak
253. *Strymon assamica* Kollar (Sensu: *Strymon sassanides* in Wynter-Blyth, 1957, *Strymonidia assamica* in D'Abrera, 1993) – White-Line Hairstreak
254. *Euaspa milionia* Hewitson – Water Hairstreak
255. *Chrysozephyrus ziha* Hewitson (Syn: *Euaspa ziha*, *Zephyrus ziha* in de Niceville, and *Neozephyrus ziha* in D'Abrera, 1993) – White-Spotted Hairstreak
256. *Chrysozephyrus ataxus* Doubleday and Hewitson (Syn: *Thecla ataxus* in Wynter-Blyth, 1957 and *Neozephyrus ataxus* in D'Abrera, 1993) – Wonderful Hairstreak
257. *Chrysozephyrus syla* Kollar (Syn: *Thecla syla* in Wynter-Blyth, 1957 and *Neozephyrus syla* in D'Abrera, 1993) – Silver Hairstreak
258. *Chaetoprocta odata* Hewitson – Walnut Blue
259. *Narathura sanesa* (Syn: *Amblypodia dodonea* Wynter-Blyth, 1957 and *Arhopola dodonea*) – Pale Himalayan Oakblue
260. *Narathura rama* Kollar (Syn: *Amblypodia rama* in Evans, 1932, and *Arhopala rama* in Marshall and de Niceville, 1890) – Dark Himalayan Oakblue
261. *Arhopola ganesa* (Syn: *Panchala ganesa* and *Amblypodia ganesa* in Wynter-Blyth, 1957) – Tailless Bushblue or Tailless Oakblue
262. *Apharitis epargyros* Eversmann – Yellow Silverline
263. *Apharitis acamas* Butler – Tawny Silverline
264. *Spindasis vulcanus* Moore – Common Silverline
265. *Spindasis ictis* Felder – Common Shot Silverline
266. *Spindasis elima* Ormiston – Scarce Shot Silverline
267. *Tajuria cippus* (Fabricius) – Peacock Royal
268. *Deudoryx epijarbus* Fruhstorfer – Cornelian
269. *Virachola isocrates* (Fruhstorfer) – Common Guava Blue
270. *Rapala manea* Moore (Syn: *Rapala schistaceae*) – Slate Flash
271. *Rapala iarbus* Kollar (Syn: *Rapala melampus* in Evans, 1932) – Indian Red Flash

272. *Rapala selira* Moore (Syn: *Hysudra selira* and *Rapala micans* in Evans, 1932) – Red Himalayan Flash
273. *Rapala extensa* Evans – Chitral Flash
274. *Rapala nissa* (Kollar) – Common Flash

Subfamily RIODININAE

275. *Dodona durga* (Kollar) – Common Punch
276. *Dodona dipoea* Hewitson – Lesser Punch
277. *Dodona eugenes* Bates – Tailed Punch

Family HESPERIIDAE

Subfamily COELIADINAE

278. *Bibasis sena* (Moore) – Orange Tail Awl or Pale Green Awlet in Haribal (1992)
279. *Hasora alexis* (Fabricius) (Syn: *Chromus alexis*) – Common Banded Awl
280. *Badamia exclamationis* (Fabricius) 1775 – Brown Awl
281. *Choaspes xanthopogon* (Guérin) (Syn: *Choaspes benjaminii*) – Indian Awlking

Subfamily PYRGINAE

282. *Celaenorrhinus leucocera* (Kollar) (Syn: *Leucocera leucocera*) – Common Spotted Flat
283. *Celaenorrhinus munda* Moore, 1884 (Syn: *Leucocera munda*) – Himalayan Spotted Flat
284. *Sarangesa purendra* (Moore) – Spotted Small Flat
285. *Seseria dohertyi* Watson, 1893 – Himalayan White Flat
286. *Lobocla bifasciatus* Moore, sub-sp. *casyapa* – Marbled Flat
287. *Caprona agama mettasuta* (Fruhstorfer) – Spotted Angle
288. *Caprona ransonetti yerburi* (Felder) – Golden Angle

Subfamily HESPERIINAE

289. *Gomalia elma* (Trimen) – African Marbled Skipper
290. *Hesperia zebra* Butler (Syn: *Syrichtus zebra* and *Spialia zebra*) – Zebra Skipper
291. *Hesperia galba* (Fabricius) 1793 (Syn: *Spialia* and *Syrichtus galba*) – Indian Skipper
292. *Spialia sertorius* (Hoffmansegg) (Syn: *Syrichtus orbifer carnea* in Evans, 1932) – Brick Skipper or Red Underwing Skipper of Europe
293. *Spialia phlomidis geron* Watson (Syn: *Syrichtus geron* or *Hesperia geron*) – Persian Skipper, formerly called Balochi Skipper
294. *Spialia geron evanidus* Butler (Syn: *Spialia doris*, Evans, 1949, Syn: *Syrichtus evanidus*) – Sind Skipper
295. *Muschampia staudingeri plurimacula*, previously called *Muschampia* (Syn: *Hesperia*) *poggei* (Lederer), and the same species as *Syrichtus poggei* Sakai, 1981, and *Syrichtus plurimacula* Evans, 1949 – Syrian Skipper or Streaked Skipper

296. *Hesperia alpina* (Syn: *Pyrgus alpina*, Evans, 1949) – Mountain Skipper
297. *Carcharodus altheae* Hübner (Syn: *Carcharodus floccifera balucha* and sub-sp. *dravira* Moore) – Tufted Marbled Skipper
298. *Carcharodus alceae alceae* (Esper) and sub-sp. *swinhoei*, Watson – Plain Marbled Skipper or Mallow Skipper of Europe
299. *Nisoniades marloyi* Boisduval (Syn: *Erynnis marloyi*) – Inky Skipper
300. *Aeromachus stigmata* (Moore) 1878 (Syn: *Aeromachus inachus*) – Veined Scrub Hopper
301. *Suastus gremius* (Fabricius) 1798 – Indian Palm Bob
302. *Notocrypta fiesthamelii* Moore – Spotted Demon
302b. *Halpe homolea* (Swinhoe) (Syn: *Halpe egena*) – Indian Ace
303. *Actinor suada radians* Moore – Veined Dart
304. *Taractrocera danna* (Moore) 1878 – Himalayan Grass Dart
305. *Taractrocera maevius* (Fabricius) – Common Grass Dart
306. *Padraona dara* Kollar – Himalayan Dart
307. *Astycus pythias bambusae* Wynter-Blyth, 1957 (Syn: *Telicota ancila bambusae* in Evans, 1949) – Dark Palm Dart
308. *Astycus augias* (Linnaeus) (Syn: *Telicota augias*) – Pale Palm Dart
309. *Hesperia comma* (Linnaeus) (Syn: *Pamphila comma*) – Chequered Darter or Silver Spotted Skipper of Europe
310. *Baoris eltola* Hewitson – Yellow Spot Swift
311. *Baoris discreta himalaya* Evans, 1932 – Himalayan Swift
312. *Pelopidas mathias* Hübner (Syn: *Baoris mathias*) – Small Branded Swift
313. *Parnara guttatus* Bremer (Syn: *Baoris guttata*) – Straight Swift
314. *Guttatus bevani* (Moore) 1878 (Syn: *Baoris bevani* and *Parnara bevani*) – Bevan's Swift
315. *Gegenes nostrodamus* Fabricius – Dingy Swift
316. *Eogenes alcides* Hers-Sch, 1869 – Torpedo
317. *Eogenes lesliei* Evans – Leslie's Hopper

ADDENDUM
Omission due to authors ignorance

*318. *Euthalia patala* (Kollar) – Grand Duchess (recorded in *Quercus incana* forest, Murree)

New records for Pakistan, by E. Kreuzberg-Mukhina and A. Kreuzberg, of Uzbekistan, collected July 1999 and specimens deposited with WWF Pakistan.

319. *Graphium sarpendon* – Common Bluebottle (from Murree foothills)

320. *Mycalesis nicotia* – Bright Eyed Bush Brown (from Gilgit District)

xxiv THE BUTTERFLIES OF PAKISTAN

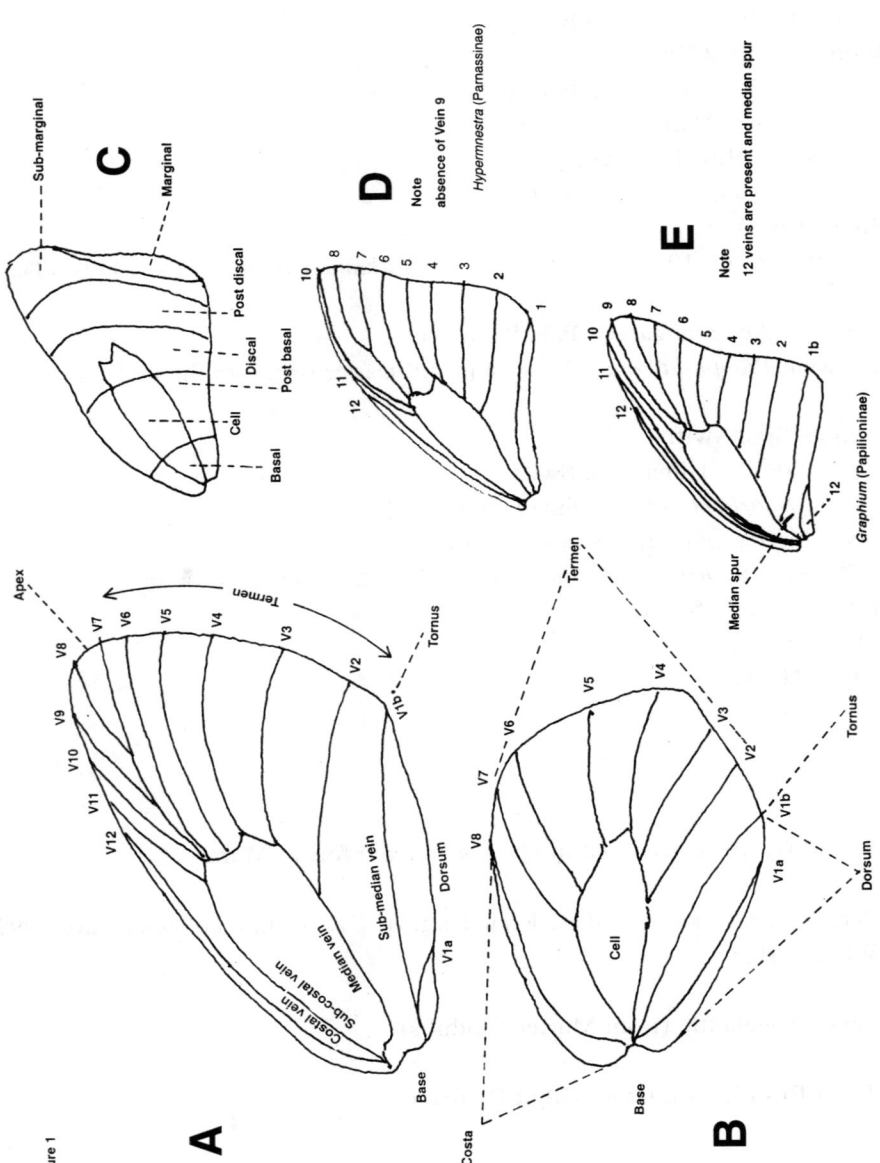

Fig. 1 Wing Venation and Generalized Areas
A. Forewing venation
B. Hindwing venation
C. Generalized wing areas
D. Parnassian venation
E. Papilionid venation

Chapter 1
Introduction

With a growing awareness worldwide of the importance of natural areas with specialist biodiversity, many countries are now devoting more resources to research upon butterflies. Not only are butterflies very sensitive to habitat degradation, but they also have a much more rapid passage of generations than higher vertebrates, and hence can be quicker to react to small changes in their environment than those declining bird and animal populations that have already been valued as sensitive indicators of environmental change and deterioration. They are therefore especially important and valuable to ecologists, environmentalists, and indirectly to government development planners, as well as to scientists unravelling the complexities of evolutionary theory and the mechanisms of genetics.

Man's ever increasing resource needs lead directly to most of our planet's problems, which are becoming especially acute in Pakistan. We need to try to mitigate, or if possible, halt these effects, which we are growing to realize, almost too late, can effect our health and will certainly give rise to a poorer quality of life for our descendants. These rather broad issues, because they occur sometimes slowly and almost imperceptibly, are nevertheless real and urgent and do require an underpinning of basic information about all of Pakistan's wildlife resources. I hope this little book will encourage some young Pakistani scientists to take up the challenge.

Readers should refer to the remaining part of this introductory chapter to be better able to evaluate the whole book. Unfortunately, there has been virtually no systematic study of butterflies since the collections made at the beginning of the century and up to the early 1930s by such stalwarts as Brigadier W. H. Evans, who spent most of his working life as an army engineer in the region which is now Pakistan. Consequently, the most perplexing and challenging problems in compiling this account have been to relate early classifications and scientific names with the many taxonomic revisions and recent changes in scientific nomenclature which have been enabled by new techniques such as studying DNA from bits of wing tissue, and mitochondrial X-ray analysis. Without access to all the more recent specialist publications, particularly by Japanese authorities on the Lycaenidae, I have had to limit myself in this book largely to the nomenclature used by the following authors. For Lycaenidae, Cantlie's *Revision,* published by the Bombay Natural History Society in 1962. For Hesperiidae, Brigadier Evans's *Catalogue of the Hesperiidae* (1949), and, where relevant, by the classification used by Meena Haribal (1992) in writing of the butterflies of Sikkim. The Hesperiidae are a very complex but often 'look-alike' family, whose classification has been largely based upon microscopic examination of the male genitalia (claspers). Evans's book is still used by museum workers despite its limitations, because nobody has yet had the patience or time to improve on his *Catalogue,* covering European and south-east Asian families. Dissection and examination of male claspers was done by him almost entirely from dry museum specimens (pers. comm., Phillip Ackery, 1996), which inevitably can lead to distortions in their appearance under a microscope, but Evans's work still stands as a major reference. Besides the above books, I have relied upon the two volumes published by Talbot (1939 and 1947) in the *Fauna of British India* series, covering Satyrinae, Pieridae, and Daninae.

As a young man working in Pakistan, I was an avid collector of butterflies during the 1950s and 1960s. When the Bombay Natural History Society published Dr Dillon Ripley's *Synopsis of the Birds of the Indian Sub-continent* in 1962, I found this reference far more comprehensive and familiar in its classification than Wynter-Blyth's book (1957) on the subcontinent's butterflies. In fact, because Pakistan has so much mountainous area and more Palaearctic faunal affinities compared with India, I had already discovered that Wynter-Blyth's book paid scant attention to many species typical of Pakistan's high altitude and western border regions. Regrettably, from the viewpoint of this book, I turned my attention to the study of Pakistan's birds and mammals, and have taken up this project too late to strengthen it with more field studies. In the species accounts, I have listed under the initials TJR those specimens from my own personal collection. The greater part of this book can therefore be ascribed to derivative investigations rather than to original fieldwork, yet my continued intense interest in the biology of butterflies, and aesthetic appreciation of them, I hope justifies this attempt to record the presently known extent and status of the region's butterfly fauna.

The first task has been to compile a checklist of what is believed to occur in Pakistan, and readers should not be discouraged by references in the text to very old collections, because there have been so few recent butterfly surveys. Use of herbicides and insecticides in agriculture and forestry is still very limited in Pakistan compared with the wealthier Western-Hemisphere nations, except for the cotton-growing regions. This means that most of the butterflies, which occurred at the turn of the century, still have a good chance of being found in the same localities. Where species are of likely, but as yet unproved, occurrence, or of doubtful occurrence, they have been listed in the checklist in parentheses and by adding the suffix 'b' to the previous checklist number.

Besides the collection of butterflies held by the Zoological Survey of Pakistan in Karachi, there are four other Institutes holding collections. A small one in the National Museum in Karachi, one in the Agricultural University, Faisalabad, one in the Pakistan Forest Institute, Peshawar, and a more recent one in the Agricultural Research Institute at Tarnab. In 1992 a complete list was published of all moths and butterflies housed in these collections as identified by the British Museum of Natural History (Hashmi and Tashfeen, 1992). The great majority of these were collected in pre-independence days and from localities outside of Pakistan (Sri Lanka, Sikkim, Assam and southern India). These specimens are nearly all without provenance (collection localities), but I have gone through them carefully and out of a total of 550 butterflies listed, only 135 species are known to have occurred in Pakistan.

Every species in the plates (painted by the author) with but several exceptions, has been chosen from specimens I collected in Pakistan, or from specimens from regions now part of Pakistan housed in the British Natural History Museum Collection. Unfortunately, because they were executed over a long time period, the reader should be warned that they are not painted to the same scale.

I conclude this chapter with another caveat. There are several distinctive groups of butterflies from different families, which have shown great variation and speciation, but are still recognizable at generic level by their very similar appearance. Examples which spring to mind include the *Erebia* (*Paralasa*) and *Maniola* Meadow Browns, the Nymphalid 'Sergeants' and 'Sailers', and particularly, in the fauna of Pakistan, the *Ypthima* 'Rings' and the Red Band Small Fritillaries (*Melitea* spp.). The *Ypthima* butterflies differ often between dry season and post monsoon adult markings, yet their classification is in large part based on the number and position of ocelli (rings) on the underwings, which can vary even within one species. Similarly, the Small Fritillaries are very similar on the upper wing surface, but with minute differences in the under hindwing pattern. I am not satisfied that I have been able to cover these two difficult groups accurately. The reader is also warned that butterflies, like all living things, vary individually in appearance, and that the specimens illustrated have been chosen as typical of those found in Pakistan, but they do not show every possible variance, though the written descriptions have tried to emphasize consistent characters enabling separation between similar species. Whilst the butterfly illustrations are not to scale, even within a single plate, their dimensions are given in the text. Thus, wingspan is measured in millimetres, from tip to tip of the forewings, when the butterfly is set with the forewing dorsum at right angles to the thorax.

Finally, I have included English or Trivial names wherever these have been used in well-known publications, but I have resisted the temptation to invent names for those few species which are only known by their scientific nomenclature.

Chapter 2
Form and Function

A knowledge of entomology is part of the curricula in most schools today, but this chapter is written mainly for the non-specialist who may lack basic information about insects, in particular the moths and butterflies assigned to the order Lepidoptera.

The life history of butterflies and moths still brings a sense of wonder even to the most knowledgeable biologist, and I hope that these introductory chapters will help to inform the reader of their complexity and many adaptations to their environment, as well as those areas of their behaviour and ecology which still need further understanding. As is well known, the brightly-coloured butterflies which visit the flowerbeds in our garden pass through four very distinct life phases or forms in the course of their development to adulthood.

The Egg

What of the egg itself? There is an extraordinary diversity both in the shape and the surface of the eggs of butterflies, varying from barrel-shaped, truncated cones, button-shaped, doughnut-shaped, to flanged domes, and even spined like Echinoderms (Sea Urchins). The shell, or chorion, as it is called, consists of chitin, a tough, non-porous protein, and its surface is patterned according to the species in a variety of reticulations, beaded ridges, indentations, and even, in some cases, spines (as, for example, the eggs of the Common Castor [*Ariadne merione*]. Generally speaking, the shape of the egg is consistent within the main families [*see*, Fig. 2], and differs between other families, so that this can be an aid to identification for the expert (Garcia-Barros and Martin, 1995). The top of the egg contains an opening, usually within a slight depression, and this is called the micropyle. Through it, the sperm enters the egg during its passage to the female's oviduct, and through it gas exchange for the developing embryo takes place. Inside the egg is also yolk upon which the embryo feeds. Initially, most eggs are pale whitish or green, often darkening with age. In most tropical species the caterpillar hatches within four or five days, especially with multiple-brooded species, but some of the high-elevation Palaearctic species may take weeks to hatch, or may be deposited at the end of the summer and not hatch until the following spring, a vital adaptation if their foodplant is herbaceous and dies back in the winter.

The female butterfly must first lay eggs, and to do so she has to be fertilized by sperm from a male of the same species. Many other families within the class Insecta can produce eggs without fertilization, especially some wasps (Hymenoptera) weevils (Curculionidae), and even some moths. As far as we know, no butterfly can do this, a phenomenon known as parthenogenesis. As will be seen in later chapters, both wing colour and scents produced by the adult butterfly play a vital role in finding a mate and in courtship. Once fertilized, the females of most species take great care in selecting suitable sites to deposit their eggs. Only certain plant families, and in many cases only one plant species, will provide the right food for the larvae (caterpillars). If the egg is laid on the wrong plant the caterpillar will not survive. The female is able to detect the correct foodplant by using her senses of sight, discerning differences in leaf shape and colour (Ilse, 1956), taste, and smell. She will often hover over the leaves of certain plants and also settle on leaves. In some cases the female can recognize suitable plants by smell picked up through her antennae, whilst other butterfly species will settle on a leaf and tap rapidly with the front pair of legs on its surface, and it is known that these legs can convey a sense of taste to the nerve centre of the female, so that if the message is correct she will be induced to lay her egg or eggs. A few species, which feed on a variety of grasses, will scatter their eggs at random on grass sward. Some of the Skippers (Hesperiidae) and some of the Browns (Satyrinae) scatter their eggs in flight. Those that lay their eggs before the Palaearctic winter, especially many *Vanessa* spp., may do so on a nearby tree trunk, where the newly-hatched larvae can shelter under bark crevices until the spring. Some species, which are gregarious in their early stages, will lay all their eggs in a batch on one leaf. Some of

THE BUTTERFLIES OF PAKISTAN

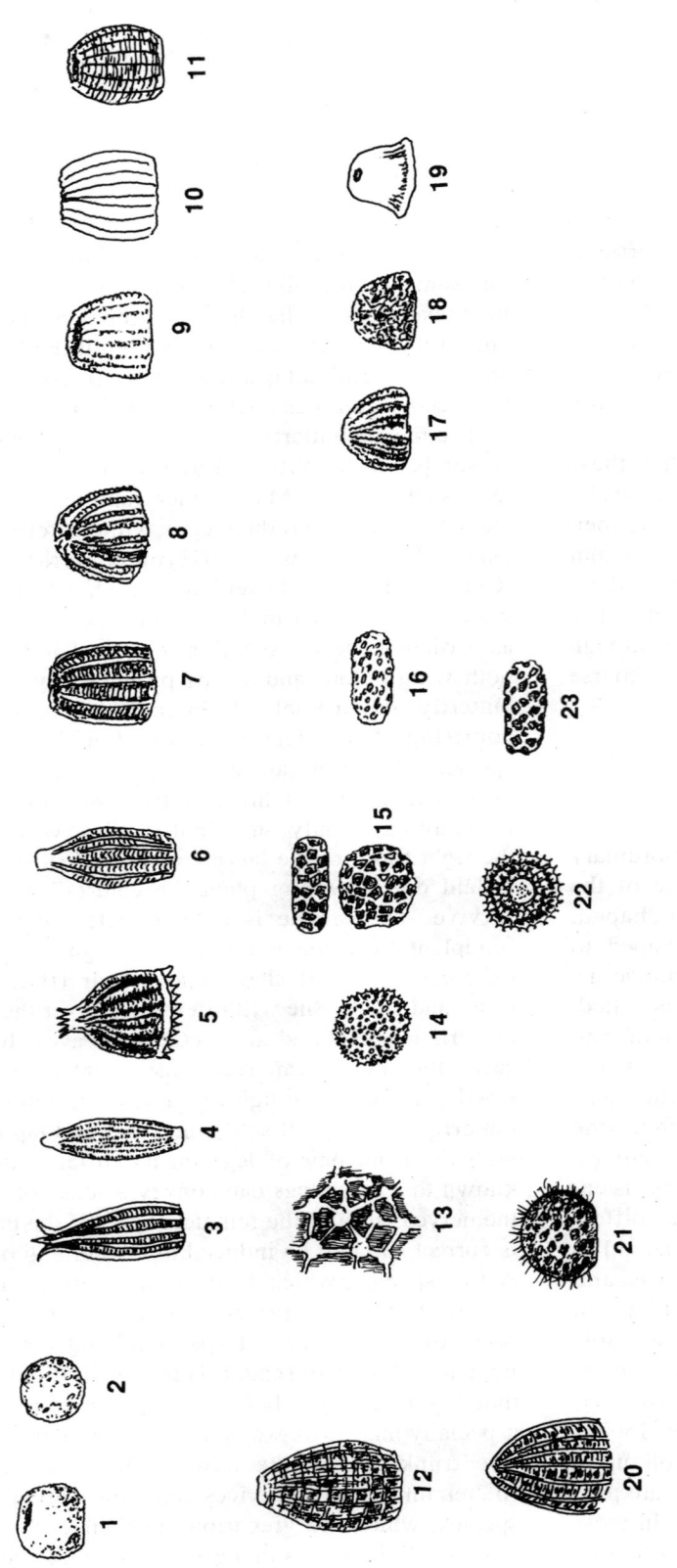

Fig. 2 Butterfly Eggs

Top Row: 1. *Papilio machaon* 2. *Pailio demoleus* 3. *Ixias marianne* 4. *Colias fieldi* 5. *Delias eucharis*
6. *Pieris rapae* 7. *Nymphalis* Tortoiseshells 8. *Argynnis* Fritiliaries 9. *Erebia nirmala* 10. *Eumenis mniszechii*
11. *Pararge schakra*

Middle Row: 12. *Danaus genutia* 13. *Nacaduba* surface sculpturing 14. *Zizeeria maha*
15. *Lycaena phlaeas*, side and top view 16. *Aphantis acamas* 17. *Badamia exclamationis*
18. *Gomalia elma* 19. *Hesperia comma*

Bottom Row: 20. *Danaus chrysippus* 21. *Limenitis trivena* 22. *Plebejus* spp. 23. *Lampides boeticus*

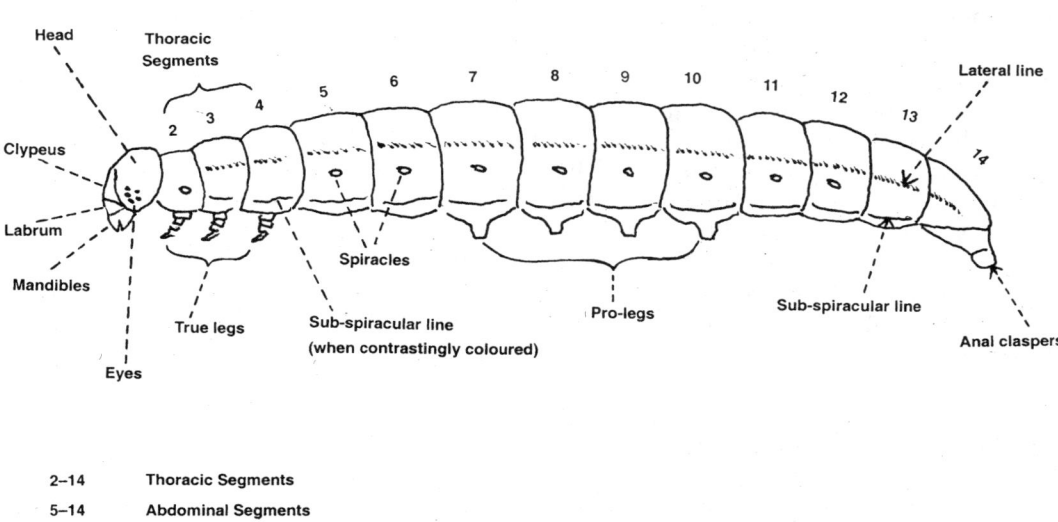

Fig. 3 External Features of Larva

the Nymphalidae, such as the Common Indian Tortoiseshell, lay all their eggs in an untidy cluster, in batches of 80 to 100 eggs, but these are an exception to the rule, and most species lay their eggs singly and carefully concealed, on the axil of a stem, or a thorn, or the tip of a leaf. Cannibalism is widespread among the larvae of certain families of butterflies, especially the Lycaenidae (Blues), which will eat any smaller caterpillar they come across, though fortunately this is not a characteristic of gregarious larval species, but may be one of the reasons why most females take such care to deposit their eggs separately and over a considerable area. Experiments have demonstrated that females can detect the presence of other butterfly eggs on their foodplant and will avoid depositing their eggs on such leaves (Chew and Robbins, in *Biology of Butterflies*, 1984). Most butterfly species lay large numbers of eggs, about 500 on average, for example, the Clouded Yellows (*Colias* spp.) (Feltwell, 1986), though a Silver Washed Fritillary (*Argynnis* sp.) in captivity (Feltwell, *op. cit.*) laid 600 eggs, and some of the Lycaenidae (Blues) lay fewer. The female extrudes a droplet of liquid with the egg which acts as a glue and sticks the bottom of the shell to the surface upon which it is deposited, though there is no such adhesive produced by those species which scatter their eggs in flight.

The Larva or Caterpillar

In general appearance, a caterpillar is familiar even to the disinterested person. One writer has aptly described this form as a 'specialized feeding tube', and indeed its two main functions are to eat incessantly until it has stored up enough energy to develop through the last two stages without major sustenance, and secondly, of course, to avoid being eaten by a predator.

The millions of different insect species on this planet share several important common features in all their life stages, despite their diversity in appearance and form. Their bodies are jointed and divided into distinct segments, usually with three thoracic segments, from each of which two pairs of jointed legs emerge. They are further protected or covered by a tough skin or shell made up of a non-porous protein compound known as chitin, within which the soft body parts are suspended without anything analogous to the skeleton of higher vertebrates. This exo-skeleton is furthered hardened and strengthened in places by special sclerotized tissue, for example, in the legs or mouthparts of adult insects. The chitinous skin is flexible to a limited extent, but cannot stretch indefinitely, so that when the sub-adult insect needs to increase in size, it must shed its skin by splitting it longitudinally and crawling out. Underneath it has a new, as yet soft, exo-skeleton which, after allowing for rapid bodily expansion, dries and hardens. Every butterfly larva goes through this moulting process, known as **ecdysis**—in the case of the sub-family Satyrinae, four times, and for the other families, five times. The growth stages in between are known as **instars**. The newly-hatched larva is minute considering the dimensions of the egg (for example, the egg of the Common Crow [*Euploea core*], a large butterfly, is about 1.4mm high and 1mm broad according to Bell [1919]), and most butterfly eggs are smaller than the proverbial pin-head. It must therefore grow rapidly to produce the adult-sized butterfly. Usually it eats its way out of the top of the eggshell and thereafter also eats the rest of the shell (chorion), before taking its first vegetable meal. The Map Butterfly (*Cyrestis thyodamus*) is a curious exception, as its egg possesses a small flapped lid which the emerging larvae can open. Quite commonly, in the early stages of growth, the larva looks completely different from the appearance of later instars. It may be cryptically coloured and with successive skin moults develop brighter colours and skin patterns. For example, the young larvae, of the Common Lime (*Princeps demoleus*) and Common Yellow Swallowtail (*Papilio machaon*) are coloured dark brown and white, with the forepart of their body thickened and closely resembling a bird dropping. Later, the full-grown caterpillar of the former is bright yellow-green in colour with a white band along the rear part of its body, and of the latter green and black striped. Many Pieridae larvae are slender in shape and leaf green in colour, so that if they lie along the stem or leaf midrib of their foodplant they are almost invisible. As a generalization, moth larvae tend to be hairy, some covered with long thick furry setae, though the large Sphingid Hawkmoth larvae are naked. In contrast, butterfly larvae appear naked, though they may often be covered with very short, semi-transparent hairs, giving the skin a shagreened effect (e.g., Pieridae larvae). Their bodies may bear scattered fleshy spines which themselves may be branched (many Nymphalidae spp.), whilst others are naked skinned but with long fleshy tentacles in pairs on the front and rear part of their bodies (e.g., Danainae larvae). The larvae of the Lycaenidae have short-humped bodies with small heads which they can retract under the first thoracic segment, whilst the larvae of Hesperiidae (Skippers) have large heads wider than their necks or first thoracic segment, and often have delicate, semi-transparent skins and an instinctive behaviour which enables them to construct a leaf shelter or cell inside of which they can hide when not feeding.

The anatomy of a caterpillar needs brief description (*see*, Fig. 3). Its body comprises a head formed by the fusion of six segments, followed by thirteen rounded body segments known as somites. The first three of these are known as the thoracic segments, and the remaining ten are the abdominal segments or somites. The head bears a pair of powerful mandibles covered with sclerotized tissue, enabling them to chew through tough vegetable matter. Often the newly-hatched caterpillar does not have strong enough mandibles to chew mature plant tissue and must confine its feeding to unopened leaf buds or even flower parts, turning to more abundant larger leaves as it grows bigger. The larvae of some of the smaller Lycaenidae feed exclusively upon and inside flowers. On the sides of the head near the mouth parts are a group of six simple eyes like tiny black specks. It is thought that they cannot see in our sense of the word, but that through these ocelli they have a perception of light and darkness. They are also equipped with a pair of rudimentary short antennae looking like withdrawn tentacles. Each thoracic segment bears a pair of segmented true legs, each made up of five segments and terminating in a claw. As will be seen below, the development of the first pair of legs in the adult butterfly varies between families, and provides an important taxonomic distinction. From segment 7 to 10 are pairs of fleshy, unjointed, false legs known as prolegs, plus a clasper leg on the last and anal segment. The internal organs include respiratory, excretory, and nerve organs, and extend throughout the length of the abdominal segments, including the tubular heart and the breathing or oxygenating organs known as trachea. The caterpillar is furnished with two other vital external organs. Along the sides of the abdominal segments are nine breathing pores, known as spiracles, which are connected to the tracheal tubes. These are often contrastingly coloured, and their location is useful in describing the position of colour stripes along the larval body. On the sides of its head are two silk-spinning glands, known as spinnerets. Frightened caterpillars of some species can drop to the ground from their foodplant using a silken 'lifeline'; the silk strands are also used by the larva to construct leaf shelters, and to weave secure beds upon which they can attach themselves before pupating. Whilst most lay persons are aware of the amazing tensile strength of silkworm silk—produced by a moth—few realize that all butterfly larvae have this ability, though it is not generally used to weave thick, tough cocoons, as is the case with many moth species.

Two other peculiar features are possessed by some butterfly families. In the Papilionidae (Swallowtails), behind the head and on top of the first thoracic segment is a brightly-coloured, extrusible organ like a forked tongue known as the osmeterium, contained in a pouch. The full-grown larva can flick this rapidly in and out and it possesses a strongly aromatic, often unpleasant, odour. It is thought that this feature may help to deter predators. In some members of the family Lycaenidae (Blues), particularly *Polyommatus, Nacaduba,* and *Spindasis* spp., there is also a small protrusible gland on the top of the tenth (abdominal) segment. This develops as the larva grows and produces droplets of a sweet substance that is greedily eaten by ants (Formicoidae). There is, in fact, a unique symbiotic relationship between certain larvae and their ant hosts, which protect them from predators in return for this sweet secretion. Colour plates 3, 4, and 5 are intended to show the diversity of colour and form between different families and species of butterfly larvae, but the reader is cautioned to realize that it is only practicable, in the species accounts and illustrations, to show the appearance of 5th instar or full-grown larvae. In their early stages, most tiny larvae appear quite different. For example, in the Lycaenidae, the egg larvae usually bear long hairs which are shed with later moults. Also, the colour of larvae, even within one species, can vary according to the foodplant.

The Pupa or Chrysalis

Most people are aware that, in the more developed and less primitive insect families, there is a larval stage before the adult insect emerges, for example, maggots in flies, or 'cutworms' in Tipulidae (Daddy Longlegs), while anyone in Pakistan who has suffered Cockroach (*Blattaria* spp.) infestation in the kitchen (unfortunately possible in the cleanest and newest of houses!) is aware that baby cockroaches look exactly like miniature adults. In the case of Lepidoptera, however, there is such a dramatic change from the appearance of the larva to that of the adult insect that there is an intermediate stage, known as pupation, during which the insect is relatively passive and immobile whilst its internal organs undergo a complete metamorphosis or transformation within the chrysalis skin. During this third pupal stage, when its internal organs are in a semi-liquid state, the body cells are reorganized into distinct groups by chemical originators known as imaginal buds (Ford, 1945), and from these the new internal and external organs are developed. As yet, attempts to fully unravel the chemistry or physiology of this process have defied investigators.

Shortly before pupating, the fully-grown caterpillar stops eating and even shrinks in size. Often it wanders widely from its foodplant until it finds a suitable place to pupate, where with its spinnerets it makes a pad or bed of silk upon which it can securely fasten its anal proleg or claspers before splitting out of its skin and emerging as a chrysalis. A few groups actually spin with their silk a complete skin or cocoon inside which they pupate. This is common among moths, but in butterflies is associated with the more primitive families such as the Skippers; it is also an evolutionary development in that specialized Alpine family, the Snow Apollos, which need to protect their pupae from sub-zero temperatures. The pupa itself has been rather inaptly described as a 'dormant stage'. In fact, many pupae species can wriggle violently if disturbed, and others possess stridulating organs situated on their lower abdominal segments which produce a hissing sound when rubbed together, both actions which might deter some predators. Moreover, the physiological transformation undergone during this stage implies profound internal activity.

The pupa that emerges is very different in shape from the caterpillar (*see,* Figs. 4 and 5), as its legs and mouthparts are shed with the larval skin, and now its outline clearly shows the main external organs, including, in some species, the encased wings. Pupae are varied in shape, being rather smooth and rounded in the Skippers, but in the *Vanessa* spp. (Admirals and Tortoiseshells) they are angular, with small humps or points along their dorsal surface, often decorated with bright metallic gold or silver spots. In the Danainae the chrysalis is rather broad and barrel-shaped, and also often brightly decorated with metallic streaks and spots. The pupa of the Pierids is rather pointed at the head end, and generally not so studded with sharp points as that of Nymphalidae.

The method of attachment of the chrysalis is consistent within families and also quite variable. For example, in the Swallowtails (Papilionidae) it is suspended horizontally or vertically, head upwards,

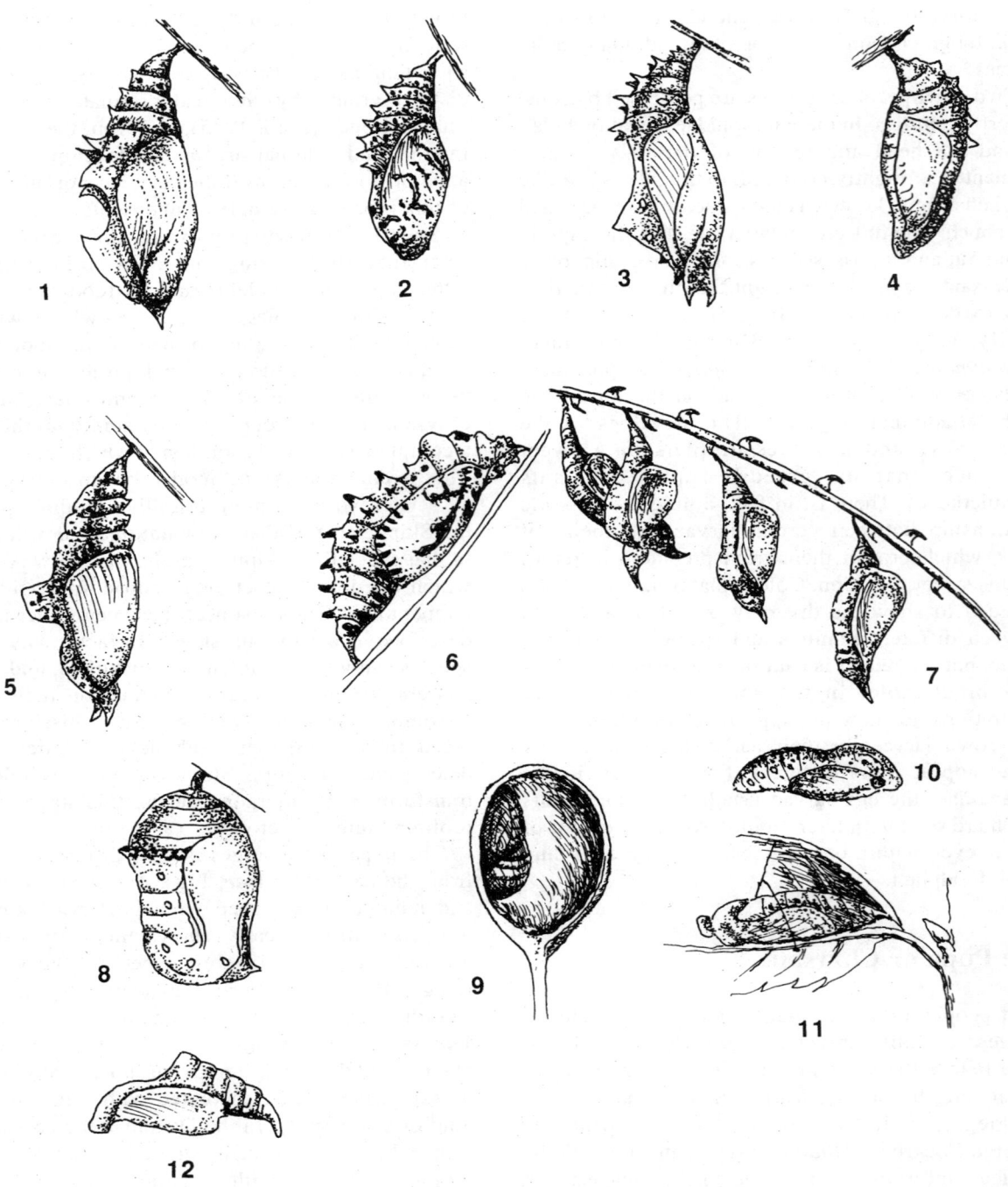

Fig. 4 Butterfly Pupae

1. *Hypolimnas missipus*
2. *Melitea didyma*
3. *Argynnis hyperbius*
4. *Junonia orithya*
5. *Limenitis trivena*
6. *Delia eucharis*
7. *Eurema hecabe*
8. *Danaus chrysippus*
9. *Deudoryx epijarbus*
10. *Rapala nissa*
11. (Below) *Badamia exclamationis*
12. *Libythea lepita*

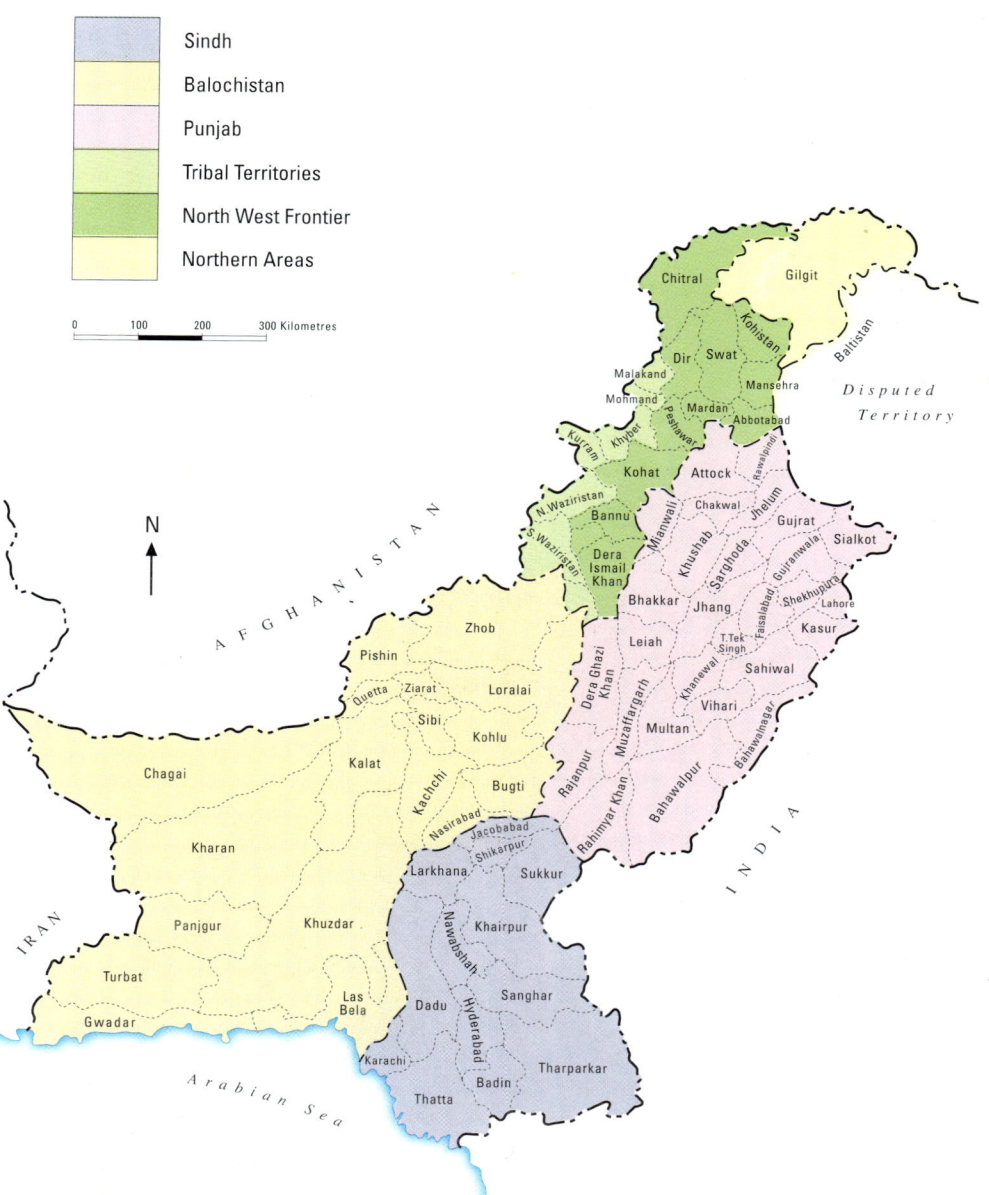

NORTHERN AREAS OF PAKISTAN

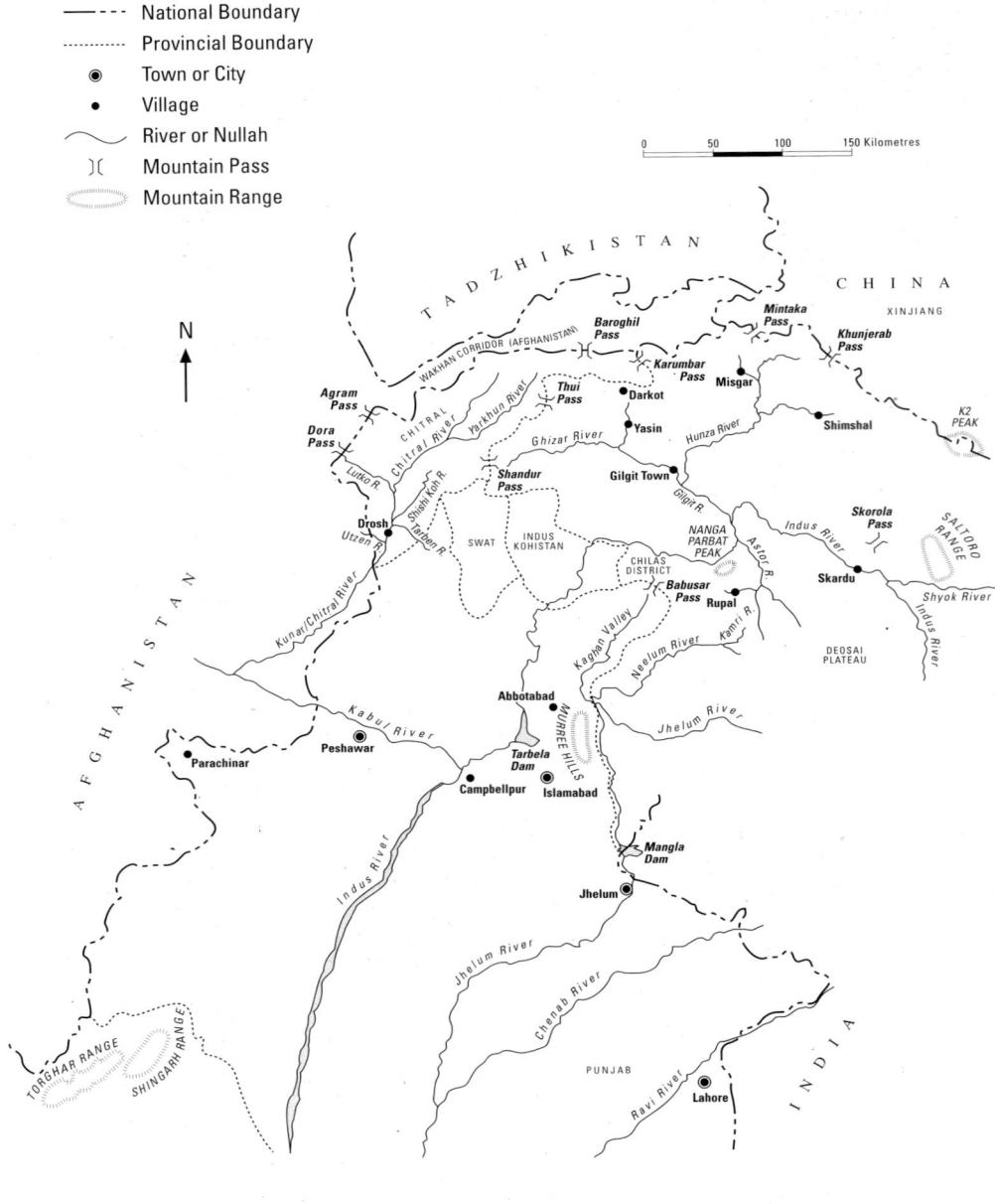

PLATE 3: Larvae - Swallowtails (Papilionidae) and Whites (Pieridae)

(a). *Hypermnestra helios* - Desert Apollo **(b).** *Pachliopta aristolochiae* - Common Rose **(c).** *Papilio machaon* - Common Yellow Swallowtail **(d).** *Papilio polyctor* - Common Peacock **(e).** *Chilasa clytia* - Common Mime **(f).** *Colias fieldi* - Dark Clouded Yellow **(g).** *Catopsila crocale* - Common Emigrant **(h).** *Eurema hecabe* - Common Grass Yellow **(i).** *Delias eucharis* - Common Jezebel **(j).** *Colotis amata* - Small Salmon Arab

PLATE 4: Larvae - Browns (Satyrinae) and Nymphalidae, including Danainae

(a). *Melanitis leda* - Common Evening Brown (b). *Parage schakra* - Common Wall (c). *Maniola pulchella* - Tawny Meadow Brown (d). *Eumenis mniszechii* - Tawny Rockbrown (e). *Junonia orithya* - Blue Pansy (f). *Melitaea didyma* - Fiery or Spotted Fritillary (g). *Argynnis childreni* - Large Silverstripe (h). *Hypolimnas misippus* - Danaid Eggfly (i). *Euthalia aconthea* - Common Baron (j). *Apatura ambica* - Indian Purple Emperor (k). *Vanessa cashmiriensis* - Indian Tortoiseshell (l). *Neptis hylas* - Common Sailer (m). *Euploea core* - Common Crow (n). *Danaus chrysippus* - Plain Tiger

PLATE 5: Larvae - Blues (Lycaenidae), Punches (Riodinidae), Beaks (Libytheinae) and Skippers (Hesperiidae)

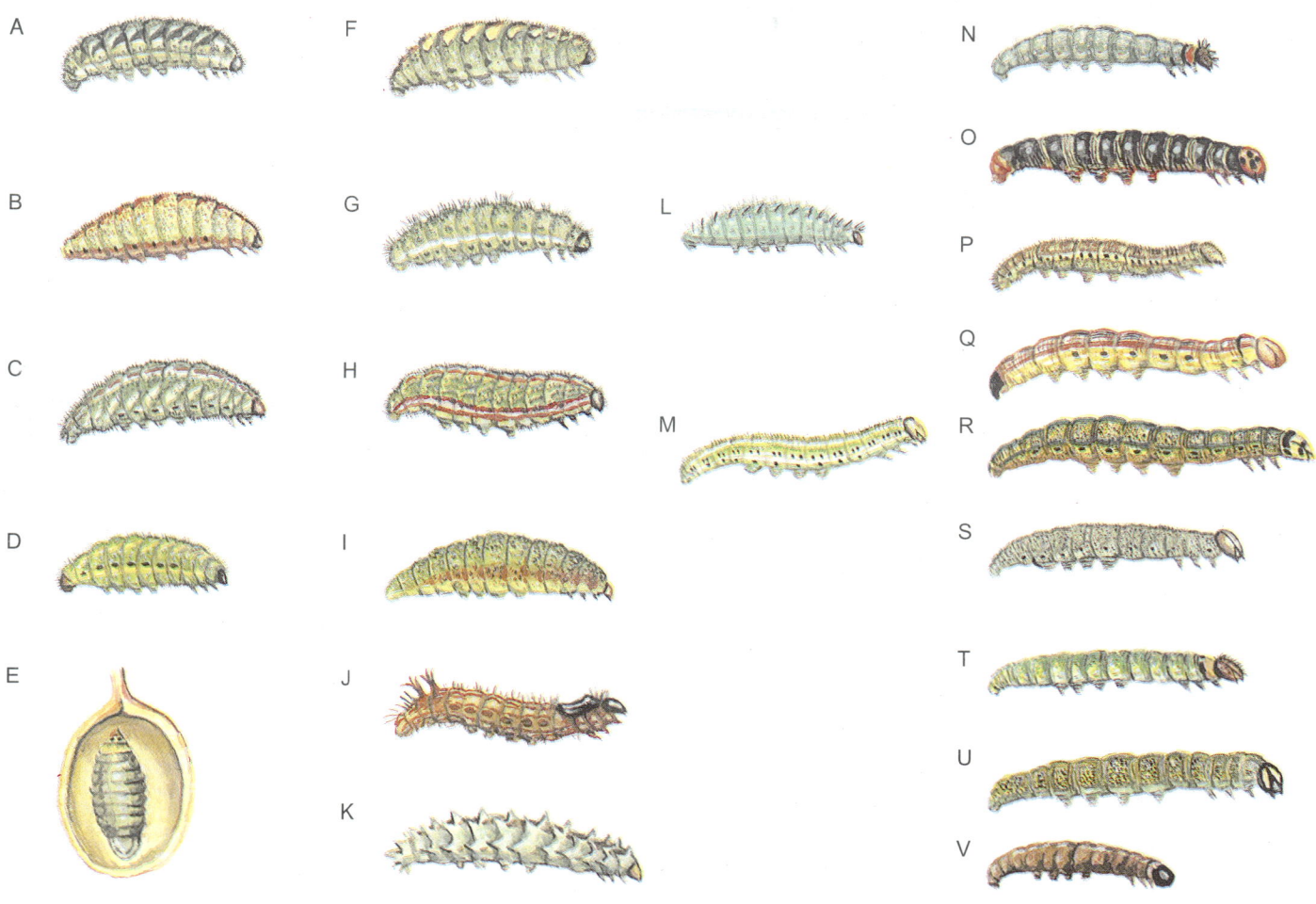

(a). *Callophrys rubi* - Green Hairstreak **(b).** *Lycaena phleas* - Common Copper **(c).** *Everes argiades* - Tailed Cupid **(d).** *Catochrysops cnejus* - Gram Blue **(e).** *Deudoryx epijarbus* - Cornelian **(f).** *Celastrina argiolus* - Hill Hedge Blue or Holly Blue of Europe **(g).** *Polyommatus eros* - Common Meadow Blue or Eros Blue of Europe **(h).** *Aricia agestis* - Orange Bordered Argus **(i).** *Nacaduba nora* - Common Lineblue **(j).** *Spindasis acamas* - Tawny Silverline **(k).** *Rapala iarbus* - Indian Red Flash **(l).** *Dodona eugenes* - Tailed Punch **(m).** *Libythia celtis* - Eurasian Beak **(n).** *Spialia galba* - Indian Skipper **(o).** *Choaspes xanthopogon* - Indian Awlking **(p).** *Muschampia staudingeri* - Streaked or Syrian Skipper **(q).** *Hasora alexis* - Common Banded Awl **(r).** *Badamia exclamationis* - Brown Awl **(s).** *Hesperia comma* - Chequered Darter or Silver Spotted Skipper of Europe **(t).** *Caprona ransonetti* - Golden Angle **(u).** *Notocrypta fiesthamelii* - Spotted Demon **(v).** *Sarangesa purendra* - Spotted Small Flat

PLATE 6A: HABITAT SCENES, INCLUDING DESERT

◀ **(a) Cholistan Desert. Typical plants,** *Tamarix aphylla, Callygonum polygonoides, Leptadenia spartium, Capparis decidua, Aristida depressa,* and *Cenchrus* spp. **Typical butterflies,** *Colotis vestalis, Colotis danae, Anaphaeus aurota, Danaus chrysippus, Syntarucus jesous, Zizeeria knysa.*

(b) Kalankot, Thatta Dist. Typical plants, *Salvadora persica, Euphorbia caducifolia, Acacia senegal.* **Typical butterflies,** *Pachliopta aristolochiae, Colotis protractus, Colotis fausta, Tarucus theophrastus, Tarucus balkanica, Catachrysops strabo, Euchrysops parhassius.* ▶

◀ **(c) Hab Valley, Las Bela. Typical plants,** *Acacia senegal, Commiphora mukul, Zizyphus nummularia, Prosopis spicigera, Acacia jacquemontii.* **Typical butterflies,** *Colotis fausta, Precis orithya, Tarucus balkanica, Tarucus theophrastus, Syrichtus evanidus.*

(d) Hazar Ganji, Chiltan. Typical plant species, *Fraxinus xanthoxiloides, Pistacia cabulica, Pistachia khinjak, Sophora molle, Artemisia brevifolia, Tulipa lehmania, Rosa lacerans.* **Typical butterflies,** *Euchloe charlonia, Euchloe ausonia, Maniola davendra, Melitaea trivia, Polyommatus loewii, Lampides boeticus, Apharitis acamas.* ▶

PLATE 6B: HABITAT SCENES, INCLUDING DESERT

◀ **(e) Kohat Dist., near Lachi. Typical plants,** *Acacia modesta, Olea cuspidata, Withania coagulans, Dodonea viscosa, Monotheca buxifolia, Ostotegia limbata, Rhazya stricta.* **Typical butterflies,** *Euchloe lucilla, Anaphaeus aurota, Hipparchia parisatis, Tarucus balkanica, Tarucus theophrastus, Polyommatus eros.*

(f) Hazara foothills, near Tarbela. Typical plants, ▶ *Acacia modesta, Dodonea viscosa, Zizyphus mauritiana, Punica granatum, Adhatoda vasica, Sagretia theezans.* **Typical butterflies,** *Colias erate, Argyreus hyperbius, Junonia hierta, Pararge schakra, Ypthima bolanica, Hipparchia parisatis, Maniola davendra, Tarucus nara, Deudoryx epijarbus.*

◀ **(g) Ziarat, Balochistan. Typical plants,** *Zuniperus excelsa, Berberis baluchistani, Prunus eburnea, Lonicera hypoleuca, Eremurus stenophyllus.* **Typical butterflies,** *Pieris rapae, Papilio alexanor, Callophrys rubi, Melitaea lutko, Polyommatus loewii, Strymon assamica, Neolycaena connae.*

(h) Khaliphat slopes, Ziarat Dist. Typical plants, ▶ *Zuniperus excelsa, Eremurus stenophyllus, Onobrychis cornuta, Prunus eburnea, Rosa moschata, Cotoneaster nummularia.* **Typical butterflies,** *Melitaea trivia, Vanessa egea, Neolycaena connae, Pararge menava, Polyommatus loewi, Polyommatus iris, Polyommatus bogra.*

PLATE 7A: HABITAT SCENES, INCLUDING FOOTHILLS

(a) Margalla Hills, in November. Typical plants, *Ficus virgata, Bauhinia variegata, Acacia modesta, Carissa opaca, Celtis eriocarpa, Pyrus pasha, Malotus philippensis, Heteropogon contortus.* **Typical butterflies**, *Papilio polyctor, Eriboea athamus, Lethe rohria, Catopsila crocale, Ariadne merione, Argynnis hyperbius, Nacaduba nora, Caprona agama.*

(b) Kao Forest, below Dunga Gali. Typical plants, *Aesculus indica, Acer caesium, Quercus incana, Populus ciliata, Juglans regia, Viburnum cotinifolium, Rosa brunnoni, Impatiens* and *Geranium* spp. **Typical butterflies**, *Atrophaneura polyeuctes, Pieris canidia, Aporia leucodice, Argynnis lathonia, Vanessa cashmiriensis, Vanessa indica, Vanessa canace, Ypthima sakra, Libythia lepita, Chrysozephyrus milionia, Arhopola ganesa, Badamia exclamationis, Chaetoprocta odata.*

(c) Margalla Hills, in March. Typical plants, *Punica granatum, Zizyphus mauritiana, Adhatoda vasica, Cassia fistula, Acacia modesta, Dodonea viscosa, Carissa opaca, Aristida cyantha, Apluda aristata, Themeda anthera.* **Typical butterflies**, *Pieris brassicae, Euthalia aconthea, Ariadne merione, Cyrestis thyodamus, Ypthima asterope, Castalius rosimon, Zizeeria maha, Parnara guttatus, Taractrocera maevius.*

(d) Lake Shandur, N.W. Gilgit. Typical plants, *Artemisia maritima, Silene viscosa, Sedum crassipes, Corydalis crassissima, Carex cruenta, Scirpus planifolius, Poa* spp. **Typical butterflies**, *Parnassius actius, Parnassius jacquemontii, Parnassius delphius, Parnassius charltonius, Baltia shawi, Pieris callidice, Colias alpheraky, Colias eogene, Argynnis jerdoni, Argynnis aglaia, Melitaea shandura, Maniola pulchella, Erebia mani, Everes buddhista, Turanana cytis, Albulina metallica, Polyommatus sarta, Lycaena solskyi.*

PLATE 7B: HABITAT SCENES, INCLUDING FOOTHILLS

(e) Shandur Plateau. Typical plants, *Artemisia maritima, Lagochilus cabulicus, Pedicularis bicornuta, Potentilla salesoviana, Potentilla peduncularis, Corydalis* and *Rhodiola* spp., *Poa* spp. **Typical butterflies,** *Parnassius actius, Parnassius charltonius, Parnassius jacquemontii, Erebia mani, Maniola pulchra, Maniola pulchella, Hesperia comma, Hesperia alpina.*

(f) Khunjerab Pass, N. Hunza. Typical plants, *Aster himalaicus, Artemisia brevifolia, Jurinea ceraocarpa, Tanacetum tibeticum, Silene gonosperma, Lepidium apetalum, Rhodiola himalensis, Rhodiola heterodonta, Androsace baltistanica, Potentilla curviseta, Potentilla microphylla, Primula buryana, Pedicularis pyramidata, Pedicularis verae.* **Typical butterflies,** *Parnassius actius, Parnassius tianschanikus, Parnassius staudingeri, Parnassius loxias, Papilio ladakensis, Baltia butleri, Pieris deota, Colias stoliczkana, Colias cocandica, Eumenis heydenreichi, Karanasa boloricus, Agriades pheretiades, Polyommatus eros.*

(h) Deosai Plateau, between Chilim and Skardu. Typical plants, as in (g), and *Juncus membranaceus, Eremurus himalaicus, Allium neopolitanum, Perovskia abrotanoides, Androsace mucronifolia, Primula macrophylla.* **Typical butterflies, as in** (g), *Parnassius* spp., *Melitaea* spp., *Agriades pheretiades, Polyommatus sarta, Iolana gigantea, Hesperia alpina, Hesperia comma.*

(g) Deosai Plateau, Nr. Chilim. Typical plants, *Saxifraga hirculus, Primula macrophylla, Allium fedtschenkohnum, Androsace mucronifolia, Corydalis clarkei, Polygonum persicaria, Pleurospermum candollei, Alopecurus aequalis, Carex alpina, Carex haematostoma, Kobresia bellardii, Scirpus* spp. **Typical butterflies,** *Parnassius epaphus, Parnassius delphius, Parnassius simo, Argynnis jerdoni, Bolares pales, Melitaea arcesia, Satyrus pimpla, Karanasa heubneri, Aricia chiron, Albullina metallica, Polyommatus sarta.*

PLATE 8A: HABITAT SCENES, INCLUDING FAR NORTHERN BORDERS

(a) Central Baroghil. Peak on right Kogo Zum, 6872m. Typical plants, *Sibbaldia cuneata, Oxytropis immersa, Cerastium cerastioides, Potentilla flabellate, Primula macrophyllum, Sedum crassipes, Potentilla anserina, Potentilla argyophylla, Juncus* and *Carex* spp. **Typical butterflies,** *Parnassius actius, Parnassius jacquemontii, Parnassius delphius, Parnassius tianschanika, Parnassius stoliczkanus, Baltia shawi, Pieris callidice, Colias wiskotti, Colias leechi, Colias marcopolo, Maniola hilaris, Eumenis heydenreichi, Albulina pheretes, Albulina metallica, Iolana gigantea, Polyommatus sarta, Halpe holmolea.*

(c) Upper Baroghil, with Zondikharam Glacier behind. Typical plants, as in (b) below, with *Anaphalis virgata, Artemisia brevifolia, Jurinea himalaica, Leontopodium himalayanum, Androsace villosa, Carex* and *Poa* spp. **Typical butterflies** as in (a) and (b) above, including *Baltia shawi, Pieris callidice, Pieris kreuperi, Colias wiskotti, Colias leechi, Melitaea minerva, Maniola hilaris, Maniola pulchra, Hipparchia heydenreichi, Polyommatus gracilis, Polyommatus eros, Hesperia alpina.*

(b) Central Baroghil, northern Chitral border. Typical plants, *Arenaria orbiculata, Rhodiola quadrifidum, Silene moorcroftiana, Rhodiola himalensis, Sedum ewersi, Saxifraga sibirica, Potentilla flabellatum, Sedum crassipes, Juncus membranaceus, Agrostis pilosula, Bromus gracillimus.* **Typical butterflies,** *Parnassius jacquemontii, Parnassius delphius, Parnassius tianschanika, Pieris callidice, Colias eogene, Maniola minerva, Melitaea hilaris, Albullina pheretes, Albulina metallica, Polyommatus sarta.*

(d) Darkot Pass, looking up Gankhush Valley. Typical plants, *Tamaricaria elegans, Hippophae rhamnoides, Anemone rupicola, Corydalis onobrychis, Capparis spinosa, Arabidopsis mollissima, Artemisia maritima, Androsace himalaica, Agrostis canina, Koeleria argentea.* **Typical butterflies,** *Parnassius delphius, Parnassius staudingeri, Parnassius epaphus, Colias wiskotti, Colias alpheraky, Erebia kalinda, Erebia mani, Albulina pheretes, Albulina metallica, Polyommatus sarta.*

PLATE 8B: HABITAT SCENES, INCLUDING FAR NORTHERN BORDERS

(e) Summit of Karumbar Pass, N.E. Chitral. Typical plants, *Juncus leucomelas, Juncus membranaceus, Carex cruenta, Carex karoi, Aegilops canaliculatum, Chesnya cuneata, Cicer songaricum, Draba altaica, Draba cachemirica, Draba oriades, Luzula spicata, Agropyron himalayanum, Agropyron repens, Corydalis crassissima, Corydalis onobrychis, Potentilla anserina, Potentilla argyrophylla, Allium chitralicum, Allium semenovi.* **Typical butterflies,** *Parnassius epaphus, Parnassius tianschanika, Parnassius delphius, Parnassius staudingeri, Pontia callidice, Colias alpheraky, Colias fieldi, Colias wiskotti, Maniola pulchra, Maniola pulchella, Maniola hilaris, Eumenis mniszechii, Eumenis heydenreichi, Karanasa heubneri, Erebia shallada, Aricia chiron, Albulina pheretes, Albulina metallica, Polyommatus sarta, Spialia sertorius, Hesperia comma.*

(f) Sawr, looking back to Darkot. Typical plants, *Hippophae rhamnoides, Juniperus communis, Artemisia maritima, Myricaria elegans, Tricholepis chaetolepis, Tanacetum longifolium, Saussurea gnaphalodes, Sibbaldia cunneata, Polygonum affine, Poa alpina, Poa pratensis, Arabidopsis mollissima.* **Typical butterflies,** *Parnassius delphius, Parnassius epaphus, Parnassius actius, Parnassius charltonius, Pieris callidice, Colias wiskotti, Colias alpheraky, Colias eogene, Clossiana hegemone, Melitaea minerva, Maniola pulchra, Maniola pulchella, Maniola hilaris, Eumenis mniszechii, Eumenis heydenreichi, Hipparchia moorei, Erebia mani, Aricia chiron, Albullina metallica, Polyommatus gracilis, Polyommatus eros.*

PLATE 9: 'Swallowtails' and 'Whites'

(a) Common Peacock Butterfly, *Papilio polyctor*

(b) Common Mormon, *Papilio polytes*

(c) Common Yellow Swallowtail, *Papilio machaon*

(d) Common Wanderer, *Parenonia anais*

(e) Dark Clouded Yellow, *Colias fieldi*

(f) Lesser Brimstone, *Gonepteryx farinosa*

FORM AND FUNCTION 9

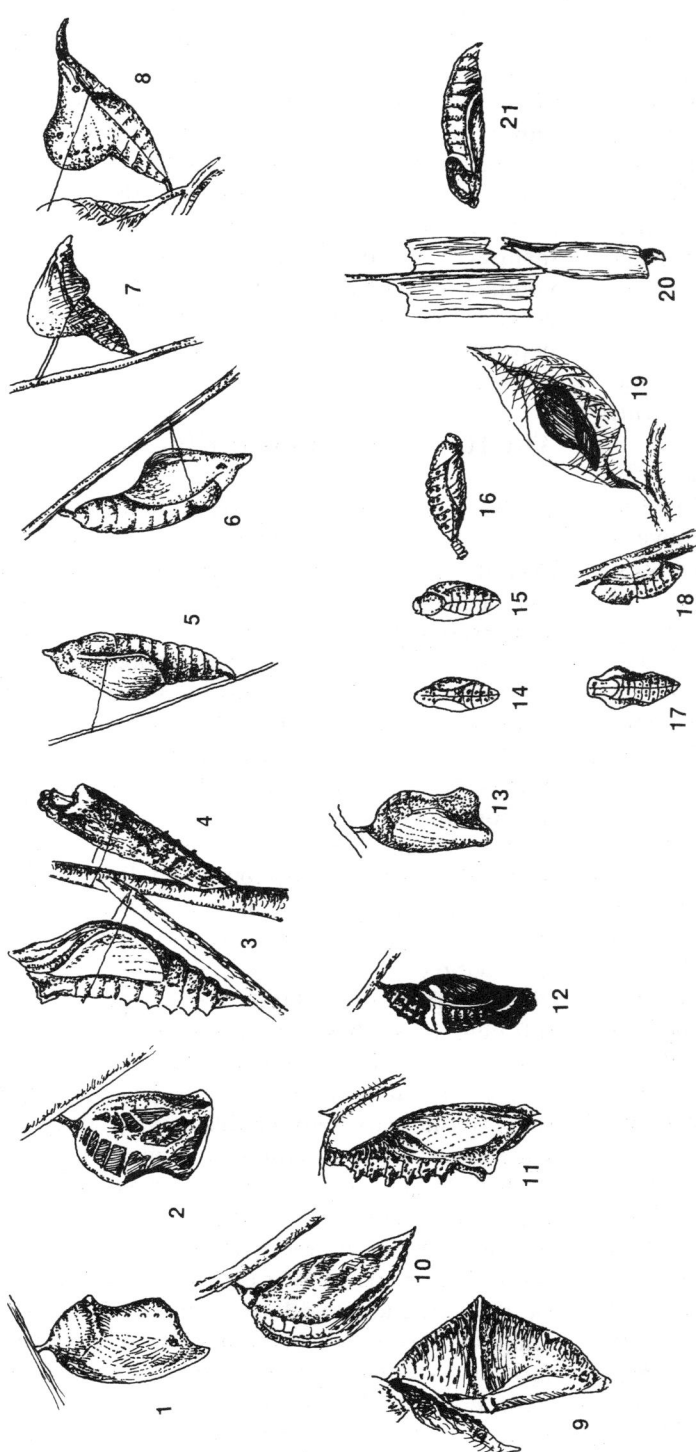

Fig. 5 Butterfly Pupae

Top Row: 1. *Danaus* sp.   2. *Euploea core*   3. *Papilio demoleus*   4. *Chilasa clytia*
5. *Colotis amata*   6. *Catopsila pyranthe*   7. *Ixias pyrene*   8. *Parenonia valeria*

Bottom Row: 9. *Eriboea athamus*   10. *Euthalia aconthea*   11. *Vanessa xanthomelas*
12. *Argynnis lathonia*   13. *Melanitis leda*   14. *Euchrysops cnejus*
15. *Polyommatus* spp.   16. *Aricia agestis* (Below)   17. *Narathura sanesa*
18. *Lampides boeticus*   19. *Apharitis acamas*   20. *Taratrocera maevius* in grass cell
21. *Hesperia comma*

along a plant stem or twig, with a stout silken band over the thorax and its tail claspers firmly embedded in a pad of silk. In the Nymphalidae, especially the Tortoiseshells and Admirals, the pupa is suspended head downwards from the tail clasper alone, hanging freely, whilst in the Pieridae the pupa is embedded in strands of silk on top of a leaf, or in a bark or rock crevice, usually head upwards, and in the Satyrinae (Browns) the pupa is suspended by the tail claspers alone, hanging head downwards, or buried in the base of plant stems. The Skippers are unique among the sub-order **Rhopalocera** (Butterflies) in being formed in a cocoon shelter of silk, within a cell made from leaves or a folded leaf, secured together with its own silk.

In many tropical species, the pupa itself may only stay in this stage for seven to twenty days before the adult emerges, because the larval food plant is always available for the female on which to deposit her eggs, whilst in those species found at high elevations and in cold conditions, the pupa may remain from late summer, throughout the winter, hidden safely among the base stalks of its withered foodplant, or the newly-hatched larva may 'hibernate' or remain dormant and not start feeding until the following spring. At all stages in its life history the butterfly is vulnerable to predators and parasites, which subject will be discussed in Chapter 3. The relatively immobile pupa is particularly vulnerable, but in the course of evolution has developed a remarkably cryptic appearance and an ability to produce a chrysalis skin which can vary in colour within the same species and even from eggs laid by the same female, according to the substrate upon which it pupates. For example, the Indian Cabbage White (*Pieris canidia*) can produce green and black speckled chrysalises when attached to a withered leaf, or grey and brown blotched chrysalises when attached to a tree trunk or rock crevice.

There are some classical examples of unique behaviour at the pupal stage which help to protect the developing butterfly. The pupa of the Common Mime (*Chilasa clytia*) is always suspended upright at a 45° angle from a woody stem, and exactly resembles, in colour and shape, a broken-off twig. The pupae of the Sailers (*Neptis* spp.) are formed inside the withered leaves of their foodplant, which are loosely held together by their larval silk. These fall to the ground either during or before pupation, and inside this leaf shelter the caterpillar first spins a thick bed of silk, sufficient to repel water. During tropical monsoon downpours the ground where it lies may often be flooded, but without any harm to the pupa, as the waxy silk keeps it dry. Unlike all other butterfly families, the Snow Apollos (Parnassiinae), protect themselves from the severe alpine winter as pupae inside a thick, parchment-like cocoon and underground, within a rock overhang or soil crevice. Likewise some of the Hesperiidae (Skippers), with their relatively delicate chrysalises, protect themselves not only inside a roughly-woven cocoon within a folded leaf shelter, but also by a white waxy powder secreted by the caterpillar just prior to pupating.

The Imago or Adult Butterfly

Unlike more primitive insect families, such as Collembola (Spring-tails) or Ephemeroptera (Mayflies), once it has emerged, the adult butterfly cannot change its size. Consequently, in some species with broods developed during the dry cold weather as well as during the humid monsoon season, there is seasonal variation in size, as well as marked differences in wing pattern. The availability of food for the larva determines their ultimate size, and adult butterflies need a minimum of food for energy, but not for growth.

Whilst butterflies are too familiar in appearance to need detailed description, the following account will enable the reader to appreciate some aspects of their behaviour, classification, and limited distribution.

Butterflies have large compound eyes, quite different from those at the larval stage, and it is known that their vision produces a multi-faceted image. The dimensions of their eyes are very large in proportion to their bodies, giving them a very wide arc or angle of vision; also, males always have slightly larger eyes than their female counterparts. Butterflies have acutely developed colour vision, including the detection of ultraviolet rays which are invisible to some of their predators. The significance of these features will be considered in the next chapter. The colours on the wing surface, which have attracted so much interest among naturalists, are entirely due to the presence of scales, which are like minute roof-tiles, loosely attached in overlapping series to the transparent wing membrane by tiny pegs. As is well known, if a butterfly is roughly handled, these scales come off in a veritable cascade of powder. In fact high-speed photography has now shown that with every wing beat, a cloud of scales is showered into the air, but it has been estimated that the average butterfly possesses about one and a half million of these scales (Feltwell, 1986), so that there are more than enough throughout adult life to retain clearly the wing pattern upon which sexual and conspecific recognition depends. If a butterfly could be deprived of most of its scales without harm, it could still fly perfectly. The Parnassiinae (Apollos), with their largely transparent and thinly-scaled wings, demonstrate this well in the harsh, windy environment where they fly. The colours of these scales are due largely to pigments partly derived from the larval foodplant. Pigments have chemical properties in their structure such that they exhibit selective absorption of light rays at different intensities, resulting in different colours and shades of brightness. A particular pigment type is often confined to one or more families (Ford, 1945). For example, in the family Pieridae (Whites), the colours are derived from uric acid compounds built up from the insects excretory products. Another important pigment is derived from melanin, which produces the black and red colours prevalent in many families; especially Swallowtails, which lack pigments, derived from uric acid (Ford, *op. cit.*). Another group, known to be derived from the larval foodplants, are flavones, present in some Skippers and Satyrid butterflies. These flavones produce ivory colours, shading to deep yellow, but curiously are not found in the sub-family Coliadinae, the yellow species of Pieridae. Besides the colours derived from pigments, iridescent colour is also produced by a process known as multiple thin film interference filters, which are present in the outer layers of the scales (Silberglied, in *Biology of Butterflies*, 1984). Some scales have an intricate cellular structure, with barriers which interfere with light transmission within the scale. Put in lay terms, these barriers act as mirrors, doubling or intensifying the colour reflection, thus producing iridescent or metallic colours, and those colours which vary when viewed from different angles (Sustare, in *The Behavioural Significance of Color*, Burtt, Ed., 1979). The best examples of this are shown in the many iridescent species of male Lycaenidae, or the purple hues in the male of Apaturninae (Indian Purple Emperor). It is thought (Silberglied, *op. cit.*) that with their capacity to see ultraviolet light, adult butterflies have especially acute vision when iridescent colours are displayed. In male butterflies of some species there are also specialized scales of different shape which have special scent-producing cells at their base, and these scales are shaped to disperse scent. They are known as **androconia**. They may be scattered randomly throughout the wing surface, as is typical of most Lycaenidae, or be borne in distinct patches known as brands, as in many Satyrinae and some Hesperiidae. These brand areas are often contrastingly coloured from the rest of the wing. In the larger Fritillaries, these scent scales may be confined within narrow black folds along the edge of forewing veins, while Danainae males have small flaps on the underwing surface from which scent can be released, as well as a tuft of hair-like cells at the tail tip, known as hair pencils (*see*, Fig. 8), for the same purpose. Some Charaxinae and Satyrinae males also possess tufts of scent-producing hairs on their hindwings which can be erected during courtship. These scents are vital in intra-specific recognition and in stimulating the female to be ready to mate. Another phenomenon associated with scales is the presence at the tip of the abdomen in females of tufts of scales, which are shed during ovipositing, so that the newly-laid egg is covered. This feature is found in some Lycaenidae and Hesperiidae (Chew and Robbins in *Biology of Butterflies*, 1984), and the best known example in Pakistan is found in the eggs of the Walnut Blue (*Chaetoprocta odata*). It is thought that this helps to protect and conceal the egg from predators.

The anatomy of the adult butterfly differs markedly from that of its larva. Not only are the antennae now long and delicate, and made up of many segments with the end part swollen, but they are now highly developed as sensory and principally as smelling organs. In the adult there are no chewing mandibles, the mouthparts being modified instead into a long flexible tube known as the proboscis, which can only be used to suck up liquids, and which is tightly curled up under the palpae when not feeding. The proboscis, technically called the **haustellum**, is actually made up of two half tubes tightly held together by a series of small hooks, rather like a zip fastener. These anatomical changes admirably fit the adult for its main objective, which is to find a mate and reproduce, the energy needed for egg or sperm production having been largely or wholly acquired by the caterpillar. Some females require additional nectar from flowers for egg production, and males do require energy to patrol suitable territories looking for a mate or driving away potential male suitors, and for this they need to suck nectar from flowers or, in some species, minerals from dirty water, sap from trees, or even salts from the urine of mammals, liquid bird droppings, and carrion. Haribal (1992) observed a skipper, the Purple Redeye (*Metapa purpurascens*), feeding on a dried bird dropping, onto which it deliberately squirted droplets of fluid from its abdomen before commencing to suck. After the male has passed its spermatophore into the female, it requires sodium salts to replace what it has lost, and this accounts for the prevalence of male butterflies of some species sucking at damp patches on the ground. This habit is known as 'puddling'. The author once counted a congregation of twenty-three Common Punch butterflies (*Dodona durga*) sucking from a small patch of damp sand in the Manga Valley. (*See also*, Plate 11, showing a congregation of Common Grass Yellow butterflies in Changa Manga Forest Plantation.)

When the butterfly first emerges from the chrysalis, which splits open along its back, its wings are damp and compressed, and it has to climb up a stem where the wings can hang freely and expand. The wings at this stage can be likened to an empty bag, and the butterfly is capable of pumping blood through the wing veins, which at the same time contract so that the two wing surfaces are drawn together as they expand. In insects, the 'blood' does not flow through special ducts as veins and arteries but generally throughout the body, and it does not carry oxygen but nutriments to the various internal organs, and is known as haemolymph. The wing 'veins' are actually channels through which both nerves and some haemolymph can pass, and are more correctly called nervures. As the wings dry and harden the nervures become stiff and rigid, acting as struts, and the adult is ready to fly. The average life span of most tropical butterflies at this stage may only be fifteen to twenty days according to Haribal (1992), although the *Vanessa* spp. often last for weeks, and those that hibernate over winter can live as long as seven or eight months.

The modified mouthparts beneath a butterfly's head, known as palpi (*see*, Fig. 8) have largely unknown sensory function, but are believed to be organs of touch. They are soft and usually very hairy, and can be used by the butterfly for grooming its forelegs and, it has also been postulated for cleaning their large compound eyes, though this function is usually performed by the epiphysis on the tips of their foreleg.

The surface area of the fore and hindwings of a butterfly can be mapped or described according to the position and number of wing veins and the general wing outline and shape. There are from ten to twelve veins, according to family, which give rigidity and strength to the wings. These are numbered in conventional sequence, and it will be seen from Figure 1 that, to give extra strength to the leading edge of the forewings, they run parallel and close to the **costa** and **dorsum**, whilst the intermediate veins radiate outwards from the oblong-shaped enclosed area known as the **cell**. Taking the outline of, say, the right forewing, we can describe each edge and corner by distinct names. Starting at the point where it leaves the body, known as the base, the leading edge is known as the **costa**, which leads to the pointed wing tip known as the **apex**. Proceeding down the outer edge, this area is known as the **termen**, or, just inside this, the **margin**; then down to the bottom right-hand corner, known as the **tornus**, and proceeding straight-lined towards the body (thorax), this wing edge is known as the **dorsum**. The hindwing bears these same names but, being slightly different in shape, the tornus is closer in towards the abdomen, and the dorsum is rather parallel to the sides of the abdomen. The terms used to denote certain areas are constantly mentioned in the following species descriptions. For this reason Figure 1 is inserted at the front of this book for easy reference. In butterflies the forewing always overlaps the hindwing and is held in place during flight by the presence of a lobe at the base of the hindwing known as the humeral lobe. This is another feature which distinguishes butterflies from most moths. The latter have a special wing-coupling organ known as the frenulum. Moths are very much more diverse in number of species than butterflies and, most being nocturnal in activity, are cryptically coloured on their upper wing surface, which helps to protect them during daylight when they are at rest. They also possess more complex antennae, often with side fronds like a fern, and they are capable of detecting from distances of several kilometres the scent produced by a female of the same species, even from a scattering of molecules of the pheromone she produces. Butterflies are also distinguished from moths by the presence of a thickened tip or club at the end of their antennae which moths lack, but their courtship behaviour suggests that they do not have such sensitive antennae. However, Haribal (1992), living in the Bombay area, records growing certain plants in pots on an upstairs verandah of her home which were the food plant of butterflies not generally found in that locality, yet females entered the verandah and attempted to lay eggs on these plants—a testimony to the sensory powers of their antennae.

The Flight of Butterflies

This is perhaps the most notable and interesting function of the adult butterfly. The arrangement of its internal organs, with those of respiration, digestion, and excretion being spread throughout the abdominal segments, allows the thorax to provide ample space for flight muscles, and, not surprisingly, there are two sets of relatively large muscles inside this area, one contracting the thorax to raise the spread wings and the other expanding the thorax to lower the wings. The direction and angle of the wings during flight are controlled by much smaller muscles located directly at the base of the wings. As has been described in the section dealing with the adult butterfly, the wings are given rigidity as well as some flexibility by the wing 'veins', similar in function to the struts inside an aeroplane wing. Because of their small size and small body weight in relation to wing surface area, there is a very favourable wing-loading ratio, with few of the problems which have to be overcome in producing flight amongst much heavier vertebrates such as birds and bats (Chiroptera). In fact, the insect has to contend more with problems associated with wind currents and air temperature than with induced lift from the difference in air speed between the upper and undersing surfaces, as in flying vertebrates. Recent research by a team of Cambridge scientists (Ellington *et al.*, in *Nature*, Vol. 384, 1996) has shown that because of the small surface area of butterflies' wings, they cannot produce the differential in air speed between upper and lower wing surfaces merely by wing shape, which produces lift in larger flying animals. However, their experiments with Sphingid moths in wind tunnels, using high-speed photography, showed that the shape of the forewing tip produces a vortex, analogous to a miniature whirlwind, through which air flows at an accelerated speed up along the wing surface to the apex, and that this greater air speed on the upper surface produces ten times the normal lift that such a small surface would otherwise create in flight.

The observer will soon notice that there are very variable flight characteristics between different butterfly species. Those with large, relatively heavy bodies in relation to their wing surface area generally being capable of much swifter flight. The Skippers particularly exemplify this principle. Their wings also beat up and down at a very much faster speed. Others, with relatively slender bodies and larger wing surface, are often not capable of such swift flight, but are much more efficient at travelling by gliding with wings held rigid, instead of by energy-expensive wing flapping. There is some need in flight to control or diminish air turbulence on the trailing edge of the wings, and particularly at the wing tips, as this produces drag and upsets the equilibrium of the flight path. Species which can glide with minimum wing flapping usually have rather rounded wing tips or forewing apex, a good example being the Sailers (*Neptis* spp.) and the Blackveins (*Aporia* spp.). Swift-flying species tend to have a more angular forewing apex, such as the larger Awl Skippers (*Badamia* spp.) and many Nymphalidae. The wings of many species, particularly the smaller ones, are bordered by a fringe of special hair-like scales known as cilia, which are often contrasting in colour from the main wing surface. These cilia are known to have a scent disseminating function in some Pierids, but must also help to reduce air turbulence in flight.

Unlike warm-blooded flying vertebrates such as birds and bats, insects can only fly when their muscles, controlling wing movements, reach an optimum temperature. Experiments by Feltwell (1986) showed that butterflies can only fly if their flight muscles have reached a temperature of 30° Centigrade. To achieve this they frequently need to bask with spread wings in the sun. For the same reasons, the flight behaviour of different species is heavily dependant upon air temperature, and the majority of species only fly when the sun is shining and the air is warm. In alpine regions the night temperatures drop rapidly, and butterflies adapted to such conditions need to bask in the early morning and to warm up their flight muscles before they can fly efficiently. In high alpine regions, many species of Apollos and Fritillaries living on north-facing slopes do not take to the wing until about 11 a.m. If disturbed in cold, cloudy conditions, many are only capable of a short, weak, floppy flight or can be picked up by the human hand, quite uncharacteristic of their normal ability on a sunny day. By contrast, a few species confined to a warm tropical climate are specially adapted to fly only in less bright sunlight conditions, like some Skippers and the Evening Brown (*Melanitis leda*), which only comes out at dusk. This butterfly with conspicuous ocelli on its upper forewings is well camouflaged like a dried leaf when its wings are closed, and only flies for very short distances, settling frequently on the ground, where it becomes hard to see.

In open, windswept areas most species, such as the Meadow Browns (*Maniola* spp.) and many alpine Fritillaries, such as Jerdon's Silverspot (*Argynnis jerdoni*), only fly close to the ground surface and avoid exposure to uncontrollable air currents by keeping close the low-growing vegetation, where friction reduces the wind velocity. I have noticed that, when flushed in windy conditions or when foraging naturally, such high-elevation species as the Common Blue Apollo (*Parnassius hardwickei*) and the High Brown Silverspot (*Fabriciana hegemone*) tend always to fly sideways with the wind and very close to the ground surface, but at the same time these species are strong and powerful flyers, presumably with better control over their flight direction.

Remembering that the main function of adult butterflies is to find a mate or to lay eggs, it is also noteworthy that there is often a very marked difference between the flight behaviour of males and females of the same species. The males tend to settle much more frequently on some prominent place such as the tip of a tree branch, from where they can look out for any butterfly which comes within sight range. If a male, it can be chased away, and if a female, it can be pursued in courtship. Females tend to spend much more time in the proximity of their larval foodplant, and in some species which lay their eggs on trees, such as the Walnut Blue (*Chaetoprocta odata*), seldom come near the ground, preferring to sail around tall walnut-tree tops, whereas the males frequently visit low-growing shrubs and flowers.

The Common Mime (*Chilasa clytia*) has a peculiarly slow-sailing flight with very little wing flapping. This is very efficient in terms of energy needs, consequently it seldom needs to settle, either to rest or to bask to absorb heat, and characteristically remains on the wing, circling lazily around treetops, for hours on end. If, however, it is frightened, or sees a potential predator, it is capable of very swift evasive flight. The Common Map (*Cyrestis thyodamus*) is another species with a very peculiar, slow, sailing flight, and this species consequently seldom settles or needs to bask, but when it does so, it plunges inside bushes and settles, often on the underside of a leaf.

There is a difference also between the feeding behaviour, of butterfly species. The Swallowtails particularly are noted for their preference to suck nectar while hovering and without alighting on a flower. Consequently they tend to visit only larger flowers borne at the end of stems and easily accessible. Such flowers often have rather deep corollas, with the nectar deep down within the flower, but the larger size and longer proboscis of Swallowtails enables them to exploit such flowers. Many Nymphalidae, such as the Tortoiseshells and Admirals, when foraging, settle on the flower or flowering raceme and crawl over it quite rapidly, visiting one flower after another, with wings partially open. By settling and using their partly-open wings for balance, they can exploit very small flowers such as the Compositae family, which Swallowtails usually avoid. Many of the small Lycaenidae (Blues) settle only on low-growing flowers, and invariably with their wings closed, before feeding, but make very frequent short foraging flights between feeding.

A different category of flight behaviour is exhibited by the noxious-tasting Danaines such as the Plain Tiger (*Danais chrysippus*). All these butterflies have a particularly slow, lazy-looking, sailing type of flight, but this gives potential predators ample time to recognize them and so avoid an unpleasant-tasting encounter.

The Systematics or Classification of Butterflies

The different species of all life forms can be grouped together in logical sequences of apparent relatedness, as well as evolutionary development, and this is the study of Phylogeneticists. Placing the many species of butterflies in some sort of logical order of supposed evolutionary development not only helps the biologist to know their expected characters and life history, but also highlights their differences. This is the basis for systematics.

Differentiation between butterfly families and sub-families depends not only upon easily recognized shared features such as egg shape, larval appearance, and the appearance of the adult butterfly, but also, at the species and even generic level, largely depends on such features as small differences in the palpi (as in Hesperiidae), the origin or place of insertion of wing venation, shape of antennae tips, and the copulatory organs in the case of males. The structure of the various modified anal appendages known as genitalia in the male butterfly do vary greatly in shape between species, and this can easily be seen under low-powered magnification after dissection.

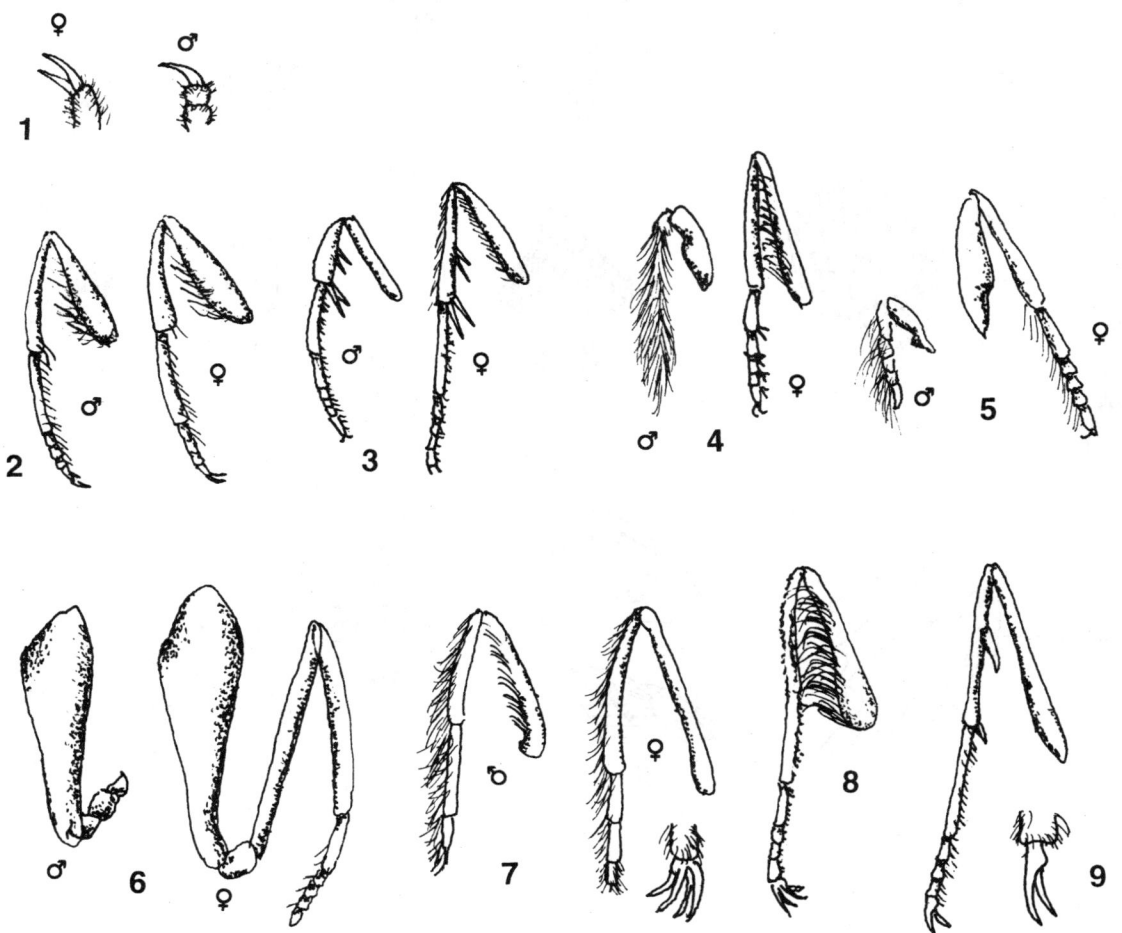

Fig. 6 Forelegs (Prothorax Pair) of Butterfly Families

Top Row (inset above):
1. Foreleg tip of male and female Lycaenidae
2. *Lycaenidae*
3. *Tagiades* (Hesperiidae)
4. *Libytheinae* (Beaks)
5. *Riodinidae* (Punches)

Bottom Row:
6. *Ypthima* (Satyrinae)
7. *Mycalesis* (Satyrinae)
8. *Pieris* (Whites)
9. *Papilio* (Swallowtails).

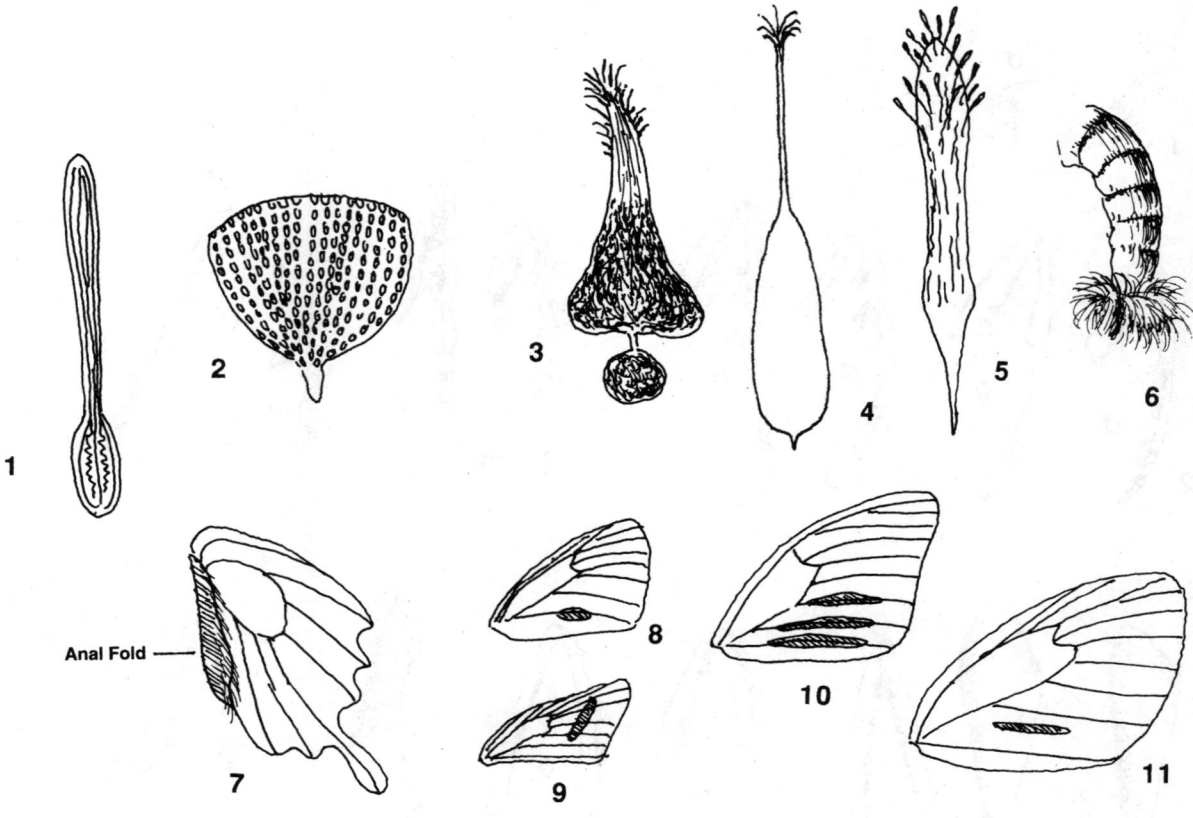

Fig. 7 Scent Scales (Androconia) and location of Brands or Stigmata
Not drawn to scale

1. *Azanus ubaldus*
2. *Nacaduba nora*
3. *Delias eucharis*
4. *Pseudochazara*
5. Unidentified Satyrinae
6. Hair Pencil on tail tip of *Eupole core*
7. Brand on *Graphium* sp. (Swallowtail)
8. (Top) Brand on *Mycalesis* sp. (Saturniane)
9. (Bottom) Brand on *Telocota* (Hesperiidae)
10. *Argynnis childreni* (Nymphalidae)
11. *Eupole core* (Daininae)

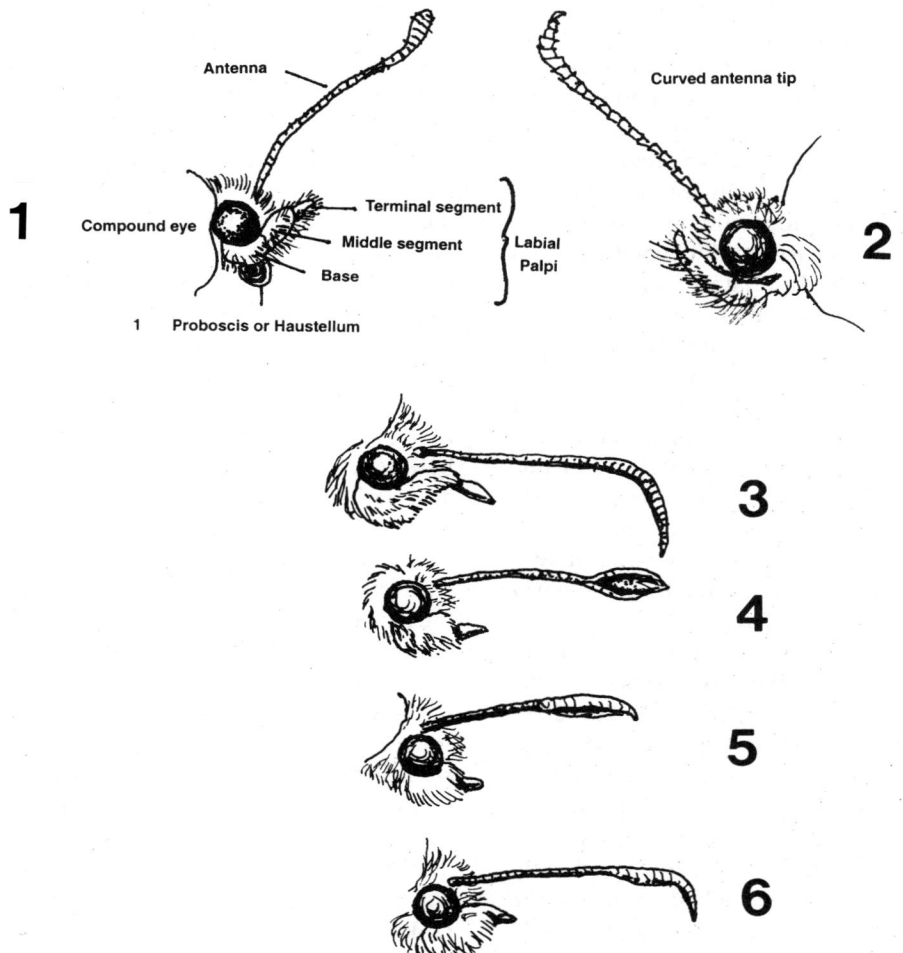

Fig. 8 Antennae and Palpi

1. Head of *Pieris* species, showing parts of palpi
2. Head of *Papilio* sp.
3. Showing antenna and palpi of *Bibasis* and *Badamia* spp.
4. Head of *Taractrocera* sp.
5. Head of *Gegenes* sp.
6. Head of *Baoris* sp.

Their peculiar shape provides the equivalent of a lock and key combination between the sexes of the same species. This difference is an evolutionary development to inhibit hybridization, and is a feature also shared by some higher vertebrate animals such as small Microtine rodents. In such a diverse and large family as the Lycaenidae, particularly those males with largely blue upper wing surface, the shape of the male genitalia are always clearly different as between species, especially sibling species. The same holds true for the many species of Hesperiidae, whose classification at the species level has largely been based on the shape of the male genitalia. The main families within the Hesperiidae are partly arranged according to the distinctive shape of their antennae tips, coupled with their palpi (*see*, Fig. 8). The antennae of Papilionidae are always curved downwards at the tip, and with a thickened club tip. The antennae of the closely-related Pieridae widen gradually, like a wedge, into the thickened tip. Those of the Hesperiidae are more gradually thickened, without rounded club-like ends, and often with the extreme tip sharply pointed (*see*, Fig. 7). Wing colour has in a few cases been the main basis for differentiating between two sibling species, but this can be a variable trait even within the same species and is probably the least reliable taxonomic feature, though the presence and position in the wing pattern of various coloured spots and ocelli are clearly significant in species separation.

Another distinctive feature separating the main families of butterflies is in the development of the first pair of legs or forelegs (*see*, Fig. 6). These are vestigial or non-functional in some families—for example, in the Danainae the forelegs are degenerate in both sexes and useless for walking. In the Nymphalidae, and especially the Satyrinae (Browns), the forelegs are also degenerate; they are thickly covered with hairs and carried tucked up under the 'chin', so that this family are sometimes called 'Brush-footed Butterflies'. By contrast, in the Riodininae (Punches and Judys) the female has well-developed forelegs, but those of the male are degenerate and useless for walking, whilst in the Lycaenidae the forelegs in both sexes are functional and fully formed, though often larger and better developed in the female. In fact, within most families the forelegs of the female are larger or possess more segments than in the male, even if non-functional for walking, as they play an important role in detecting suitable sites for egg laying. In some species of adult butterfly (e.g., Parnassiinae), the claw at the foreleg tip is bi-fid (Ackery, 1975), and in female Lycaenidae, but not the males, the forelegs end in two claws. In the Pieridae, the forelegs in both sexes are fully formed and used for walking, and the tips end in four claws. In the Papilionidae, the first pair of forelegs are also fully formed, which is one of the reasons that these two families are considered to be more closely related to each other than to others.

Scientific nomenclature follows internationally agreed rules and conventions, so that every user, whatever his native language and script, can recognize the same species, yet in the case of butterflies there seems to be a bewildering variety of alternative generic names, especially among the Satyrinae. In this book the author has not necessarily followed conventional taxonomic rules, by listing the earliest accepted nomenclature, but has tried to give those most often used in earlier references to the butterflies of the subcontinent. These synonyms may no longer be accepted today, because advances in our knowledge have enabled taxonomists to arrange species in more realistic generic order. Also more significantly, advances in molecular genetics, and the study of mitochondrial tissue, enables biologists not only to determine relatedness but often to discern differences which have been interpreted as warranting the raising of sub-species to full specific rank.

Chapter 3
Senses and Sensibility

Despite being much less complex organisms than higher forms of animal life, butterflies possess all of the senses that humans have with the exception of hearing in the conventional human sense. It is a curious fact that, within the class of insects, closely-related moths (Heterocera) have been shown by dissection to possesses well-developed hearing systems in their head (Ford, 1945), but no equivalent has been detected in butterflies. They can, however, detect vibrations caused by sound waves by other sensory means, which in physical terms is equivalent to hearing. Butterflies have vision, smell, tactile or sensory feeling, and taste or gustatory sensibility. But in some cases these senses are so different in specialization from that of humans that we have difficulty in understanding or appreciating their mechanism or effectiveness. Reference to the location and form of these sensory organs has already been made in chapter one, but to understand the behaviour and ecology of butterflies, this chapter will attempt to enlarge upon their capabilities.

Vision or Sight

It has long been recognized that adult butterflies possess well-developed colour vision, enabling them to preferentially select nectar-bearing flowers according to their colour as well as scent (Eltringham, 1933), and to recognize the colour and wing pattern of other butterflies, pursuing conspecifics, but shortly after approach, leaving other species alone.

The feature which has attracted mankind's most widespread interest and admiration for butterflies lies in the bright colours and patterns on their wings. These colours must serve two functions. They must as far as possible serve to identify the wearer for sexual selection and recognition of conspecifics, and they must help as far as possible to camouflage (crypsis) and protect from visually hunting predators. As we shall see, they may be conspicuously patterned, enabling a potential predator to easily recognize their distastefulness. These apparently contradictory forces acting on succeeding generations have lead to the development of more striking colours, and particularly of iridescent colours, in male butterflies, and to more diffused or drab colours in females where there is such sexual dimorphism. In the same manner, the underside of their wings have tended to produce cryptic effects, for example, the wonderful dried leaf-like appearance of the Common Beak (*Libythia lepita*), or the Evening Brown *(Melanitis leda)* when they settle and close their wings. The selective evolutionary pressures on males to develop different bright colours and wing patterns, which enable them quickly to recognize rival males of the same species and which must also serve to attract females, are therefore understandable. Females, by contrast, with smaller eyes and less well-developed vision, may appear more similar to each other, even between different species. This is particularly evident in the Lycaenidae. Another piece of evidence for such selective evolutionary development lies in the identical underwing pattern needed for crypsis, which is a valuable key to identification in the Lycaenidae, and which is displayed by both sexes even in some polymorphous species. It is possible to draw an evolutionary comparison with the relative homogeneity in plumage pattern of female Birds of Paradise (Paradisaeidae), even between different genera, whilst the males exhibit such a striking variation in spectacular plumage colour and shape.

While flying, the butterfly can flash colour signals, particularly in the UV band, to other butterflies from a considerable distance, but as soon as it settles and closes its wings it becomes hard to see. Thus there is evolutionary selection for bright colours in males, but not to the same extent in females, which experiments have now shown do not possess such acute visual powers as their male counterparts (Silberglied, in *Biology of Butterflies*, 1984). As will be seen from this chapter, males use colour vision largely to identify other males and then avoid conflict, or to maintain their territory by chasing away competitor males, in addition to finding females from some distance. The later stages of sexual recognition and courtship depend not on vision but on olfactory messages. Vision is reinforced by other sensory perceptions, as is the case with higher life forms.

As has been mentioned in Chapter 1, the eyes of butterflies are not only very large and spherical in shape, but they comprise a series of compound lenses or visual cones such that they see a multi-faceted image. They are also capable of detecting movement over very small degrees of arc, as well as seeing movement across a very wide field. Anyone who tries to approach closely to some of the shyer species will be aware of how difficult it is not to seen. Experiments have demonstrated that, though they have poor spatial resolution and cannot see the whole shape or bulk of objects with the clarity of human vision (Silberglied, *bp. cit.*), this is compensated for by their high sensitivity to motion, especially to flicker. Their eyes are able to see colour over a much broader spectral range than any other animal form, from red to ultraviolet. It is their UV vision which makes them particularly well able to see the iridescent colours of other males, and to discern the flicker of wings from small objects such as fast-flying butterflies a considerable distance away. Ultra-violet light cannot be seen with the human eye, but research has shown that butterfly wings exhibit a striking variety of UV patterns, especially in groups or families that appear concolorous to our eyes, such as the white Pierids and brown Satyrids. They thus pass 'secret messages' to other butterflies in the sense that visually orientated predators cannot detect. It has been demonstrated by Silberglied and Taylor (1978) that Clouded Yellows (*Colias* spp.) can detect very small differences between the under hindwing greenish-yellow of males and yellowish-green of females. Other experiments, where male *Colias* spp. were dyed with red, blue, and even black, showed that females of *C. Philodice* would mate indiscriminately with all these colours but that one species *C. eurytheme* was sensitive only to the Ultra-violet signals of one particular male species which, if masked, did not successfully mate with conspecific females.

In conclusion, we know that egg-laying females are able to identify host plants from leaf shape, and that nectar-seeking butterflies can discriminate between colours. Experiments by Eltringham (1933) demonstrated that *Vanessa* spp. (Admirals and Tortoiseshells) preferred yellow and violet coloured flowers, whilst Pierids (Whites, such as Cabbage Whites) were more attracted to red and purple flowers, all these colours being preferable to white flowers. In mate-seeking and courtship, colour plays a role of varying importance between different species, with olfactory cues being in most cases essential to the successful completion of courtship.

Ultraviolet light reflections are mainly developed by males and mainly on the upper wing surface, and these can be important recognition signals for both females and males. The Danaid Eggfly Butterfly (*Hypolimnas missipus*) is well known for the prominent circular white wing patches of the male (lacking in the female). These have been shown to reflect intense UV light from both the white and the black areas, but especially in the intermediate zone between the two, which shows a purple sheen to our human vision. Males spend a lot of their time perched on the top of bushes or leaves with their wings held horizontally, exposing the UV pattern to the maximum. This is probably in order to send important signals to other males and females, as well as purely basking to absorb heat for the flight muscles.

Smell or Olfactory Sense and its Role in Courtship

It was long ago recognized that butterflies could produce scents which were sexually attractive to females of the same species, (DeGeer, 1752, quoted in Boppré, in *Biology of Butterflies*, 1984), and that butterflies possessed differently-shaped scales on their wings which were variable both within and between different species. These specialized scales are called androconia (*see*, Fig. 5). These have been described in Chapter 1. Their existence reflects the importance of scent production in the biology of butterflies. Only males possess androconia, and the 'sexual' or epigamic scents they disseminate are used by males for recognizing other males and females, for inhibiting flight escape, and for encouraging acceptance by females. As has been mentioned in Chapter 1, males of Danaid species also possess a brush of special modified scales at the tail tip which is extrusible and is used for disseminating volatile scents. By contrast, many butterfly species possess diffuse smells not confined to one sex, and these are derived from chemicals sequestered from the larval food plants. Such odours are called aposematic and are particularly well developed in some families like the Danainae and Acreaeidae (Costers). Their function is to warn potential predators of their unpalatability or even poisonous taste. Reverting to the function of androconia, pioneering studies by Tinbergen (1953) showed that a male Rockbrown (*Hipparchia* spp.) in the final stages of courtship approached the female on the ground, walking up to her face to face, and then clasped the tips of her antennae between his forewings which were then rubbed together, releasing powerful scents from the androconia which helped to make her ready to mate. Courtship in the weak-flying Psyche (*Leptosia nina*), a Pierid, involves similar behaviour, with the two sexes approaching each other on a leaf, face to face, with the male flapping his wings over the female's antennae. Similar studies by Barth (1944), and later by Magnus (1950), on the powerful-flying Silver-washed Fritillary (*Argynnis paphi*) showed that courtship by a flight dance was an essential preliminary, with the males swooping under and over the flying female, during which, it is presumed, he signals to the female from his forewing androconial wing folds. Then both sexes settle on the ground and approach head to head as in *Hipparchia* spp., with the male flapping his wings over the female's antennae tips. Courtship in some of the Danainae is more complex, with the male in rising circling flight regularly dispersing scents over the female's head from the hair pencil at his tail tip; if the female is receptive, she alights on the ground or a leaf surface, and then he approaches her still in flight and again alternately protrudes and expands his hair-pencils around her head. If ready to mate, she closes her wings and allows him to settle and copulate. Feltwell (1986) notes that when a female settled on the ground is approached by a courting male, she raises her abdomen as a signal if she is not ready to mate or has already been fertilized. It is important to note that the antennae are receptive to chemical scent messages, and that androconia may be located all over a wing surface and not just in localized brand patches, so that rubbing the wings together may make the whole wing surface disperse the pheromones which are intra-specific chemical signals. In the Danainae the pocket or flap on the hindwing undersurface may help to protect the male's aromatic scents from volatilizing, or it may contain chemicals which can combine with those produced from the hair pencil to result in more potent pheromones. Boppré, (*op. cit.*) discusses this possibility and records that *Danais chrysippus* regularly makes mechanical contact between its hindwing pockets and the abdominal hair-pencil tip. It is also important to remember that olfactory cues are used in conjunction with visual ones in varying degree between different butterfly species, and that our understanding of these phenomena is still imperfect. Before considering the remaining senses of butterflies, mention must be made of behaviour exhibited by some Danainae, especially the Common Crow (*Euploea core*). Sevastopulo (1944) has described encountering flocks of this butterfly sailing around in a forest clearing with their hair pencils everted until the whole air was scented with their peculiar musky aroma, whilst there did not appear to be any females in this congregation. We still do not have a satisfactory explanation for such behaviour.

Tactile or Sense of Touch

In butterflies, and indeed many other insect orders, the body is covered with flexible hairs called setae which are actually different from mammal hairs, but are connected with sensory nerve cells at their base. They convey a very discerning perception of touch to the butterfly, and are most conspicuous around the wing bases and along the thorax and abdomen. E.B. Ford (1945) has described how a butterfly settled on a blade of grass will not be disturbed by the sweeping touch on its body of adjacent grass blades in a breeze, yet the lightest of touches from an unfamiliar body such as a slender rod will at once put it to flight.

Female butterflies have been observed, when searching for the right or optimum ovipositor sites, settling on leaves and drumming on the surface with their front pair of forelegs. For example, neotropical Satyrids (Singer, in *Biology of Butterflies*, 1984), and many Oriental spp. of Nymphalidae (Bell, 1919) do this when searching for the right larval foodplant. However, the sensory capacity of female butterflies' forelegs appears to involve both tactile and something equivalent to gustatory ability, as there is a growing body of evidence that the forelegs and even other thoracic legs possess taste discrimination abilities. Chew and Robbins (*op. cit.*) refer to specialized receptors in the forelegs of *Pieris brassicae* which are sensitive to glucosinolates (mustard oils) contained in certain Cruciferous plants which play an important role in the female's detection of suitable egg-laying sites. Experiments by Ford (*op. cit.*) were conducted with captive Nymphalidae, in which the last pair of thoracic legs was immersed in apple juice; this immediately elicited a proboscis uncoiling reaction. When the butterflies were exposed only to the scent of the apple juice, without immersing the hind legs, there was only occasional similar reaction, the message received through the tips of their tarsi being much stronger in eliciting feeding response.

Taste or Gustatory Sense

This has already been discussed above in reference to the peculiar chemo-sensory capacity of the adult insect tarsi. Studies backed up by experimental evidence have shown that butterflies, like nearly all higher animals, have four basic taste senses (Brower, in *Biology of Butterflies*, 1984). They can detect substances which are sweet and therefore likely to be palatable and energy-supplying, substances that are salty and therefore capable of supplying minerals which might be necessary for ionic balance, and substances that are sour—experiments have shown that liquids which have sourness generally elicit aversion. Finally, they can taste bitterness, which invariably leads to aversion. It is a human experience also that nearly all bitter-tasting substances are poisonous, and similar taste responses have been elicited from birds in laboratory and field experiments. This shared sensory perception is an important factor in the development of chemical defences against vertebrate predators, a widely occurring phenomenon which will be discussed in chapter four.

Food Seeking

In the larval stage of a butterfly's development, as we have seen, an efficient feeding physiology enables them to store most, if not all of the energy requirements for females to produce eggs and for males to produce spermatophore packages, so that the adult butterfly can devote its energies to finding a mate and to reproduction.

For the female, the finding of a suitable host plant for depositing her eggs is crucial. We have seen that females utilize visual, chemo-sensory, and especially olfactory cues for searching out the best host plants. Species whose larvae can utilize a wide variety of host plants are called polyphagous, and those with specialist requirements confined to one suitable host plant species are called monophagous. Obviously there are intermediate relationships, with many species of butterfly larvae being able to feed successfully on a variety of plant species belonging to one particular family; these are termed oligophagous (Courtney, 1984). The distribution and abundance of certain butterflies must be dependent on the availability of such host plants. If the host plant is stringent in its habitat requirements, then the distribution of the butterfly which utilizes that plant will tend to be erratic and limited in abundance. Studies by Michael Singer (1984) on host plant/butterfly relationships in the tropics show that many female butterflies whose larval host are grasses will search for the right place where the host grass species is growing, but will quite frequently lay their eggs on some other plant or a stick nearby. Other studies show that some species preferentially select plants which are growing in clumps rather than isolated individuals of the right host plant. Singer (*op. cit.*) suggests, that if the larva eats all the leaves of its original host plant, it can migrate to another nearby plant without difficulty, whereas if that plant is isolated, the risks of predation to the larva as it searches for another foodplant are greater. Many studies have also shown that butterflies make ovipositing mistakes, laying their egg or eggs on unsuitable host plants. It has been postulated that, with many short-lived species of adult butterfly, especially those living in regions of uncertain weather conditions, there is a time constraint in finding suitable ovipositing sites. This may partly account for some eggs being laid on less suitable host plants.

Mention has earlier been made of the relationship between the larvae of certain Lycaenidae species and ants, which protect the larvae from predators in return for rewards of a sweet liquid excreted by the larva from a dorsal gland. Female butterflies with this larval relationship have been seen to select those sites where the suitable ant species are present and active, rather than choosing the right host plant. In many instances of such Lycaenidae, for example the Babul Blues (*Azanus* spp.), the host plants are the flowers of *Acacia* spp., and the ants will even carry the newly-hatched larva onto an appropriate *Acacia* twig. Some of the Pierrots (*Tarucus* spp.) are closely attended by tree ants, which even make shelters for the small larvae during moulting. Consequently the females of these species will lay their eggs on a variety of plants such as 'Ber' trees, *Zizyphus* spp., and Siris (*Albizzia* spp.), taking more care to choose sites where the tree ants are present.

Some species of butterfly are noted for the wide range of different foodplants which are suitable and used by the larvae. Quite often such species are also migratory, with a tendency in the adult stage to travel in a continuous direction from one brood to the next. It is obviously an advantage in such species that a variety of plants from different families are suitable for egg-laying. The Painted Lady (*Vanessa cardui*) is the classic example of this polyphagous character the Pea Blue (*Lampides boeticus*) is another. It is no coincidence that both these species are strongly migratory.

Mention was made earlier of the adult butterfly's need to maintain ionic balance, particularly to replace lost sodium salts in the case of males, which are capable of mating with a number of different females. Each time they do so and pass a spermatophore packet to the female, they lose such salts, and to replace these they have the habit of frequently seeking out damp places and seepage, tree sap, and the faeces or urine of large vertebrates. This habit is known as puddling, and often the insect can be seen to suck up moisture and almost immediately excrete droplets of water from the tip of its abdomen. It is presumed that the butterfly can rapidly extract soluble salts from this sucked-up moisture and void unwanted water. Often quite large numbers of exclusively male specimens can be seen gathered at a suitable damp spot puddling together, an important behavioural adaptation because such male species, when seeking females and courting, will not tolerate the approach of other conspecific males.

Chapter 4
Co-evolutionary Competition—Predators and Defence

In general terms, the struggle for survival in both the plant and animal kingdoms has led over the course of millenniums to the extraordinary diversity in occurrence, appearance, and behaviour of living species, and this is one of the more fascinating attributes shared by the family of butterflies, one which has led biologists into many fields of research.

Plants, in order to produce the optimum amounts of fruit and seed, have to defend themselves from herbivores, whether vertebrate or insect. Herbivores again, if they are to reproduce, have to protect themselves from predators which are carnivorous, whether vertebrates or arthropods, as well as from pathogens which might attack their internal physiology. Butterflies have developed so many defences against predators that it is fair to assume that the latter have exerted a profound influence on their ecological and evolutionary development.

This chapter can only look briefly at this co-evolutionary 'arms race' between plants, the source of all organic energy, and predators versus prey.

In the plant kingdom the development of thorns, spines, hairy leaves, tough, silicate-embedded cell structure, and, most widely, the manufacture of toxic chemicals within the leaf tissue, all help to make such edible parts difficult to ingest or distasteful, and therefore avoided by herbivores. Research has shown that the larvae of many butterflies can not only develop immunity to chemical toxins in leaf tissue, but that they are capable of storing or sequestering such chemicals in their body tissues, with the end result that they themselves become distasteful to potential predators.

The reader may ask, what of the interdependence of many plants for pollination by insects? This complex question is answered in part by the diversity of insect species which are specialist feeders on pollen or nectar, which unwittingly act as pollinators, and which do not eat the vegetative parts. A most interesting study in South Africa throws some light on this subject. Compared with other insects such as hymenoptera, butterflies are not usually important pollinating agents, but in the South African study (Johnson, 1994), a Satyrinae species of butterfly, *Meneris tulbaghia*, is the main pollinator of an orchid, *Disa ferruginea*, which actually contains no nectar reward, but which commonly grows alongside stands of a nectar-producing gladiolus (*Tritoniopsis triticea*, family Liliaceae), as well as another nectar-producing plant commonly called the Red Hot Poker (*Kniphofia uvaria*, family Asphodelaceae). It was found that the orchid produced either orange-coloured flowers closely resembling *Kniphofia,* or scarlet flowers closely resembling *Tritoniopsis*, in different regions and that fruit setting was better where these plant species grew in mixed stands, where the butterfly was unable to discriminate between the unrelated but look-alike flowers. This not only reveals the importance of colour and shape to nectar-seeking butterflies, but in Johnson's opinion was a nice example of Batesian mimicry on the part of the orchid. Batesian mimicry will be explained later in this chapter.

Another example of pollination by butterflies is shown by the sub-tropical family of mainly perennial woody shrubs belonging to the genus *Clerodendron* in the plant family Verbenaceae. A study in Andhra Pradesh, Central India, (Reddy and Reddi, 1995) showed that the shrub *Clerodendron infortunatum,* which produces white flowers with long-protruding stamens and a deep corolla, was pollinated by regular visits from five species of Swallowtails. These species all possessed relatively long proboscises and typically sought nectar from large, deep-corolla flowers which were accessible by flight hovering. The visiting Papilionids grasped the flower with their legs while hovering and, in doing so, their fluttering wings stroked the anthers, picking up pollen grains, which were then transferred to the female stigma of other flowers. In botanical terms, the flowers anthesed by slowly opening to protrude the male anthers first, which later curled outwards and backwards exposing, on the second morning, the female stigma, so that butterflies visiting flowers at different stages would be forced to transfer the adhering pollen onto the stigma of flowers which opened a day earlier. In a more recent study, (Reddy and Reddi, 1996) on the pollination of *Duranta repens* (family Verbenaceae), it was also found that butterflies were attracted to the blue flowers of this woody shrub, which produced nectar for only one day after flower opening. It was found in the study area, Andhra Pradesh, that seventeen species of butterflies visited this plant and that all of them picked up pollen grains in the process, and that Lepidoptera comprised over 50 per cent of pollinating insects, with bees about 34 per cent, and wasps only 1 per cent. *Duranta*, though of South American origin, is a widely planted and most popular hedge plant in Pakistan.

Now let us consider the dangers from predators faced by butterflies at all stages of their life cycle, and some of the protective strategies which they have evolved.

The Egg

Though we have seen in Chapter 1 that the eggs of butterflies are minute, there are Micro-ichneumonidae wasps which lay their eggs inside those of butterflies; their larvae feed on the egg yolk, pupating inside the egg and hatching out as tiny wasps from the dead and empty eggshell. These Ichneumon wasps actually follow females searching for ovipositor sites and can account for a major part of the butterflies' egg production. Besides wasps, there are mites (*Acari* spp.) which feed on butterfly eggs (Ehrlich, in *Biology of Butterflies*, 1984), and some insects which search out and eat butterfly eggs. For example, cockroaches have been observed doing this, and some birds (Brower, in *Biology of Butterflies*, 1984). There is evidence that some butterfly eggs may be protected by containing poisonous chemicals within their tissue, especially those species which lay their eggs in clusters and produce brightly-coloured eggs (Chew and Robbins, in *Biology of Butterflies*, 1984). The eggs of the Plain Tiger (*Danaus plexippus*) have been shown to contain toxins distasteful to other insect predators (Brower, *op. cit.*), as also the eggs of *Pieris brassicae* and *P. rapae*.

The Larva

Two very large families within the Hymenoptera are potential predators upon butterflies, being parasitoids, and these are Ichneumon wasps and Braconid wasps. Also Chalcid wasps and Tachinid flies are in some of their species, parasitoids of caterpillars. By definition a parasitoid eventually kills its host, in contrast to the normal life cycle of parasites, which merely feed upon their host but do not kill it. Chalcid wasps (e.g. *Cleonymidae* spp., super-family Chalcidoidea) eventually crawl out of the dying caterpillar and pupate as tiny white cottony cocoons alongside its body. Ichneumonidae pupate inside the caterpillar or, in some species, eventually pupate inside the butterfly pupa. Tachinid flies (thickset, hairy-bodied Diptera), fall to the ground when fully grown and pupate there, away from the now dead caterpillar. To add to this depressing list of parasitoids there are Mason wasps (Eumenidae), which paralyse caterpillars with their sting and then provision their eggs with the caterpillar as food for the larva when it hatches, and Social wasps, such as *Vespa* spp., and Paper wasps, *Polistes* spp. which masticate caterpillars and then provision their grubs with this semi-digested food. Many other arthropods which are not parasitoids, such as spiders and praying mantises, attack and eat caterpillars. Nocturnal arthropods tend to eat early instars or small caterpillars, while day-flying insects such as Hymenoptera attack larger, later instars. A Tachinid fly has been recorded (Bell, *op. cit.,* in many life history accounts) as parasitoiding whole colonies of the Indian Tortoiseshell (*Vanessa cachmiriensis*). *Apanteles* spp. of Braconidae are specialist parasitoids on Large White Cabbage Butterflies (*Pieris brassicae*). All these hymenopterous insects lay their eggs inside the body of the young caterpillar and, after hatching feed, only on the soft fatty tissues, avoiding the vital organs, which enables the caterpillar host to continue surviving.

There are a variety of vertebrate predators which readily take caterpillars when available and on an opportunistic basis. Among these may be mentioned lizards, tree frogs, birds, and small arboreal rodents and insectivores. Numerous studies of foliage-foraging bird species have shown that they consume an enormous number of caterpillars, and the Tits (*Paridae* spp.) and Warblers (*Phylloscopus* and *Sylvia* spp.) specialize in such food hunting.

What are the defences and devices used by caterpillars to protect themselves?

Many caterpillars are coloured green like the leaves upon which they feed, and, when not foraging, lie on the underside of leaves or along the midrib, where they are difficult to see. Colonial-feeding caterpillars often construct protective silk webs and, if disturbed, can jerk their bodies in a way that startles some predators. Examples of colonial larvae with these habits are the Indian Tortoiseshell (*Vanessa cashmiriensis*) and the Large Tortoiseshell (*Vanessa xanthomelas*). The best known example of camouflage is that provided by the larva of the Common Baron (*Euthalia aconthea*), which, besides being leaf green in body colour, has a paler longitudinal line down its back and paler yellowish-green branched spines angled out from the sides of its body. It always takes up a position along the midrib of a leaf, and the paler branched spines exactly blend in with the leaf venation. As has already been indicated, many caterpillars can sequester toxic chemicals from their host plant and so become distasteful to predators (*see*, for example, Brower, in *Biology of Butterflies*, 1984). This is particularly so in the case of many Pieridae which feed on Cruciferae, particularly *Brassica* spp., containing mustard oil glucosides, which are irritant to the digestion. The main larval host plants of the desert-loving Arabs (*Colotis* spp.) are Capparidaceae (False Capers), which contain irritant glucosinolutes (Brower, *op. cit.*) Morshead (1925), describing a mass movement of the Himalayan Blackvein, another noxious-tasting species, noted that it was not attacked by any birds despite appearing in such huge congregations. The Danainae feed on plants from the family Asclepiadaceae (Milkweed spp.), which plants contain cardenolides which cause vomiting, and may effect the heart. Many Danainae also feed on plants of the Apocynaceae (Oleander) family which contain pyrrolizidine alkaloids (Rothschild *et al.*, 1975), and the Nymphalid genus *Hypolimnas*, which includes the Danaid Eggfly and Great Eggfly, feeds on Moraceae (Fig family) and Asclepiadaceae families, both of which contain cardenolides. Many Papilionidae (Swallowtails), such as the Common Windmill (*Atrophaneura polyeuctes*) and Common Rose (*Pachliopta aristolochiae*), feed on plants of the family Aristolochiaceae (Birthwort vines), which contain aristolochic acids distasteful to vertebrate predators. The Acraenae (Costers) feed on Passifloraceae (Passion Flower vines) which sometimes contain cyanogenic glycosides. Other plant families which may contain poisons include the Euphorbiaceae which contain carcinogens (Marsh, Rothschild, and Evans, in *Biology of Butterflies*, 1984).

Besides the ability in some species to sequester toxic substances ingested from their foodplant, the Swallowtail larvae (*Papilio* spp.) are characterized by the possession of an organ known as the osmeterium, which is a brilliantly-coloured forked tentacle possessing strong aromatic odours. This is normally contained in a prothoracic pouch on the dorsum behind the neck, but can be extruded suddenly when the larva is startled, and this may deter many potential insect predators. Many Nymphalidae larvae, such as the Pansies (*Junonia* spp.), possess branched spines along their body and, when alarmed, will twist up their fore and hind body segments, presenting a dense armoury of bristles which can also deter some predators. Finally, many larvae are secretive and feed only at night, when day-hunting predators are absent. The delicate larvae of many Skippers (Hesperiidae) commonly forage only at night; during the day they protect themselves inside a shelter constructed from a folded-over segment of leaf firmly held together with their silk.

The Pupa

Though it is relatively immobile, the pupa suffers comparatively less predation than any of the other life stages of the butterfly. Provided with a tough leathery skin, resistant to the ovipositor organ of many insects, and with a remarkable capacity to pupate on a background which blends with its colour, the pupa or chrysalis depends mostly on escaping attention by camouflage. However, there are plenty of tree surface foraging birds such as Tree Creepers (*Certhia* spp.), Woodpeckers (family Picidae), and Nuthatches (*Sitta* spp.) which specialize in searching out cocoons and pupae on tree bark and even rock crevices (for example, *Sitta tephronota*, the Rock Nuthatch, common in Balochistan).

Pupae of some families, for example the Riodinidae and some Lycaenidae, are capable of producing audible or ultra-sonic sounds from stridulation of organs in their 11th and 12th segments (Brower, *op. cit.*), and this is believed to deter some predators, particularly ants. In a study in England on Large Whites (*Pieris brassicae*) and Common Cabbage Whites (*Pieris rapae*) by Baker (1970), it was found that wild birds preferred the pupae of *P. napi* to *P. brassicae*, and this was probably attributable to the presence of higher concentrations of glucosides in the latter. There are a few parasitoidal hymenopterous insect which specialize in attacking pupae. *Pteromalus puparum* is a tiny Chalcid wasp which lays its eggs on the freshly-formed pupae of Cabbage Whites, before their skin has time to harden, and the larvae burrow inside the pupa, eventually emerging as adult wasps just before the host butterfly pupa dies.

The Adult Butterfly

The enemies of butterflies, with their power of flight, are fewer than is the case with their larvae, and are mostly confined to visual hunters, particularly vertebrates. Some species possess the ability to fly very swiftly if attacked, or jerkily and in such a manner as to confuse a pursuing predator. The main non-vertebrate predators on butterflies are also visual hunters such as spiders, praying mantises, and Robber flies (family Asilidae) (*see*, Plate 11). Vertebrate enemies include many species of lizards, birds, and arboreal mammalian insectivores. Bell (1919) recounts watching a Robber fly in the Sindh desert returning to the same twig perch from which it flew down and seized about six Grass Jewel butterflies (*Zizeeria trochilus*) in the space of one hour. Crab spiders (Thomisids) so, called because they move sideways like crabs, hide in flowers and specialize in feeding upon insects which visit flowers. These spiders are usually coloured pink or orange-yellow, matching the flowers within which they lie in wait. In Pakistan the Common Garden Lizard (*Calotes versicolor*) is an agile arboreal hunter, which I have watched in my garden in Malir successfully capturing Grass Yellow (*Eurema* spp.) and Blue Pansy (*Junonia orithya*) butterflies. Bell (1919), in Sukkur district, describes watching the desert-loving Fringe-toed Lizard (*Acanthodactylus cantoris*) visually spotting an overhead flying Painted Lady (*Vanessa cardui*), and pursuing it until it settled and was captured; quite a feat for such a strong-flying butterfly. The birds which capture butterflies are aerial hunters such as Flycatchers, Drongoes, and some Chats. I have watched in Ziarat, Balochistan, the Spotted Flycatcher (*Muscicapa striata*) capture *Melitea* Desert Fritillaries, and in northern Hunza, Güldenstädt's Redstart (*Phoenicurus erythrogaster*) capture a number of high-altitude *Colias* spp., while the Paradise Flycatcher (*Terpsiphone paradisi*), watched for two days in the Kaghan valley, where it had fledglings to feed, preyed upon any large insect that came within vision, including the Common Mormon (*Papilio polytes*) and the Large Cabbage White (*Pieris brassicae*). Surprisingly, the Common Drongo (*Dicrurus adsimilis*), (*see*, Plate 11) which bred in my garden in Malir, was seen by me several times catching and consuming the Plain Tiger (*Danaus chrisippus*), which is one of the noxious species normally avoided by birds.

The defensive mechanisms employed by butterflies include, as well as flight escape, those of camouflage and, in some species, the retaining within their bodies of noxious or distasteful chemicals sequestered by their larvae, and, paradoxically, the wearing of contrasting, easily recognized wing patterns and colours, which are called **aposematic**. These serve as warnings or memory aids to visual predators that the subject it distasteful. Experiments with captive-reared birds (Jacamars, *Galbula* spp.) which were specialist predators on butterflies, in a neotropical rain forest environment (Chai, 1996), showed that even young, inexperienced individuals quickly learned to recognize those butterflies which were distasteful and to reject them or avoid trying to catch them. In this experiment spanning two years, with many palatable butterfly species also being presented to adult and young birds, it was conclusively demonstrated that specialized predators quickly learned the visual cues which indicated that the butterfly prey possessed chemical defences. However, a further series of experiments (Pinheiro, 1996), with another specialized butterfly predator, the South American Flycatcher known as the Kingbird (*Tyrannus melanochilus*), revealed the surprising result that this bird frequently captured and ate aposematically coloured butterflies which were known to be distasteful, perhaps indicating that there is a high mortality price to be paid by distasteful butterfly species in the process whereby the Flycatcher learns to avoid such species. It was also revealed in these experiments that some aposematically coloured butterflies which were known to be palatable, but were skilled at evading capture, could indicate an evolutionary development of aposematic wing pattern in species which were not worth the effort of trying to capture, a co-evolutionary development of warning signals which in this latter case indicates skilled evasion ability.

Earlier in this chapter, the case was cited of orchid flowers imitating lilies with a rich nectar source to deceive pollinating butterflies; a process known as **Batesian Mimicry**. In the butterfly family, and in all major continents, there are truly fascinating and amazing examples of wholly palatable non-related species of butterflies which closely resemble those species which possess noxious chemicals within their bodies. In Asia and our region, the families whose adult butterflies are avoided by predators are Acraeninae, Daininae, some Pieridae, and some red-bodied Swallowtails (*Atrophaneura* spp. and *Pachliopta* spp.). Because of the plants upon which these larvae feed, adult butterflies belonging to these families can contain harmful chemicals, noxious to predators. There are examples of non-related butterflies which are wholly palatable which imitate them in wing pattern and even flight habits. This form of mimicry was first described by H.W. Bates in the 1860s. A second, more complex form of protective mimicry, known as **Müllerian Mimicry**, entails the look-alike appearance of different species, often closely related, all of which may be unpleasant or poisonous-tasting to predators. This form of mimicry was first described by F. Müller in the late 1870s. The whole subject has excited much research among Lepidopterists, but the complexities are beyond the scope of this book. Suffice it to add that different plant communities of the same species or family may not always manufacture toxic chemicals within their tissues, so that there is variation in toxicity or palatability even within one species of butterfly dependant on that plant host. In many butterflies there is stronger evolutionary or ecological pressure for the male of the species to retain its original specific wing patterns, while the female alone, who can thus recognize conspecific mates but needs to be avoided by predators, has evolved Batesian mimicry. To be successful, Batesian mimicry must evolve with a model such as the Common Crow (*Euploea core*) which is plentiful and which is mimicked by a palatable species which is comparatively uncommon, such as the female of the Great Eggfly (*Hypolimnas bolina*), otherwise vertebrate predators, which are long lived and have some learning ability, will take too long to discover that the model, i.e., the Common Crow, is distasteful. Though experimental evidence is still lacking, there is a growing body of opinion that look-alike families such as the Sailers (*Neptini* spp.) and Sergeants (*Parathyma* spp.) are Müllerian mimics, and the similarity between many species of Danaid, such as the Blue Tigers, all of which are noxious, is another example of Müllerian mimicry. It is, of course not essential in Müllerian mimicry for either of two or more species to be common; they can both be equally rare. In the ongoing evolutionary development of Batesian mimicry, it is to be expected that there are examples of slighter and greater resemblance to the model.

Given below are the examples of mimicry found in species occurring in Pakistan, and readers are asked to examine the colour plates in which they are figured. Being in systematic order, they are not usually on the same plate as their models.

Camouflage is an important protection for all palatable butterfly species, and there are many examples of species with bold patterns on their upper wing surface which, when they settle, have their underwings cryptically patterned so that they are very difficult to see. The Common Beak (*Libythia lepita*) has already been mentioned in this respect, and most of the Rock Browns (*Hipparchia*, and *Eumenis* spp.) resemble, on their undersides, lichen-covered rocks typical of their habitat. The Evening Brown (*Melanitis leda*) is also cryptically marked and almost invisible when it settles with closed wings among dead leaves on the ground.

Many smaller species, especially the Hairstreaks (Theclinae) and Tailed Blues (*Euchrysops* and *Jamides* spp.), when settling immediately close their wings, which bear on the hind wing a tornal lobe with an eye spot and beyond this a thin-haired like tail. Frequently these species also rub their hindwings together slightly when settled. This gives the impression that the lower edge of the hindwing may be a head with waving antennae, and often deceives attacking birds and lizards, which seize a part which the butterfly can lose without fatal injury. In India there are many examples of species which, with closed wings, so closely resemble dried leaves as to be mimics, even to the point of bearing realistic venation and fungal blemishes on their surface, yet

Model	Mimic
Common Windmill (*Atrophaneura polyeuctes*)	A day flying moth, *Epicopia polydorus*
Common Rose (*Pachliopta aristolochiae*)	Common Mormon (*Princeps polytes*)
Blue Tiger (*Tirumala limniace*)	Common Mime (*Chilasa clytia*, form *dissimilis*)
Common Crow (*Euploea core*)	Great Eggfly female (*Hypolimnas bolina*)
Plain Tiger (*Danaus chrisippus*)	Indian Fritillary (*Argyreus hyperbius*)
	and Danaid Eggfly female (*Hypolimnas misippus*)
Pioneer (*Anaphaeis aurota*)	Common Gull (*Huphina nerissa*)
Common Wanderer (*Parenonia valeria*)	Striped Albatros (*Appias libythea*)

their upper wing surface is brightly coloured. The Great Orange Tip (*Hebomia glaucippe*) and Oakleaf (*Kallima* spp.) butterflies which do not occur in Pakistan, are classical examples of leaf mimics.

In the introductory chapter I referred to the value of scientific studies on butterfly populations as revealing the genetic and evolutionary forces at work in shaping species variation. Though it is a moth and not a butterfly, the classic study of the European Peppered Moth (*Biston betularia*), which is polymorphic with dark (melanistic) forms and light, whiter patterned forms, gave new insights into the action of these evolutionary forces. Due to sooty emissions during the Industrial Revolution in the nineteenth century, which darkened all tree surfaces in many regions, it was noticed that the whole population of Peppered Moths in such areas became correspondingly dark and consequently harder for predators to see. This happened over a comparatively short space of time on the evolutionary scale, but revealed how natural selective pressures could rapidly act upon critical polymorphic genes.

In contrast to the above, mention has already been made of aposematic colouration. The slow sailing flight with conspicuous spotted or striped wing patterns of the Danainae are a good example of those species which are unpalatable and need to give warning to potential predators that they are not worth attacking.

Migration

The majority of butterfly species in Pakistan rarely travel more than a few hundred metres from their birthplace in all stages of their life history, particularly the Satyrids, such as *Maniola* Meadow Browns, and *Erebia* Arguses. The more sedentary species, particularly those living in mountainous regions, tend to develop a multitude of different forms or sub-species due to the gradual effects of genetic mutation within a relatively small, isolated population without the opportunity for widespread gene exchange which would give homogeneity over a wider population. This explains, in part at least, the great variety and number of forms encountered in Pakistan within such genera as *Erebia* and *Maniola*, and indeed *Hipparchia* and *Karanasa*, all belonging to the sub-family Satyrinae and adapted to mountainous habitat.

However, there are some species which are notable for both short-distance local migration and long-distance migration. The practical problems for any researcher in tracking the flight path or migration route of butterflies have meant that our knowledge of this phenomenon still remains very scanty. There are techniques for marking adult butterflies with dyes and even small adhesive paper discs, but, their wings being so fragile and delicate, such marks not only risk affecting the individual's ability to migrate normally, but also present great difficulties in making adequate recoveries any distance from the point of release. With bird migration, recoveries have always depended upon public or volunteer response for recovery of individuals previously identified with aluminium leg rings. These are long-lasting and, since a huge range of Palaearctic species are migrants, a picture has gradually emerged about their flight paths and distances travelled. Butterfly migration studies would also entail large scale co-operation from the public, even if only limited to interested volunteers, and sadly no such programme has ever been attempted on the Indian subcontinent, where much successful bird ringing has been carried out for decades.

Despite these limitations, it is known, from many on-the-spot observations of absence and then seasonal abundance of specimens all travelling in one direction, that certain butterfly species are regular migrants. Because their life history as adult butterflies is brief, migration over long distances is usually achieved not by one individual, but by successive broods. After one hatching of adults migrates in a certain direction, they may settle and mate in one region and the next generation, after emerging as adults, will continue migrating in the same general direction. In Pakistan this enables many species to produce broods during the cold weather along the northern plains regions and foothills, while in the spring adults tend to travel northwards into the Himalaya, returning southwards in the late summer or autumn. There is circumstantial evidence that this phenomenon is widespread amongst the very strong flying Emigrant butterflies (*Catopsila* spp.), and C.B. Williams (1938) recorded regular movements in one direction in northern India of individuals of *Catopsila crocale* and *Catopsila pyranthe* in late June and July southwards into the foothills. Similarly, the Large Cabbage White (*Pieris brassicae*) has been observed with numbers moving northwards in the spring in the same general compass direction and again many individuals moving southwards into the foothills in late summer. This species lays its eggs on different Cruciferous plants in the hills in summer and more commonly on cultivated rape or mustard crops (*Brassica campestris* cultivars) in the plains in winter. Brigadier Evans recorded a north-westward movement of large numbers of the Pea Blue (*Lampides boeticus*) from Rawalpindi district into the hills from February lasting up to April, and elsewhere in India this species is well known for its migratory tendency. Morshead (1925) recorded a mass congregation of the Himalayan Blackvein, *Leucodice soracte*, near Dalhousie, with the 'air above looking as though filled with snowflakes', though this single-brooded species does not actually show any evidence of long-distance migration. Observations have also been made (Williams, 1938) of the Pale Lemon White (*Euchloe lucilla*) travelling in one direction through the Khyber Pass, and of the Painted Lady crossing over the Shandur Pass in northern Chitral in August in a south-easterly direction. The author has noted a general south-eastwards migration of individual Blue Tigers (*Tirumala limniace*) from the northern Punjab Salt Range, post monsoon. In parts of southern India the Common Crow (*Euploe core*) has also been observed (Williams, *op. cit.*, and Haribal, 1992) migrating in a definite compass direction. There are certain species which seem to be capable of travelling continuously during their adult lifetime, but in various different directions, and this migratory tendency results in their very widespread occurrence at certain seasons only, but over an increasingly wide area. This certainly applies to the Painted Lady (*Vanessa cardui*) wherever it occurs, in Europe as well as the Indian subcontinent. Adult butterflies of this species are widespread throughout the Indus plains during the cold weather, from Karachi up to Lahore, whilst they have been seen in the summer from the Murree foothills up to the highest Karakoram passes, with adults purposefully flying in one particular direction. The same phenomenon has been observed for some *Colias* spp. (Clouded Yellow Butterflies), particularly the Eastern Pale Clouded Yellow (*Colias erate*), which, when abundant in some years through the drier mountain regions from Balochistan up to the NWFP, will subsequently turn up in small numbers in lower Sindh and the Punjab plains in winter.

R.R. Baker, who has made a lifetime study of butterfly migration, has postulated that all butterflies have a 'lifetime track' which determines where they will die in relation to their birthplace, and that for many species this may cover a distance, in one continuous direction, of many kilometres, even though such behaviour may not, strictly speaking, be defined as migration. In one study on *Pieris rapae* (Small Cabbage Whites), Baker (1978) discovered that throughout adult life, on sunny days, individuals had an average cross-country movement of 0.8 km. per hour, and that they might have a typical lifetime track of as much as 100–200 km. This Cabbage White was introduced to Australia in the Melbourne area in 1939, and, unaided by man, reached the west coast of that continent after three years. This would mean that over a maximum of twenty-five generations, this butterfly had travelled 4,000 km. The most well documented and spectacular example of migration is that provided by the North American Monarch (*Danaus plexippus*), which travels some 3,000 km. from its breeding grounds in the north-east of the USA and eastern Canada, south-westwards, to overwinter in Mexico. Many of these individuals also return north the following spring. The Indian and Pakistani population of this same species does not evince any such long-distance migratory tendencies. In fact, many species exhibit migratory tendencies in some populations, whilst other populations of the same species have a sedentary life history. This is true of the European Small Tortoiseshell (*Aglais urticae*). In those cases known from the Indian subcontinent, there is as yet no evidence that any 'migrant' species return to their original birthplace as individuals, but that most of them return post-monsoon over successive generations. Technically speaking, therefore they are not true migrants like some birds, but rather each population is either immigrant or emigrant from its birthplace.

According to Feltwell (1986), butterflies on migration navigate by the position of the sun, and only do so in favourable weather, flying between 11 a.m. and 4 p.m. During such regular mass movements, they try to take advantage of following winds, but to gain height and lift, will fly into the wind. The Large Cabbage White (*Pieris brassica*) has been recorded as flying at 15 km. per hour in good sunny weather.

In concluding this brief account, it can be assumed that there must be some ecological advantage in such dispersal or migration; that it may be triggered by increased population density of that species when conditions for its breeding have been favourable; or even that it enables more successive broods to be hatched by exploiting the favourable climate and food availability found in the northern areas in summer. It is also evident that those species with wide dispersal tendencies have a very uniform or homogenous phenotype, or, to phrase it in layman's terms, 'all look uniformly alike'. The Painted Lady and Large Cabbage White both illustrate this point, with no significant races or sub-species having been described over a very cosmopolitan world distribution. As we saw in the beginning of this discussion, those species which are very sedentary, or by reason of geographic and physical barriers are unable to disperse, are much more inclined to produce, over time, a variety of different-looking forms, ultimately leading to speciation, as evidenced by the *Maniola* Meadow Browns.

Hibernation

Butterflies have evolved diapause at every stage of their life cycle. Diapause can be defined as a resting period or status during which the insect has no physiological or developmental activity except that which is minimally necessary to maintain viability.

The species accounts which follow will show that many Satyrid species breeding in northern alpine regions of Pakistan lay eggs which hatch out at the very end of the summer, after which the tiny larva in its first instar goes into hibernation, hidden deep down within the base stems of some plant tussock, preferably of its foodplant. It does not awake and start feeding until the following spring, and if there is an unseasonable snowfall, many at this stage which survived sub-zero temperatures for weeks on end, will within minutes die from hypothermia.

The majority of instances of diapause in fact occur during the larval stage, though in some species the larvae may feed for some time and pass through several instars before hibernating. Some of the high-elevation Fritillaries, such as Jerdon's Fritillary (*Argynnis jerdoni*), hibernate as partly-grown larvae throughout the winter, recommencing to feed the following April. The Common Snow Flat (*Tagiades atticus*) found in mountainous regions of India, hibernates as a fully-grown larva. Many of the mountain-dwelling Skippers, such as *Erynnis* spp. occurring in Europe, hibernate throughout the winter as fully-grown larvae, but do not pupate until the following spring. There are fewer examples of hibernation at the egg stage, but the High Brown Fritillary (*Argynnis adippe*) is reported by Ford (1954) to hibernate as an egg, hatching in the following spring, and all the *Parnassius* Snow Apollos pass diapause in the egg (Kreuzberg and Pljushch, 1989), but after the little larva inside has fully developed. The Silver Spotted Skipper or Chequered Darter (*Hesperia comma*) and the European Silver Studded Blue (*Plebejus argus*) also pass the winter in the egg stage. Many tree-feeding species, especially the Hairstreaks (*Thecla* and *Strymonidea* spp.), spend the winter as eggs, hatching the following spring. Their eggs are typically laid singly at the base of next year's leaf buds, where they remain until the following spring, hatching as the new buds unfurl.

There are also some species which spend the winter in northern regions as hibernating pupae. It is believed that some of the Wall butterflies (*Pararge* spp.) pupate at the end of the summer and spend the winter as pupae, emerging as adults after five or six months.

Only a few species hibernate as adult butterflies, and the best examples of this are amongst the *Vanessa* species such as the Eastern Comma (*Vanessa egea*) and Indian Tortoiseshell (*Vanessa cashmiriensis*).

Generally speaking, those species which are truly Oriental in distribution and have a continuous larval food supply do not need to hibernate at any stage of their life cycle, whereas most of the species which breed in Pakistan's northern Palaearctic climate hibernate through the winter months, and, in some of the higher-elevation species, may spend part of two alpine summer seasons as developing larvae, producing only one adult hatching after two years, but on an annual basis.

Chapter 5
The Distribution of Butterflies in Pakistan

It is no coincidence that the distribution of butterfly species, both with regard to habitat adaptation and to zoogeographic affinities, follows closely that of higher animal forms, particularly the vertebrate kingdoms of Mammals and Birds. Readers may refer to discussions concerning their distribution in the opening chapters of *Birds of Pakistan* (Roberts, 1991) and *The Mammals of Pakistan* (Roberts, 1997). Because approximately 60 per cent of Pakistan's land surface comprises mountainous areas, including parts of some of the planet's most dramatic high-altitude ranges such as the Hindu Kush, Pamir, and Karakoram, it is not surprising that a high percentage are alpine or montane species. Those species adapted to pass the early larval stage in diapause, or whose pupa are sheltered underground or beneath stones, can survive sub-zero temperature without being killed. Such families as the Satyridae are well represented amongst alpine-adapted species, and the extreme example of such adaptation is among the 'Snow Apollos' (Parnassiinae), whose caterpillars shelter by night under stones and whose pupa develop underground. Pakistan is further characterized by having a dry steppic type of climate over a large part of these mountainous regions, resulting in the evolution of a rather distinctive Central Asian fauna, not well represented in the more mesic regions of the Western Palaearctic.

Taking the country as a whole, species adapted to more mesic conditions and natural deciduous woodland are poorly represented, and out of 317 species known to occur, rather less than 34 per cent are of strictly Oriental zoogeographic origin, whilst 66 per cent are of Temperate Palaearctic affinity, including species typical of the Sino-Japanese phytogeographical zone (the Himalayas).

As outlined in my previous books (*see above*), Oriental faunal species (in a zoogeographic sense), have found the easiest invasion or colonizing routes through the south-east coastal belt (southern Sindh) and the north-eastern Sub-Himalayan region, through Kasur, Lahore, and Sialkot districts (*see*, Plate 1). The area in between, along Pakistan's eastern borders, presents a formidable barrier to the spread of most forms of wildlife not adapted to extreme desert conditions, because of the existence of the Great Indian Desert of Rajasthan, represented within Pakistan by the Cholistan desert in Punjab and the Thar desert in Sindh. This may be illustrated by listing a few truly Oriental faunal species of butterfly usually found only within these two widely separated but restricted areas:

Oriental faunal species confined to extreme south-eastern and north-eastern regions

Papilio polytes, Pachliopta aristolochiae, Leptosia nina, Delias belladona, Delias eucharis, Parenonia anais, Appias libythea, Eurema laeta, Junonia iphita, Hypolimnas bolina, Ypthima inica, Mycalesis perseus, Euploe core, Castalius rosimon, Chilades laius, Zizula hylax, Catochrysops strabo, Jamides celeno, Tajuria cippus, Virachola isocrates, Rapala manea, Hasora chromus, Badamia exclamationis, Sarangesa purendra, Saustus gremius, Astycus pythias, Astycus augias.

From an ecological perspective, the greatest biodiversity of both plant and animal species usually occurs in the transition zone between major biomes. In Pakistan, the richest area both botanically and for all animal forms lies in a narrow belt along the Himalayan foothills characterized by having a sub-tropical climate, with annual rainfall fairly well distributed throughout the year, totalling about 40 inches (1,000mm), and having a rich and varied plant fauna with Indo-Malaysian affinities, which provides an abundance of suitable larval food plants. This area, restricted to the eastern part of the Himalayan foothills in Pakistan, includes such localities as the Margalla Hills, the Manga, Lehtrar, and Kahuta Valleys, Chaprot, and parts of Poonch (*see*, Plate 7). Here may be found some of the more spectacular Oriental faunal species such as the Glassy Bluebottle (*Graphium cloanthus*), the Common Nawab (*Eriboea athamus*), the Common Castor (*Ariadne merione*), the Map (*Cyrestis thyodamus*), the Common Sailer (*Neptis hylas*), and the Baron (*Euthalia aconthea*).

As has been recorded for mammalian and avian species, there are also a few butterflies which appear to have colonized the western mountain regions of Balochistan from the Ethiopian zoogeographic region. Examples are the Greenish Black-tip (*Euchloe charlonia*), the Green Striped White (*Synchloe belemia*), the Small Bath White (*Pieris chloridice*), the Black Spotted Pierrot (*Tarucus balkanaica*), the Grass Jewel (*Zizeeria trochilus*), the Red Copper (*Thestor callimachus*), the African Marbled Skipper (*Gomalia elba*), and the Streaked or Syrian Skipper (*Muschampia staudingeri*).

To the north, and at higher elevations than the sub-tropical scrub forest zone described above, there is a broader ecological zone characterized by Himalayan Moist Temperate Forest, which has a mixture of tall tree deciduous (*Acer, Aesculus,* and *Prunus* spp.), and coniferous forest (*Abies, Pinus,* and *Picea* spp.), with predominantly coniferous forest on the dryer southern slopes or higher north-facing aspects. This zone occurs between elevations of 6,000–10,500ft (1,800–3,200m), and receives more monsoon rains and winter snowfall than the much dryer inner mountain ranges further north. Typical butterflies of this zone include *Papilio arcturus, Atrophaneura polyuectes,* and Pieridae such as *Pieris brassica, Pieris canidia,* and *Pieris ajaka, Aporia leucodice, Gonepteryx rhamni,* and *Colias fieldi.* The larger Fritillaries (*Argynnis* spp.), 'Admirals and Tortoiseshell' (*Vanessa* spp.), and many 'Blues' such as the Hedge Blues (*Arletta* and *Celastrina* spp.), as well as some *Polyommatus* spp., are well represented. This zone also includes some of the larger Skippers, such as *Badamia exclamationis,* and *Hasora alexis,* and shade loving Satyridae such as the 'Tree Browns' (*Lethe* spp.), 'Satyrs' (*Aulocera* spp.), and the Himalayan Fivering, *Ypthima sakra.*

Higher up mountain slopes there is a Himalayan Moist Alpine Zone starting above the tree line, which may vary from 10,000 to 11,000ft, (3,050–3,350m), up to the permanent snow line at 17,000ft (5,180m).

This zone is characterized by the smaller Fritillaries such as *Clossiana, Fabriciana,* and *Bolaria* spp., Meadow Browns (*Maniola* spp.), and, amongst the Lycaenidae, 'Meadow Blues' (*Polyommatus* spp.), 'Arguses' (*Aricia* spp.) and the 'Green Underwings' (*Agriades, Albulina,* and *Glaucopsyche* spp.)

In the much dryer inner mountain ranges, beyond the monsoon influence, typified by Gilgit, Baltistan, and Hunza, the Alpine Zone is more restricted and has some high-altitude species especially adapted to severe conditions, such as the Parnassius 'Snow Apollos', and some Pieridae such as *Baltia, Pontia* spp., and the Alpine Large White (*Pieris deota*), as well as many species of 'Clouded Yellows' (*Colias* spp.).

In more southern latitudes, characterized by a Steppic Montane Zone, typified by parts of the NWFP and Balochistan, there is Juniper (*Juniperus excelsa*) or 'Chilghoza' (*Pinus gerardiana*) forest at higher altitudes, and *Pistacia,* with *Fraxinus* scrub forest at lower altitudes. Here may be found a number of Satyridae, 'Rockbrowns' (*Hipparchia, Karanasa,* and *Eumenis* spp.), as well as the smaller 'Red-band Fritillaries', (*Melitea* spp.). Amongst the Blues are typically some of the smaller 'Pierrots' (*Tarucus* spp.), *Plebejus sephyrus, Polyommatus bogra,* and some of the 'Coppers' (*Lycaena* spp.). Among the Hepseriidae are some of the smaller Skippers (typically *Gomalia, Spialia,* and *Carcharodus* spp.).

The major sand-dune desert regions such as the Thal, Sibi, Cholistan, and Thar deserts are, not surprisingly, poorly represented by butterfly species. The 'Arabs' (*Colotis* spp.) are the most typical and abundant, whilst the Pioneer (*Belenois aurota*) and Spotted and Black Spotted Pierrots (*Tarucus callinara,* and *T. balkanica*) are also quite often encountered.

The three plates illustrating different habitats have not been chosen to represent every major ecological zone, and the cultivated areas of the Indus plains, which do harbour many butterfly species, have been omitted, whilst readers interested in studying all the distinctive ecological zones are referred to the habitat plates in Chapter 3 of *Birds of Pakistan*, Vol. 1 (Roberts, 1991). Because many of the less well known and rare high-altitude species have only been found in Pakistan's far northern borders, photographs of these still remote and inaccessible regions are included instead.

Chapter 6
Papilionidae—Swallowtails

Parnassiinae—Apollos

Classification

There is wide variation in the markings of butterflies in this sub-family, even within single species, partly attributable to their habitat, which tends to result in small, isolated, disjunct populations with little genetic exchange. Consequently, there has been much disagreement in their classification, but current expert views now split them into about fifty-four distinct bi-nomial taxa, with at least twenty-two more doubtful species or valid sub-species (Häuser, 1993).

The Apollos belong to the super-family Papilionidae, or Swallowtails, but the sub-family Parnassiinae is very different in appearance from the sub-family Papilioninae, both as adults and in their early stages. They are included in the Papilionidae because they have antennae ending in a definite club, and the larvae possess a peculiar forked extrusible organ just behind the head, known as the osmeterium, which is common to all Swallowtails as well. This organ is always brightly coloured and, when extruded, produces powerful aromatic scents, derived from the foodplant, which are thought to deter some potential predators. The adult butterfly differs from the sub-family Papilioninae, which are usually large in size and with the hindwing tailed. In the Parnassiinae, adults are medium to small sized, the forewing is rounded at the tip (known as the apex) and roughly triangular in shape, whilst the hindwing (which is not tailed) is very characteristic, being pyrate or pear-shaped, strongly incurved along the inner margin, and lacking any anal fold in the wing, which many species of Swallowtails do possess. The forewing venation differs from the Swallowtails in lacking vein 9 (see, Fig. 1), so that the total number of veins are eleven compared with twelve for the Papilioninae and all other butterfly families. Also, at the base of the forewing in both Apollos and Swallowtails there are two veins numbered 1a and 1b, but 1a is always very short, usually extending down to the wing dorsum margin rather than reaching to the outer termen margin, as in other families. The Parnassiinae are distinguished in another aspect of their forewing venation, as they lack a median spur. This spur extends outwards from the median vein which bounds the lower edge of the cell, arising close to the base of vein 1b (see, Fig. 1). Papilioninae always possess this spur. Compared with Swallowtails, their antennae are short, being only about one-third of forewing length and they are often ringed black and white along the shaft. The body including the head of Parnassians is very hairy, and in males also the abdomen. Females are relatively hairless on their abdomen, but in both sexes the body is tough and leathery, whilst the wings are either white or pale creamy in ground colour, with a very thin covering of scales and therefore of translucent appearance. As a generalization, those species which live at lower altitudes and in more mesic conditions are more heavily marked with dark scaling and red ocelli, whilst the highest-altitude dwellers are paler, often lacking much black scaling on their wings. Females after fertilization possess a horny pouch at the tip of the abdomen known as the phragis, (see, Fig. 9) which is formed from a quickly-hardening substance discharged by the male after fertilization and obviously prevents any other male from passing on his genes to that female. The phragis differs in shape between species, especially the 'Keeled Apollos', and is a valuable diagnostic feature. Because of their adaptation to a most inhospitable environment with strong cold winds, they are powerful flyers, able to keep close to the ground surface and to fly sideways with the wind when on the wing. Adults freely visit flowers; the larvae are often gregarious in their early stages and the pupa; are formed underground, sheltered from the elements. Their eggs are spherical when viewed from above, but turban-shaped from the side, and are normally dull white in colour with a slightly rugose surface. The larvae are stout-bodied and usually hairless or bearing sparse short bristles, and are cylindrical in shape. With the exception of the Desert Apollo, they are usually black or dark grey in colour, with pale yellow, red, or green tubercles along the sides. Their foodplants belong to the families Zygophyllaceae, Saxifragaceae, Aristolochiaceae, Fumariaceae, and

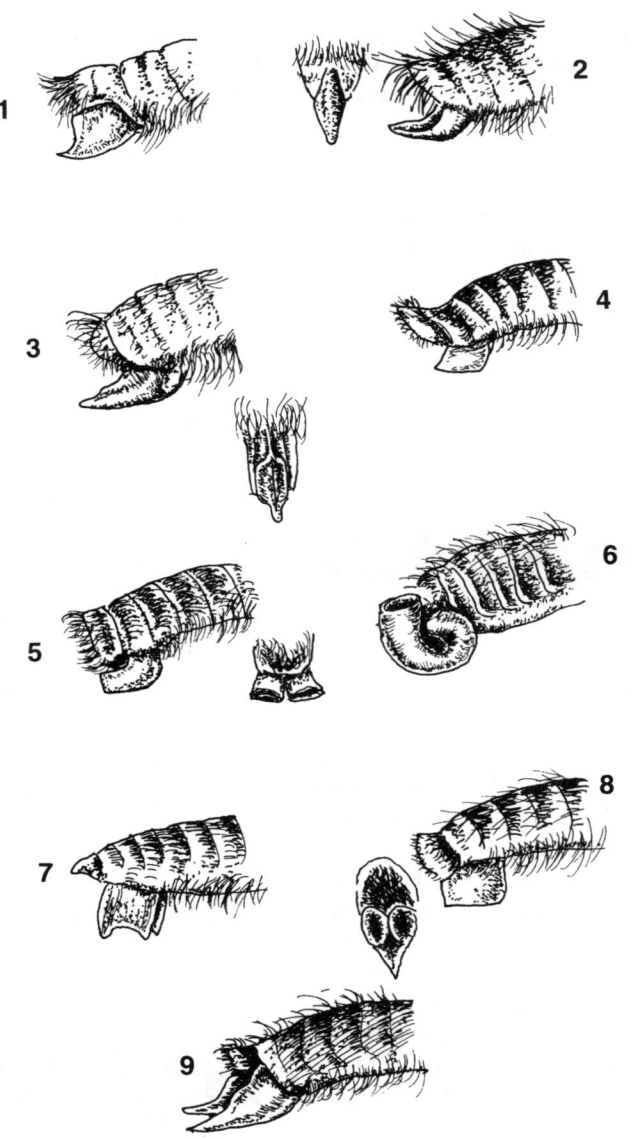

Fig. 9 Phragis of fertilized female Parnassians

    Top Row: 1. *Parnassius hardwickei*   2. *P. epaphus*, ventral and side views

    Second Row: 3. *P. jacquemonti*, side and ventral views   4. *P. stoliczkanus*

    Third Row: 5. *P. inopinatus*, side and ventral views   6. *P. charltonius*

    Fourth Row: 7. *P. acco*   8. *P. loxia*, ventral and side view

    9. *P. delphius*, side view.

Crassulaceae. Because Pakistan has such a wide area of magnificent mountainous country, Parnassiinae are well represented, with sixteen species definitely recorded, in contrast to the comparative paucity of Oriental Swallowtail species at lower elevations. Very little is known about the life history and indeed the distribution of the majority of Apollo species, because of the very high elevations where most live. Many of the specimens in major museum collections came from mountaineering expeditions, collected by persons with skills and interests other than entomology! Young Pakistani biologists are well placed to make significant contributions to our knowledge about this very unusual group of butterflies.

Genus HYPERMNESTRA Ménétriés, 1848

1. The Desert Apollo

Hypermnestra helios
(Syn: *Helios balucha*)
Wingspan 45–55mm

Description: Though this species has been put in its own separate genus, it is, like the *Parnassius* spp., a distinctive butterfly in having only very thin scaling on the wings, a very hairy tough body, and the hindwing rather pear-shaped, being incurved on its inner margin, and lacking any anal fold. It is pale straw yellow on the forewings, which bear a square black spot mid cell and another narrow spot end cell, with a black-bordered red transverse patch just beyond the apex of the cell. The wing margin has a narrow band of black scaling, fading out towards the lower or dorsal end. The hindwing bears small black spots in the upper discal and upper costal areas. The upper surface appears less yellow than the forewing, but shows the greyish shadows of the underwing pattern. The under hindwing has dark yellowish-green transverse bands in the basal, median, and sub-marginal areas of irregular curving outline. These areas are actually black, thickly dusted with yellow scales. In this genus the female never has a horny anal plug (phragis). The antennae are short and, in contrast to the genus *Parnassius*, have the club at the tip rounded, whilst it is elongated in the latter. The thorax and abdomen are black and covered with silver and black hairs.

Status: Believed to be very rare and only a few specimens have ever been collected from Balochistan, two from the Nushki area by Ollenbach, and two in the Indian Museum, Calcutta, probably collected by Stoliczka from Kach. It occurs more widely in Turkestan, Xinjiang, and extending into western Iran. Sakai (1981) collected it from Seistan province, around Girishk in southern Afghanistan, where it was on the wing during April. There are records of its occurrence from every province within Afghanistan and also from Tajikistan to the north. Probably in Balochistan, being slightly further south, it may be encountered from mid-March. Ollenbach's specimens from Nushki, both of which are in the collection of the Indian Forest Institute at Dehra Dun, were collected on 5 April 1905.

Habits: The eggs are described as pale green, and are laid singly on the underside of a leaf (Talbot, 1939). The larvae when full grown are thick-set somewhat humped, and broader at the front part, grass green in ground colour, with concentric rows of small black tubercles around each segment. The lower sides bear an indistinct whitish line, above which the spiracles are framed by yellow rings. There are two rows of yellow spots down each side of the dorsum. In Central Asia, they have been found feeding upon *Zygophyllum turcomanicum*, and in Afghanistan upon *Zygophyllum miniatum* (family Zygophyllaceae). Several spp. of succulent, desert-adapted *Zygophyllum* occur in the Chagai. The pupae are described as greenish-grey or maroon-brown in colour, and smooth and thick and obtuse in shape. They are formed underground.

Genus PARNASSIUS Latreille, 1804

2. *Parnassius actius*

Wingspan 50–60mm

Description: This rather small Apollo is not heavily marked with black scaling and its marginal borders are thin and rather obscure pale grey, the inner sub-marginal band barely extending below the apex on the forewing. Usually both male and female lack any red-centred spots on the forewing, but have two red spots ringed with black between veins 7 and 8 and veins 4 and 5. The hindwing varies in the prominence of the sub-marginal ring of black spots, these being reduced to grey scaling in some males. Females are more heavily marked. In northern Chitral, specimens have the spots in the forewing costal area red-centred, and show heavier black scaling over the wing surface. The female after fertilization has a carinate pouch similar to *P. jacquemontii*.

Status: This is a very high-elevation Apollo which has been collected in Hunza from the Shimshal valley, and from the Baroghil Pass in the Hindu Kush range in extreme northern Chitral. It has also been collected from the Doubanni mountains in Gilgit.

Habits: The larval foodplants in the Altai (Kreuzberg, 1987) are Crassulaceae, and they have been found feeding on *Rhodiola heterodonta, Rhodiola poamiroalaica*, and *Rhodiola gelida, Rhodiola coccinea, Rhodiola kaschgarica*, and *Sedum ewersii*. This latter succulent is common and widespread throughout the far northern areas of Pakistan.

3. The Keeled Apollo

Parnassius jacquemontii
Wingspan 55–70mm

Description: This is a comparatively small species, but similar in general appearance to *Parnassius tianschanicus*, having broad vitreous-grey marginal bands, with a second inner looped band slightly darker and much thinner, and broad black bars across the mid cell, and end cell of the forewing. There are usually two small sub-apical red-centred ocelli, and a larger red-centred spot between veins 1b and 2. The hindwing has two large red-centred ocelli between veins 7 and 8, and another between veins 4 and 5, the lower one white-centred. There is also a conspicuous post-discal ring of black, arrow-shaped spots, the points inwards. The wings are also speckled with black scales. The underside is more greenish along the hindwing margin, and with fewer scales, looks glassy. There is a red-centred, black, rimmed anal bar, and the body is stout with a covering of white hairs. The female after fertilization has a small keeled plug (*see*, Fig. 9).

Status: This is a medium-elevation Apollo. It has been collected on the Shandur Plateau in Chitral, though Sanders (1955) in early August, did not encounter it among four other species of Apollos he was able to collect. In the third week of August, however, he did find it plentiful in the Baroghil valley, between 3,800–4,100m (12,500–13,500ft) in company with *P. delphius*.

Habits: In Afghanistan the larval foodplant is a *Sedum* sp. (Sakai, 1981).

4. The Larger Keeled Apollo

Parnassius tianschanicus
(Syn: *Parnassius discobolus* of Tytler, 1926)
Wingspan 65–80mm

Description: This is one of the larger Apollos with bold markings and a handsome appearance. Again in this family, there is much variation between populations and many forms have been described. The forewing usually bears a very broad, vitreous, grey margin, with an inner, darker, looped grey band, and broad black spots mid cell, end cell, and sub-apical. The costa and post-discal areas are speckled with black scales and there is a red-centred spot between veins 1a and 1b. The hindwing bears two large red-centred spots between veins 7 and 8, and between veins 4 and 5, and the sub-marginal band of spots are arrow-shaped, the points inwards. The underside is rather glassy with greenish dusting around the hindwing margin, and a red-centred anal spot on the hindwing underside. As its name suggests, the female after fertilization has a prominent keeled abdominal plug.

Status: This is one of the medium-elevation Apollos, generally flying between 3,200–3,800m (11,000–12,500ft), which occurs from Chitral through the Karakorams. It has been collected from the Baroghil valley (Sanders, 1955) at 3,660m (12,000ft), and the Shandur Plateau at 3,200m (11,000ft). Also in the Turikho valley and the Yarkhun valley, all in Chitral. Also from Misgar in Hunza. It is reported to be very common in favoured localities, such as the Baroghil valley and Misgar. On the Shandur Plateau it flies from June to August, where it is sympatric with *P. epaphus, P. jacquemontii*, and *P. delphius*.

5. The Common Red Apollo

Parnassius epaphus
Wingspan 50–60mm

Description: This species is comparatively small and not heavily marked with black when compared with *Parnassius hardwickei*. Like all the Parnassians, it can be very variable in markings, but its ground colour is always white with rather narrow greyish marginal band on the forewing, and a second, sub-marginal, thinner band of grey dots, with a round black spot mid cell, a second, thicker and longer, spot in the distal end of the cell, and a small costal spot red-centred and ringed with black. The hindwing is black basally, with two red spots in the upper costal and upper discal areas. The margin bears a ring of narrow, lunar-shaped black spots, this being the main distinguishing feature between *P. epaphus* and *P. hardwickei*. The latter bears circular black spots around the hindwing margin. The under hindwing also has two faint red-centred anal spots in addition to the upper costal and distal spots. The antennae are ringed black and white along the shaft. The body is stout and very hairy, and the female after fertilization has a carinate anal plug, not keeled.

Status: This is one of the high-elevation Apollos, and, though widespread, is less often encountered than the Blue Apollo (*P. hardwickei*). It has been collected from extreme northern Chitral at 4,470m (14,700ft), and on the Karumbar Pass on the Chitral-Gilgit border, in the Hindu Kush Mountains. Sanders (1955) collected it in Astor district, both on the Burzil Pass and further north up the valley, between 4,350m–4,877m (14,500–16,000ft), where it was freshly emerged in June and sympatric with *Parnassius simo* and *P. charltonius*. It has also been collected from Hunza.

Habits: Very little has been recorded about the early stages of any of the Parnassians, and this is so in the case of *P. epaphus*. It is known that most females within the genus scatter their eggs on the ground near their foodplant, and that the larvae are often gregarious in their early stages, dispersing when they grow bigger. The larval foodplants are Crassulaceae. The pupae are formed after the larva creeps into an underground crevice or shelter beneath a stone and makes a rough cocoon, securing itself to this with silk girdles. The pupae are generally more or less cylindrical in shape, and quite thick with blunt ends. The larvae are cylindrical, with relatively small heads, and decorated with bristly tubercles along their sides, with black being the usual ground colour, variously decorated with white or red or both-coloured spots. Just behind the head there is a forked fleshy organ known as the osmeterium, which can be extruded at will. The Red Apollo butterfly has been described as frequently visiting flowers or settling on rocks to bask (Haribal, 1992), when it has its wing spread flat or the hindwings partly concealed under the forewings. In dull or stormy weather they do not fly but retreat into rock crevices. Sanders (1955) found that this species hatched about a month earlier than *P. simo* and preferred grassy slopes.

6. The Common Blue Apollo

Parnassius hardwickei
Wingspan 50–65mm

Description: Slightly larger than the Common Red Apollo, this species frequenting lower elevations is much more heavily marked with black. It can be rather variable in appearance, but usually on the forewing there are two red-centred spots, one between veins 1a and 1b, and a thinner band beyond the end of the cell. The upper wing apex is bordered by grey with a sub-apical band of black dividing into separate spots along the lower wing. The hindwing usually bears a large crimson-centred spot between veins 7 and 8, and another between veins 4 and 5, and its most conspicuous, character serving to separate it from other species, is a curving row of three or four white-centred black circular spots in the post-discal region, which on the underside of the hindwing have a bluish tinge. The underside wings look glassy, having a very thin covering of scales. The body is stout and very hairy. Females have a small balloon-shaped pouch with a longitudinal groove when fertilized, and they are more heavily marked than males.

Status: Occurring as low as 3,000m (10,000ft) and on alpine meadows at the edge of the tree line, this is the most likely Apollo to be encountered. TJR has collected or observed it on the summit of Miranjani in the Murree Hills, at Kadir Gali, and on the slopes of Makhrah mountain in the Kaghan valley, all locations between 3,000–3,350m (10,000 and 11,000ft). It has also been collected from Chitral at Ishpadrog, Gilgit around Rupal, Astor, and Baltistan on the Deosai Plateau.

Habits: The adult has a strong, fast flight which I would describe as rather fluttering, and always keeping close to the ground contours. On Miranjani, adults frequently settled on *Taraxacum* flowers and *Cynoglossum glochidiatum*. The larvae feed on *Saxifraga* spp. typical of their more mesic habitat.

7. Varnished Apollo

Parnassius acco
Wingspan 40–60mm

Description: This is one of the smallest Apollos, notable in having yellowish or tawny cilia to the wing margins. The forewings also show three fairly-well marked black bars distally from the end of the cell, and there are two well-defined dark grey marginal borders. The forewings are margined by white or yellowish-white cilia. The hindwing is margined with black and bears two red-centred spots between veins 7 and 8 and veins 4 and 5. The female after fertilization has a conspicuous phragis comprising two parallel down-curved points (*see*, Fig. 9).

Status: This Apollo has been collected from the Rindi nullah in the Baltoro glacier region in Baltistan and from western Ladakh, India. Talbot (1939) also lists it as having been collected from the Shigar valley in Baltistan near the Baltoro mountains. It is a very high-elevation species, and the Baltoro specimens were collected at 5,000m (16,500ft), where it seems to be restricted in distribution and uncommon.

8. The Kafir Banded or Delphi Apollo

Parnassius delphius
Wingspan 55–65mm

Description: This is a medium to high-elevation Apollo which is quite heavily marked, but very variable in pattern between different populations. The forewing marginal and sub-marginal borders are glassy grey, and often there are no red-centred spots, but thick black bands across mid cell and the end of the cell. The hindwing has rather small black or grey spots between veins 7 and 8 and between veins 4 and 5. These are black in *P. delphius hunza*, but red-centred in *P. delphius kafir*. The chief distinguishing feature of *P. delphius* is the pattern of spots in the post-discal region of the hindwing. Generally there are only two in the anal region, those above being indistinct and grey. The female after fertilization forms a belt-like pouch around the abdomen, prolonged below into a bi-fid lobe.

Status: This is another Apollo adapted to a wide elevation range from 3,900–4,570m (13,000–15,00ft). It occurs rather uncommonly from the Safed Koh, where it is very rare, to the western side of the main Chitral valley in Kafiristan, also in the Baroghil valley, and on the Bangol Pass. Also on the Shandur Plateau, where Sanders (*op. cit.*) found it common in early August. Further east, in Gilgit, it occurs in Yasin and the Rupal valley, Astor, and the Farsat Pass on the western border of Chilas district, also in Baltistan below the Saltoro Glacier, and as far as Burzil on the eastern boundary of the Lower Deosai Plateau. In the Baroghil valley, it is sympatric with *P. tianschanicus* and *P. jacquemontii*, though the latter species generally flies at higher altitudes and in more xeric habitat.

Habits: Sakai (1981) records the larvae as feeding on *Corydalis fedtschenkoana* in Afghanistan. There are many spp. of high-altitude *Corydalis* recorded from Gilgit and Chitral.

9. *Parnassius cardinalis*

sub-species *cardinalina*
Wingspan 65–72mm

Description: This large species is included within the *Delphius* group, all of which are distinguished by having a row of distinct dark sub-marginal spots on the hindwing upper surface which are centred with blue. The forewing bears fairly broad grey discal and sub-marginal bands, whilst the hindwing bears three red ocelli, one costal, two conjoined in the post-discal area. These ocelli are ringed with black, and in most specimens are white-centred.

Status: This was described as *P. delphius cardinalina*, being considered as part of the *Delphius* complex of species, in Talbot (1939); in Ackery's treatment of the genera (1975), as part of the *P. delphius nicevillei* group. It was elevated to full specific status by Häuser (1993) on the basis of its occurrence sympatrically with another *Delphius* group taxon in the Central Asian Republic of Tajikistan. It is included in the Pakistan checklist on the basis of Andre Avinoff's original description of a specimen collected by him on the Burzil Pass in 1916. The type specimen was described by Grum-Grshimailo in 1887 from Tajikistan. It flies between 3,000 and 3,200m (10,000–10,500ft) elevation.

Habits: In the Altai Mountains the larval foodplants are Fumariaceae (Kreuzberg, 1987), whilst Sakai

(1981) gives the larval foodplant as *Corydalis ledebouriana* in Afghanistan.

10. *Parnassius staudingeri*

Wingspan 50–55mm

Description: This species is part of the *Delphius* group also, but was separated from *P. delphius* by Kreuzberg (1985) on the basis of differences in both male and female genitalia, as well as wing pattern and larval host plants. I have not seen the original publication, in Russian. The species is striking by the presence of large crimson ocelli on the hindwing which are thickly margined with black and conjoined by a black bar between the costal and sub-discal ocelli. The post-discal spot extends well into the area above vein 5. The hindwing sub-marginal spots are not very well developed being, confined to the areas between veins 1 and 2, but generally are blue centred in the more heavily-marked females. These spots peter out towards the hindwing apex into an angular dark-grey band. There are no tornal dark spots on the upper hindwing as in *P. stoliczkanus*.

Status: This species was formerly treated as *P. delphius chitralica*, but Kreuzberg (1985) states that the two are strictly allopatric. It occurs widely in the Karakorams, having been collected from the Saltoro Mountains in Baltistan, from Hunza near Misgar, and on the Baroghil Pass in Chitral. Also from western Yasin near the Chitral border (Tytler, 1926).

11. Pir Panjal Banded Apollo

Parnassius stoliczkanus
sub-species *P. stoliczkanus atkinsoni*
sub-species *P. stoliczkanus chitralica*, and
sub-species *P. stoliczkanus tytlerianus*
Wingspan 50–60mm

Description: This very variable species has proved difficult for taxonomists to place. It was treated as a sub-species of *P. delphius* by Ackery (1975), but later separated as a full species, *P. stoliczkanus*, by Collins and Morris (1985), following Talbot (1939), who considered that the hindwing pattern, with five well-marked, dark-rimmed sub-marginal spots, made it distinctive from *P. delphius*. It has also been treated as a full species by Häuser (1993). It is very variable both in size and the prominence of wing markings, so that a number of sub-species have been placed within this taxon. The smallest is *P. stoliczkanus tytlerianus*, which is heavily marked with black scaling but lacks any red ocelli, these areas being replaced by a rather quadrate black spot between veins 4 and 5. Also, the sub-marginal spots on the hindwing are poorly developed, being solid black without any blue centres and degenerating into a wavy dark grey line towards the apex. A larger form, very similar in markings to *tytlerianus*, but with much paler, less distinct dark scaling, has been ascribed to *P. stoliczkanus chitralica*. A third sub-species, *P. stoliczkanus atkinsoni*, has also been collected from Pakistan, but this is a large, striking form, with a prominent, pale crimson ocellus beyond the cell, but only a small black (not red) spot in the hindwing costal region. There are two hindwing tornal spots, which are red-centred and rimmed black, and a well-marked, prominent sub-marginal row of blue-centred spots. It can be separated from the rather similar *Parnassius staudingeri* by these tornal red-centred spots, which the latter lacks. Females of *P. stoliczkanus* and *P. staudingeri* have a small bi-lobed phragis after fertilization.

Status: The large sub-species of the Banded Apollos, *P. stoliczkanus atkinsoni*, has been collected from the Burzil Pass and from the Haramosh Range in eastern Gilgit. The sub-species *P. stoliczkanus tytlerianus* has been collected from Yasin in north-western Gilgit, and the sub-species *P. stoliczkanus chitralica* has been collected from the Baroghil Pass.

12. Black-edged Apollo

Parnassius simo
Wingspan 40–55mm

Description: This is a rather small Apollo, heavily scaled with black on the upper wing surface. The forewing has the apex region rather pointed and a broad dark area in the post-discal region. The hindwing bears only one red spot basally in the costal region, and is margined with white cilia, whilst the forewing is margined with black cilia. The very similar *P. acco* has two costal red spots on the hindwing and has creamy white cilia bordering the forewings. The antennae of *P. simo* are all black. The female after fertilization has a narrow tubular phragis ending in two points.

PLATE 10: Nymphalids 'Punches', a Tree Brown and a 'Blue'

(a) Painted Lady, *Cynthia cardui*

(b) Indian Red Admiral, *Vanessa indica*

(c) Large or Yellow Leg Tortoiseshell, *Vanessa xanthomelos*

(d) Blue Admiral, *Vanessa canace* sipping tree sap

(e) Common Punch, *Dodona durga*

(f) *Dodona durga* sipping ground moisture

(g) Common Tree Brown, *Lethe rohria*

(h) Western Blue Sapphire, *Heliophorus bakeri*

PLATE 11: Butterfly Predators and Behaviour

(a) Robber Fly, *Aselia* sp.

(b) Common Garden Lizard, *Calotes versicolor*

(c) King Crow, *Dicrurus adsimilis*

(d) Blue Peacock, *Papilio arcturus,* with wing mites, an external parasite

(e) Common Grass Yellow, *Eurema hecabe,* males 'Puddling'

(f) Larva of Common Baron, *Euthalia aconthea*

PLATE 12: APOLLOS

A. *Hypermnestra helios* - Desert Apollo **B.** *Parnassius acco* - Varnished Apollo **C.** *Parnassius actius* **C-1.** Male **C-2.** Female
D. *Parnassius actius tianshanicus* **E.** *Parnassius jacquemontii* - Keeled Apollo **F.** *Parnassius tianschanicus baroghila* -
Larger Keeled Apollo **G.** *Parnassius hardwickei* - Common Blue Apollo **H.** *Parnassius epaphus* - Common Red Apollo
I. *Parnassius delphius* - Kafir Banded Apollo **I-1.** var. *stoliczkanus* **I-2.** var. *chitralica*

PLATE 13: APOLLOS

A-1. *Parnassius stoliczkanus* - Pir Panjal Banded Apollo, male var. *chitralica* **A-2.** Female var. *atkinsoni* **B.** *Parnassius stoliczkanus tytlerianus,* male **C.** *Parnassius charltonius* - Regal Apollo **C-1.** Male **C-2.** Female **D.** *Parnassius simo* - Black Edged Apollo **E.** *Parnassius staudingeri* **E-1.** Male **E-2.** Female **F.** *Parnassius cardinalis,* male **G.** *Parnassius loxias,* male

PLATE 14: 'YELLOW SWALLOWTAILS' AND APOLLOS

A. *Graphium cloanthus* - Glassy Bluebottle **B.** *Papilio alexanor* - Southern Swallowtail **C.** *Papilio machaon* - Common Yellow Swallowtail **D.** *Papilio ladakensis* - Ladak Swallowtail **E.** *Papilio demoleus* - Lime Butterfly **F-1.** *Parnassius boedromius*, male **F-2.** Female **G-1.** *Parnassius inopinatus*, male **G-2.** Female

PLATE 15: BLACK SWALLOWTAILS AND PEACOCKS

A. *Papilio polytes* - Common Mormon, form *stychius*, female **B.** *Papilio polytes romulus*, male
C. *Pachliopta aristolochiae* - Common Rose **D.** *Atrophaneura polyeuctes* - Common Windmill **E.** *Papilio polyctor*
- Common Peacock, male **F.** *Papilio arcturus* - Blue Peacock **G.** *Chilasa clytia* - Common Mime

PLATE 16A: 'SMALL WHITES', EUCHLOE, BALTIA AND PONTIA

A. *Leptosia psyche* - Psyche, male **B.** *Euchloe ausonia* - Pearl White **C.** *Baltia butleri* - Butler's Dwarf **C-1.** Male **C-2.** Female **D.** *Baltia shawi* - Shaw's Dwarf **D-1.** Male **D-2.** Female

PLATE 16B: 'SMALL WHITES', EUCHLOE, BALTIA AND PONTIA

E. *Euchloe lucilla* - Pale Lemon White **E-1.** Male **E-2.** Female **F.** *Euchloe charlonia* - Lemon White **F-1.** Male **F-2.** Female **G.** *Euchloe belemia* - Striped White **G-1.** Male **G-2.** Female **H.** *Pontia chloridice* - Small Bath White **H-1.** Male **H-2.** Female

PLATE 17 A: CABBAGE WHITES AND BATH WHITES

A. *Pieris napi* - Green Veined White **A-1.** Male **B-1.** *Pieris brassica* - Large Cabbage White, male **B-2.** Female **C.** *Pieris canidia* - Indian Cabbage White **C-1.** Female **D.** *Pontia daplidice* - Bath White **D-1.** Male **E-1.** *Pieris devta* - Chitral Banded White, from Chitral **E-2.** From Baltistan **E-3.** From Balochistan **F.** *Pieris ajaka* - Murree Green Veined White **F-1.** Male **F-2.** Female

PLATE 17 B: CABBAGE WHITES AND BATH WHITES

G-1. *Pieris kreuperi* - Kreuper's Small White, male **G-2.** Female **H.** *Pieris glauconome* - Desert Bath White
H-1. Male **H-2.** Female **I.** *Pieris callidice* - Lofty Bath White **I-1.** Male **I-2.** Female

PLATE 18: BLACKVEINS AND JEZEBELS

A. *Pieris rapae iranica* - Small Cabbage White **A-1.** Male **A-2.** Female **B.** *Pieris deota* - Kashmir or Alpine Large White **B-1.** Male **B-2.** Female **C.** *Aporia nabellica* - Dusky Blackvein **C-1.** Male **C-2.** Female **D.** *Aporia leucodyce* - Himalayan Blackvein **E.** *Delias Eucharis* - Common Jezebel **E-1.** Female **F.** *Delias belladona* - Hill Jezebel **F-1.** Male **G.** *Delias sanaca* - Pale Jezebel **G-1.** Male **G-2.** Female

PLATE 19: PIONEER, GULL, WANDERER, YELLOW ORANGE TIP AND EMIGRANTS

A. *Belenois aurota* - Pioneer **A-1.** Male **B.** *Cepora nerissa* - Common Gull **B-1.** Male **C.** *Appias libythea* - Striped Albatross **C-1.** Male **C-2.** Female **D.** *Pareronia anais* - Common Wanderer **D-1.** Male **D-2.** Female **E-1.** *Ixias pyrene* - Yellow Orange Tip, males **E-2.** Female **F.** *Catopsila pomona* - Lemon Emigrant **F-1.** Male, monsoon brood **F-2.** Male, dry season brood **G.** *Catopsila crocale* - Common Emigrant, female

Status: This is another very high-elevation Apollo, collected mostly from the summits of passes in the Karakoram between 4,600–5,100m (14,500ft and 16,800ft). It has been collected from the upper region of Khapalu valley, the Saltoro mountain range, the summit of Katichula pass, and at Haldi, all in Baltistan. Sanders (1955) found it not uncommon in the surrounding mountains of the Burzil Pass in eastern Astor district, where it preferred to keep to shale and scree slopes, avoiding grassy areas, where he found *P. epaphus* in the same localities. It is more widespread and commoner than *P. acco* in northern Baltistan.

13. *Parnassius boedromius*

Wingspan 45–50mm

Description: This small Apollo is very similar to *Parnassius simo* except for having paler, less distinct markings. The upper forewing bears only a rather broad, pale grey band along the sub-marginal region, and lacks the distinct discal band which is prominent in *P. simo*. The hindwing upper surface has a sagittate row of dark grey spots sub-marginally, and the sub-marginal black spot between veins 5 and 6 is more prominent than in *P. simo*, whilst the post-discal spot is wholly black, and not red-centred as in *P. simo*.

Status: *Parnassius boedromius* was originally described by Avinoff from specimens collected in the Alexandra Mountains in Central Asia. Later it was regarded as a sub-species of *P. simo boedromius* (Ackery, 1975). Häuser (1993) treats it as a full species, following Kreuzberg (1985), who found *P. simo* and *P. boedromius* flying sympatrically as well as being morphologically different. Two specimens collected by Colonel Lorimer from Gilgit and presented to Tytler were described by him (Tytler, 1926) as *P. simo nov.* sub-species *lorimeri*, though he wrote at the time that it was very like *P. boedromius*. This is now recognized as being synonymous with *P. boedromius*, and the Gilgit specimen forms part of Tytler's collection in the British Natural History Museum. It was collected from the Kine-Chish pass, south-west of Gilgit town, at 4,200m (14,000ft) (Tytler, *op. cit.*).

Habits: In the Altai Mountains the larval foodplants belong to the family Scrophulariaceae (Kreuzberg, 1987), and they have been found feeding on *Lagotis globosa* and *Lagotis decumbens*. The former grows at higher altitudes in the Karakorams of Pakistan and the latter grows at lower altitudes in more mesic conditions, having been collected from the Deosai Plateau.

14. The Regal Apollo

Parnassius charltonius
Wingspan 80–90mm

Description: As its name suggests, this is one of the finest Apollos, with a very large red ocellus on the hindwing, usually white-centred, and extending from vein 4 up to 6. The forewing bears two marginal and sub-marginal vitreous grey bands and thick black bars across mid cell, end cell, and disco-cellular. There is a dusting of black scales mainly around the junction of veins, and no red spots on the forewing. The hindwing, besides the disco-cellular spot, has an anal red bar and a small red ocellus on the outer end between veins 7 and 8; also a sub-marginal ring of bluish-white-centred black spots, surrounded by a greyish band. There is also a narrow black marginal line, interrupted at the veins. The female is larger than the male and after fertilization has a very distinctive, rounded abdominal plug, heliciform, or snail-shell-shaped, when viewed from the side.

Status: This is a medium elevation Apollo frequenting less barren mountain passes. Collected from Shandur Plateau between 3,900–4,370m (13,000–14,500ft), the Shishi Koh valley, and Bangol, all in Chitral. Sanders (1955) collected it from the Burzil Pass in eastern Astor district in July, where it was sympatric with *P. simo, P. stoliczkanus*, and *P. epaphus*. Also recorded from the Karakorams and the Babusar Pass, between Chilas and the Kaghan valley.

Habits: Sakai (1981) records the larvae in Afghanistan as feeding on *Corydalis thyrsiflora*, and *Corydalis gortschakovii*, both species which are widespread in Chitral and Baltistan, growing up to 4,400m (14,500ft). Studies in the Karakorams, in Baltistan, and from the Babusar Pass (Häuser, *et al.*, 1985) revealed that their eggs are rounded in dorsal view and matt white in colour, with the surface minutely pitted with reticulate indentations. The larva in its fifth or final instar reaches 40–50mm in length and is greyish-black in colour, covered with short black bristle-like setae. Around each segment are six spots, whitish-green in colour. When disturbed, the caterpillar typically rolls into a ball and, according to the terrain, rolls down into a crevice. The pupa is dark brown in colour, smooth on its surface, stout and plum-shaped, about 25mm long and 9mm broad with a rounded posterior. It is housed in an oval, pale yellowish-green cocoon made of densely spun larval silk. It was observed that in cloudy or wet weather the larvae sheltered under thick stone slabs, where the ambient temperature would remain fairly constant. Larvae were found at 4,400m (14,400ft) on the Shandur Plateau, and at 4,100m (13,450ft) on the Babusar. They undergo diapause, and spend the winter either as fully-developed larvae inside the egg, or as newly-hatched larvae. It is not known whether the life cycle is completed within one year, or whether some pupae also undergo diapause if developed late in the summer. It was observed that numbers of adult male butterflies would spend the midday hours patrolling along a cliff face where the updraughts enabled them to fly over a high trajectory giving maximum view. Surprisingly, females were found only rarely visiting these male colonies, and wandered quite widely. One female was seen as low as 3,100m (10,200ft) near the village of Babusar, below the Pass.

15. *Parnassius loxias*

Wingspan 55–70mm

Description: This Parnassian is slightly smaller in size than the closely-related *P. inopinatus,* and is much smaller that *P. charltonius*. It is, however, similar to *P. charltonius* in general wing pattern, but all markings are thinner and less heavily black-scaled. The upper hindwing bears two red ocelli like *P. charltonius*, but the lower one between veins 3, 4, and 5 is much smaller in area. The male does not show any ocelli, either black or red-centred in the hindwing tornal area as does *P. charltonius,* and the female has only traces of black spots in this area. However, like the Regal Apollo, *P. loxias* bears a row of five blue-centred sub-marginal black spots on the hind wing.

Status: Originally described as a species by Püngeler in 1901, it was later considered as a sub-species of *P. charltonius* (Bang-Haas, 1927). Later fieldwork on its ecology and breeding biology (Kreuzberg and Pljushch in Häuser, 1993) established the fact that it is a distinct species. Formerly only known from the Tien Shan in Khirgizia and Xinjiang (China), it was collected from the Khunjerab in the mid-1980s, the specimen now housed in the Copenhagen Museum (Harish Goankar, pers. comm., 1996).

Habits. This species has been well studied by Russian biologists in the Central Tien Shan Mountains (Kreuzberg and Pljushch, 1989). It undergoes diapause during the winter as a fully-developed larva in the egg, and there is only one generation per year. The larvae hatch during late April and early May. Their foodplants are *Corydalis* spp. of small Alpine plants growing in rock crevices. In the Tien Shan, foodplants identified were *Corydalis krasnovi, Corydalis hindukushensis,* and *Corydalis fimbrilifera.* When disturbed, like all Parnassian caterpillars, the larvae curl up into a tight ball and often roll down the slope. They will only evert their osmeterium, which is coloured yellow, under extreme irritation. The larva is greyish-blue in ground colour when fully grown, with rings of spots around each segment alternating velvety black, yellow, grey, and light orange. There are short bristly setae on each segment also. The pupa is formed in June, sheltered under slates or in rock crevices. The first adults emerge in the first week of June, with the last fresh adults seen on the wing in late July, some surviving in worn condition up to the end of August. They fly only in good weather, from about 8 a.m. up to 2–3 p.m., and within an altitude range of 2,500–4,100m (8,300–13,500ft). By late afternoon the adults descend to 2,750–3,100m (9,000–10,200ft) and then hide in rock crevices for the night. After first identifying the larval foodplant, females lay their eggs singly, some 40cm away from the foodplant, usually on a rock. Their eggs are circular when viewed from the top, and turban-shaped from the side. They are coloured white with a matt surface. Well-formed larvae can be seen inside the egg after two months.

16. *Parnassius inopinatus*

Wingspan 70–80mm

Description: Slightly smaller than *P. charltonius*, it is otherwise very similar in appearance, with a large prominent red-centred ocellus in the post-discal area of the hindwing, and two conjoined, red-centred hindwing tornal spots. In the male, the end cell ocellus is usually white-centred. It differs from *P. charltonius* in the phragis of the female being straight, not curving like a snail's shell. The male averages smaller than the female; both sexes have blue-black sub-marginal spots, but these only extend up to between veins 4 and 5, whilst in *P. charltonius* and *P. loxias*, these blue-centred spots extend into the area above vein 5.

Status: First described as a distinct species by Kotzsch in 1940, this was collected from the Safed Koh Range within Pakistan by O.C. Ollenbach in 1902. It has also been collected by Sakai (1981) from Nangarhar Province in western Afghanistan from areas bordering Pakistan on the Safed Koh Range. Ackery (1975) also lists the Firus-Kuhi Range and Koh-i-Baba Range as collection localities.

Habits: The larval foodplant in Afghanistan (Sakai, 1981) is *Fumaria vaillantii* which has also been collected (Nasir and Ali, 1972) from the Upper Kurram Agency, in the Safed Koh region of Pakistan.

Papilioninae—Swallowtails, Peacocks, and Mimes

Classification

Taking the subcontinent as a whole, the Papilioninae family contains many of the largest and most beautiful of all butterflies in the world, including the huge Birdwings of the tropical high rainfall areas, genus *Troides*. Compared with families such as the 'Blues' or 'Skippers', the total number of species within the Papilioninae is not great, with only an estimated 700 worldwide (Ackery, 1984). Pakistan, with much of its area Palaearctic and arid, has only very few representatives of this family, with eleven species out of about ninety-four recorded for the whole subcontinent. Papilioninae adults are very variable in wing shape. Many species are large in size, with the hindwing tailed—this feature always extends from vein 4 on the hindwing—and the tail is often spatulate in shape. The forewings are often rather pointed and narrow. Many species are predominantly black in colour. The closest family to the Papilionidae is thought to be the Pieridae or 'Whites'; they can be differentiated by the fact that all the Papilionidae possess fully-developed forelegs in both sexes, and only eleven veins on the hindwing, vein 1b being absent while the Pieridae which always possess veins 1a and 1b on the hindwing. The forewing always bears twelve veins, with a median spur on vein 1b, lacking in the Parnassiinae (*see*,

explanation under the account of Parnassiinae, and Fig. 1). There is usually not much difference in appearance between the sexes, but males often possess specialized scent scales in a fold along the hindwing inner margin.

The eggs of the Papilioninae are generally small for the size of the adult butterfly, spherical in shape when viewed from above, and often yellow or turning orange in colour after being laid. The larvae are spindle-shaped, with a large head, which is retractable under the first thoracic segment. They are never hairy but often possess long fleshy horns at either end, and are banded in conspicuous colours as they develop in size. Some species, particularly the genus *Papilio* have protective brown and white markings resembling bird droppings at the early larval stage, and all the larvae possess osmeteriums, which can be extruded at will and are contrastingly and brightly coloured (*see*, account of Parnassians). They are usually rather sluggish in habit and, in some of the larger species, there is a peculiar thickened saddle-like process covering the forepart of the body, and sometimes ocelli or eye-like markings on the side of the forepart of the body which may serve to startle predators. In their early stages, many larvae possess bristle-like fleshy protuberances along the sides of their body which gradually disappear with subsequent moults.

All the Swallowtail larvae feed on plant families containing strongly aromatic and even poisonous substances in their leaves, and these chemicals are often sequestered by the larvae and used to disperse scent through the osmeterium, as well as making the caterpillars themselves distasteful to certain predators. Hence the conspicuous pattern of many species larvae, making them easily recognizable to predators. The foodplants of the Mimes (*Chilasia* spp.) are *Lauraceae* and *Anonaceae*, of the black-bodied Swallowtails, *Rutaceae*, and of the Yellow Swallowtails, *Umbelliferae*. There is much variation in the shape of the pupae, but they are generally supported upright along a stem or twig by a silken girdle around the thorax and by the tail claspers holding onto a silken pad. The chrysalis of the Common Mime, *Chilasia clytia*, is unique in resembling a broken-off twig.

Genus GRAPHIUM—Kite Swallowtails

The members of this genus generally have quite long, slender tails on the hind wing, and males have a patch of specialized scent wool borne in a fold on the hindwing inner margin, which itself is fringed with hairs. The forewings in the costal region have very thin scaling, and most species are marked with bluish or green windows against a black background.

17. The Glassy Bluebottle

Graphium cloanthus
(Syn: *Zetides cloanthus*)
Wingspan 82–95mm

Description: *Graphium cloanthus* belongs to the genus of 'Kite Swallowtails' or 'Jays'. They are characterized by having their forewings long and pointed at the apex, and only thinly covered with scales, hence the common name Glassy Bluebottle given to this species. Both sexes have a conspicuous tail on the tornal margin of the hindwing extending from vein 4. Their 'windows' are a pale blue-green, with the costa and wing margins broadly bordered with velvet black. There are four pale-green semi-transparent spots on the termen of the hindwing in the black border, and the anal angle on the hindwing often bears a small crimson spot. The under hindwing bears a thin red curved line at its base between the pre-costal area and veins 7 and 8, and a series of crimson-rimmed black spots from the base of the cell down to the anal angle. The inner margin of the hindwing bends inwards and bears many fine grey hairs, important in scent distribution by the male. The sides of the abdomen are striped longitudinally with pale yellow and black.

Status: Collected TJR July, from summit of Mukhshpuri, Murree Hills. It frequents open grassy glades up to 2,700m (9,000ft), and has also been seen (author) in the Margalla Hills down to 600m (2,000ft). Butler (1886) records a worn specimen taken at Murree on 10 September. Rather rare even within its limited distribution. Can be seen on the wing during and after the monsoon.

Habits: The larvae are green, with a yellow line along the sides, below which is a white line. The head is greenish-yellow with two black and white tubercles. The 4th segment has a yellow saddle-like band, and the 13th segment is bluish-green. The pupa is bright green with yellow bumps. The larval foodplants are *Miliusa velutina* (a small deciduous tree growing in the Murree foothills), *Machilus duthiei, Machilus odoratissima* (family Lauraceae), and *Michelia* spp. The butterflies are strong in flight and, where common, congregate at damp patches.

18. The Common Mormon

Papilio polytes
(Syn: *Princeps polytes*)
female forms *stichius* and *romulus*
Wingspan 90–100mm

Description: This species belongs to a group of Swallowtails called 'Helens' and 'Mormons', which have prominent tails on the hindwing and are mainly black, with a discal band of elongated white spots on the hindwing, and usually red lunules along the hindwing post-marginal area. *Princeps polytes* is interesting in that the females have developed several forms which are distinct from the males and which mimic the Common Rose, *Tros aristolochiae*, and the Crimson Rose, *Tros hector*. The form known as *P. polytes romulus* has been collected by the author in lower Sindh, and is distinguished by its close imitation of *T. hector* in having pale greyish-white bands across the forewing, both in the end cell and post-discal regions, these bands divided by narrow black lines along the veins and mid vane. The hindwing has the discal white band of elongated spots replaced by crimson spots as well as all the sub-marginal lunules being crimson. The form *P. polytes stichius*, which the author found common all over the Punjab, imitates the Common Rose, and on the hindwing these discal spots are red in the lower area adjacent to the body, becoming white as they ascend to the cell. Unlike *Tros* spp., however, there are no red markings on the body. *P. polytes stichius* also has greyish-white bands on the forewing post-discal area, and is the form illustrated. The males, and also some females, have a more extensive discal band of white spots on the hindwing which extends up to vein 7, and this band is often quite creamy yellow. There is no trace of any crimson, and the forewing bears small marginal white spots which peter out towards the apex.

Status: Widespread and common during and after the monsoon in Sindh, Punjab, and the NWFP, in the plains, occurring up to Islamabad.

Habits: The egg is spherical in shape, pale orange or clear lemon-yellow coloured, smudged with brown, and is laid singly on the upper or underside of the leaves of its foodplant. The larva is velvety dark grass green with a white lateral line bordered yellowish. The head is yellow, and there are raised crests on segments 4 and 5, and white bands on segments 7 to 10. The osmeterium is scarlet. Sevastopulo (1940) describes a full-grown larva with ocelli on either side of the 3rd segment, encircled yellow, and the pupa as having a bi-fid head with small horns, the thorax keeled, and suspended by a girdle of silk and a tail pad of greyish silk. Its colour is variable, usually green, sometimes mottled with brown and grey. The foodplant of the caterpillars belong to the family Rutaceae, such as *Murraya paniculata* and *Murraya koenegii*, *Tripahasia* spp., *Zanthoxyllum* spp., and *Armatum* spp. The first is sometimes grown in gardens for its scented flowers, and the others are all plants of Pakistan's foothill zone. The larva will also feed on limes and oranges, which belong to the same family. The butterflies regularly visit flowers, and are strong flyers.

19. The Common Rose

Pachliopta aristolochiae
(Syn: *Tros aristolochiae*)
Wingspan 80–110mm

Description: This species belongs to a large group of black Swallowtails with red on their bodies which sequester poison from the larval host plants, making them distasteful to potential predators. The forewings are very elongated and black, with paler greyish stripes inside the distal part of the cell and between the veins. The hindwing is also narrow and elongated, with a series of white spots around the lower end of the cell, the lowest one near the body being tinged with pink. There is also a series of crescentic pinkish-crimson spots along the margin. The abdomen is red distally and along its sides and also the head. The hindwing margin is scalloped and bears a spatulate tail.

Status: Widespread in Sindh, and Punjab in the plains, but commoner in the south, especially Sindh, and Rahimyar Khan district in southern Punjab; absent from northern Punjab.

Habits: The larva is a rich blackish brown with two rows of red or pink tubercles down each side of the body. The osmeterium is orange. In some specimens the whole of the 7th segment is white. The pupa is fawn coloured and has the appearance of a dried curled leaf. The thorax is keeled with five thin white lines, and is suspended by a girdle around the body and a tail pad of brown silk. The foodplant of the larvae is *Aristolochia* spp.

20. The Common Windmill

Atrophaneura polyeuctes
(Syn: *Tros philoxenus*)
Wingspan 110–130mm

Description: This species is another red-bodied Swallowtail whose larvae feed on poisonous plants, from which they are able to sequester toxins which protect them from predators. Even to humans they smell unpleasant. As can be seen, this is a large butterfly with very long and narrow hindwings, deeply scalloped along the margin. The forewings are black with greyish streaks, but otherwise unmarked, whilst the hindwing bears a large white square-shaped patch between veins 4 and 5, just below the cell, and a small white patch below this between veins 3 and 4. There are three broad pinkish-red lunules in the post-discal region from the tornus outwards. There is a broadly spatulate tail which also bears a terminal red spot. The abdomen is crimson with black inter-segmental bands and small black spots along the sides. The head is also red. The sexes are alike.

Status: This is a Himalayan species, and is common in forested areas in the Murree Hills, also in the lower Kaghan valley, and in Azad Kashmir especially Machiara. Flying from 1,500–2,700m (5,000–9,000ft).

Habits: It is interesting in that the mimics of other red-bodied Swallowtails are mainly plains-dwelling butterflies, but this hill butterfly is mimicked by a large black day-flying moth, *Epicopia polydorus*, which the author has collected in Dunga Gali, where it appeared uncommon. The full-grown larva is 2.5 inches long, and pale purple-brown in colour, with a small rugose head, which is covered with black hairs. There are short tubercles on the 2nd segment, and a row of smaller tubercles on each segment from 3rd to 13th. These are tipped red, and there are dark purple spots between the tubercles along the back. The larvae feed on species of *Aristolochia*, probably *A. punjabensis*, which grows in the Murree Hills. In the early stages they feed gregariously. The pupa is reddish-ochraceous with a broad head slightly bifurcate, the wing cases swollen laterally, and the abdominal segments with lateral foliaceous appendages. It is capable of making a squeaking sound when touched. The butterfly is especially fond of the flowers of the Horse Chestnut (*Aesculus indica*) and, later in the summer, of Buddleia. The adult insect itself has an unpleasant smell, and has a rather lazy, floating sort of flight.

21. The Southern Swallowtail or Balochi Swallowtail

Papilio alexanor
Wingspan 75–90mm

Description: This Swallowtail has a similar colouration and body pattern to the more widespread and less rare Yellow Swallowtail, *Papilio machaon*. It differs from the latter mainly in lacking black lines along the veins, and is also without black areas at the base of fore and upper wings, this area being clear yellow. There is a broad black diagonal band across both fore and hindwings, a short black bar across the end of the cell, and another just beyond the end of the cell. The hindwing is tailed and deeply scalloped, with a broad sub-marginal band frosted with blue and green, and bordered black. There is a small scarlet tornal spot, and the margins of both wings are edged with black. The underside is a paler creamy yellow, and the body is black with two thin longitudinal yellow lines on top of the thorax and another lateral line along the sides of the abdomen.

Status: Very rare in Pakistan, occurring mainly in the extreme border areas of the Chagai in Balochistan and in juniper forest around Ziarat and the upper Uruk valley. Recently Raza Abbas, filming for Pakistan Television in the Toba Khakar Range of Zhob District at Torghar, filmed at close range a fully-grown larva feeding on an unidentified Umbellifer, identified by the author when shown the film. Its range may therefore be more extensive than the few museum specimens indicate.

Habits: The larvae are described as whitish-green, with a black band on the anterior part of each segment, and nine black and yellow dots. The pupa is like that of *P. machaon* but more flattened on the dorso-ventral side, with the head square, not bifurcate. It is coloured light or dark grey and is fastened upon rock faces, with which it blends perfectly. Sakai (1981) records the larval foodplant in Afghanistan as members of the Umbelliferae. In southern Europe, the main foodplant is the Mountain Meadow Seseli, *Seseli montanum*, and *Seseli dioicum*, also *Trinia vulgaris*, all members of the Umbelliferae. In Balochistan there are many representatives of the *Ferula* genus which are extremely xerophytic, and it is likely that in the Chagai one of these would be the larval foodplant.

22. The Common Yellow Swallowtail

Papilio machaon
Wingspan 75–90mm

Description: This handsome butterfly is primrose yellow above with the base of both wings black, dusted with yellow scales. The cell and veins are outlined in black, and there is a broad black bar mid cell, with a second bar curving round the disco-cellular region. The forewing is margined black and contains a row of yellow spots, inside this a sub-marginal band of black, scalloped along its inner edge and thickly dusted with blue and green scales. The hindwing at vein 4 is produced to a slender tail and is scalloped along the margin, with larger yellow spots inside the margin and a broad sub-marginal band, dusted with blue and green scales. There is a large red tornal spot, and the inner margin of the hindwing is also narrowly bordered black. The underside is a paler creamy yellow, and inside the sub-marginal band of green and blue scales there are russet red spots between veins 2, 3, and 4, as well as a black-ringed tornal spot of red. The abdomen is largely yellow with narrow longitudinal black lines. The sexes are very similar, though females tend to be larger.

Status: Widespread throughout the northern mountain areas from about 1,800–3,000m (6,000–10,000ft) from Chitral and Swat through to Gilgit. Evans (1950) records it in Balochistan from May to September in the higher valleys such as Uruk, Ziarat, and on the Shingar Range, where it was uncommon. The author did not encounter it in the Murree Hills, but Butler (1886) records it as quite common in August around Murree, with a specimen being taken from Campbellpur on 9 July. Occurs up to 4,270m (14,000ft) in alpine meadows. Collected TJR in Swat at Miandam and observed in Indus Kohistan above 2,700m (9,000ft).

Habits: The eggs are spherical and yellow at first, developing red-brown markings. They are usually laid singly on the flowers of the foodplant, but sometimes in batches of 4 or 5. The larvae when young are black with numerous tubercles and a white band round the middle. In later instars green bands develop around each segment. In the last instar the larva is green with black bands separating each segment, which can be very variable in width. Along the sides are small red or yellow tubercles. The osmeterium is bright orange. The pupa is variable in colour according to the substrate to which it is attached. It is held by a silken band across the thorax and by the tail claspers attached to a silken pad; the commonest colour is green. The head is bifurcate, with two rows of dorsal tubercles which are black with red tips. The larval foodplants are the Common Carrot (*Daucus carrota*) which has naturalized in many hill areas, as well as *Heracleum* spp., *Selenium* spp., and *Angelicine* spp. The butterfly is strong on the wing and fond of visiting flowers, over which it generally hovers without actually settling.

23. The Ladaki Swallowtail

Papilio ladakensis
Wingspan 70–80mm

Description: This Yellow Swallowtail is very similar to the Common Swallowtail, *P. machaon*, but differs in having a very short tail on the hindwing. It is also more hairy on the body, with a slightly paler yellow wing colour if a good series are compared, and the hindwing tornal red spot is smaller but rounder in shape. The sub-marginal spots on both fore and hindwings are shorter and broader than in the Common Yellow Swallowtail.

Status: *P. ladakensis* was treated by Talbot (1939) as a sub-species *P. machaon ladakensis*, but was in fact described as a distinct species by Moore in 1884. More recent studies have shown that the genitalia differ from *P. machaon*, from which it is also generally allopatric, being confined to higher elevations in the extreme northern mountain ranges. It has been collected from the Pamirs in northern Chitral and is also not uncommon on the Khunjerab in northern Hunza. It flies between 3,600–4,200m (12,000 and 14,000ft).

24. The Lime Butterfly

Papilio (syn: *Princeps*) *demoleus*
Wingspan 80–100mm

Description: The Lime Butterfly lacks any tail on the hindwing, and is more heavily patterned with black than *Papilio machaon*. The upper forewing is largely black in the cell and outer wing margin, with a series of irregular yellow spots in a discal band, the lower spots being the largest, and a second, much smaller, outer band of sub-marginal spots. There are two yellow spots at the upper end of the cell and several scattered yellow spots in the apical region. The upper hindwing bears a red tornal spot; the discal black band is not interrupted with yellow spots but is dusted with yellow scales, and has a series of sub-marginal yellow spots, with a prominent black circular spot in the apical region dusted with blue. The underside is paler yellow, with the black areas more heavily dusted with yellow, and on the hindwing a series of ochraceous red spots in an irregular discal band, each rimmed with black. When freshly emerged, the yellow colour is pale creamy on the upper side, but this darkens to deeper yellow in older specimens. The body is yellow with thin black longitudinal lines, and the sexes are similar.

Status: One of the commonest butterflies in the plains in all four provinces, and can be seen on the wing in every month, but most frequently after the monsoon from August to October. It also occurs rarely in the major lower valleys of Chitral and Swat.

Habits: The egg is spherical and pale yellow or greenish-yellow, with a pebbled surface. The larvae in their early stages are blackish-brown with a white 'V' mark and creamy white diagonal lines at the anal end, and some rows of bristles. They resemble bird droppings, an effective camouflage. When full-grown, the larvae have a brown head and their bodies vary from yellow-green to deep green with a broad greasy-white sub-lateral stripe, and small spines on the 11th segment. There are black-centred, fawn- or white-ringed ocelli on the sides of the 3rd segment. Sevastopulo (1939) describes the osmeterium as deep orange in colour, whilst Talbot (1939), describes it as flesh-coloured. At full growth the caterpillar is 33mm long. The pupa is generally green if formed against a leaf or twig, but can be some shade of yellow-brown against other backgrounds. The head has two frontal projections, and there are sub-dorsal tubercles on each abdominal segment. It is generally fixed with quite a short silk band around the body. The larval foodplants are mainly *Citrus* spp., limes and pomelos being preferred, but they will feed on *Zizyphus* and other members of the Rutaceae family, such as *Ruta graveolus*, and *Glycosmis pentaphylla*. The butterfly has a swift, rather jerky flight, and frequently visits flowers; females can often be seen patrolling through citrus orchards.

25. The Common Peacock

Papilio polyctor
Wingspan 90–130mm

Description: The 'peacock' butterflies are aptly named, having iridescent blue patches on their wings, with bands of golden-green dusting on a black background—colours reflected in the neck and train plumage of the peacock bird. *Papilio polyctor* differs from the only other species of this group occurring in Pakistan in that the male has broad black stripes of woolly scent scales on the veins of the forewing, and with the blue patch on the hindwing not entering the cell as it does in *Papilio arcturus*. In both sexes the forewing costa is arched and the outer margin is slightly concave, and there is a broad sub-marginal band of golden-green scales, with more scattered green scales basally. The hindwing has a long spatulate tail protruding from vein 3, and a series of purplish-red lunules encircling the top of black spots along the wing margin. The iridescent blue-green post-discal patch varies in size and shape and in blue or blue-green intensity according to the angle of light, and extends from the hindwing apex down to between veins 1a and 1b. The under hindwing bears much more prominent violet and red lunules along the entire wing margin, with a thinner scattering of golden-green scales, and the forewing apex and discal areas are rather greyish. The body is black with a scattering of green scales on the dorsal surface of the abdomen.

Status: This 'Peacock' occupies lower elevations than the similar Blue Peacock (*P. arcturus*), and is not uncommon in the lower Murree Hills, being recorded as high as Lower Topa, also throughout Hazara district, around Abbotabad, and the main vale of Swat, but rare in the lower valleys of Chitral. It is on the wing from mid-April to the end of September, from about 600–2,100m (2,000–7,000ft).

Habits: The full-grown larva has both head and body green, the body being speckled with paler yellowish-green. The forepart of the body (thorax) bears a thick shield-like covering which can project over the head, and is marked with slender black involute lines. There are four oblique darker lateral stripes and bright blue sub-dorsal spots from 7th to 10th segments. The osmeterium is yellow. Early larval stages look quite different, being coffee brown with spines and looking like a bird dropping. The pupa is suspended by a girdle, with the tail fastened to a pad of white silk on the stem or underside of the foodplant. It is pinkish-brown in colour with darker, purplish-brown markings (Sevastopulo, 1947), but some pupae have been described as bluish-green in colour (Talbot, 1939). Its head is produced to two triangular points, and the lateral keel is very distinct, with broad, low, thoracic horns. The larval foodplants belong to the family Rutaceae, including *Xanthoxylum armatum*, very common at lower altitudes, up to 1,500m (5,000ft) in the Murree Hills and Swat, also *Zanthoxylum acanthipodium* and *Clausena* spp. Mohyuddin (1987) found the larvae feeding on *Rosa* sp. leaves in Rawalpindi in April. The butterfly is a fast, strong flyer, typically hovering over flowers when feeding, and fond of open sunny glades away from trees.

26. The Blue Peacock

Papilio arcturus
Wingspan 110–130mm

Description: Similar in general appearance to *Papilio polyctor*, but in this species, the male lacks any black elliptical bars of woolly scent scales on the veins of the forewing, and both sexes have a deeper blue post-discal patch on the hindwing, which enters the lower part of the cell. There is a rather narrower band of golden-green scales along the sub-marginal area of the forewing than in *P. polyctor*, and on the hindwing a black tornal spot, ringed with violet and mauve, and two similarly-coloured lunules between veins 3 and 4, and 4 and 5. These lunules are more prominent than in *P. polyctor*. As in the former species, the hindwing margin is scalloped, with a spatulate tail protruding from vein 3. The underside is blacker with hardly any green scales on the under forewing, which has a more prominent grey post-discal band than in the Common Peacock. The under hindwing, besides five marginal lunules, has two tornal spots completely ringed with violet and red, turning to orange on the outer margin.

Status: Common on hilltops and at higher forested elevations in open glades, from the Murree Hills across to the Kaghan valley. Not recorded from Chitral.

Habits: The full-grown larva is green in body and head, with a thick fleshy thoracic shield, margined with yellow, covering the first four segments. The rest of its body is heavily speckled with yellow. The 3rd segment bears ocelli on its sides, with brick red centres outlined above in white. There are 4 mauve spots on the upper part of the shield, and yellow diagonal stripes on the hind end of the body. The osmeterium is yellow. The pupa is suspended by a girdle and tail pad, and is coloured purple-brown with a darker dorsal stripe, the wing cases greyish-brown, and suffused orange on the ventral side of the abdomen. The adult or imago is, like others of the genus, a strong swift flyer, fond of visiting flowers, especially Buddleia in late summer. The larval foodplants are Rutaceae of various spp. Males are territorial, often frequenting small hilltops or bush-dotted ridges, their habit being to perch and sally forth to intercept visiting butterflies until they are identified.

27. The Common Mime

Chilasa clytia
form *dissimilis*
Wingspan 90–100mm

Description: This species belongs to a genus of five species, of which only *C. clytia* occurs in Pakistan. All the genus mimic *Danaid* species, and the form *C. clytia dissimilis* which occurs in our area mimics the Blue Tiger, *Danais lymniacae*. Form *dissimilis* is a pale yellowish-green, with each vein broadly margined in black and along the costa and wing margins. The forewing upper side has the base of the cell yellow, dividing into three or four separate streaks, and the yellow areas between veins 2, 3, 4, and 5 are split into two to three arrow-shaped spots. The upper hindwing cell is wholly yellow, and there is a marginal row of small yellow lunar spots. The hindwings are a paler colour, almost white, with a series of orange-yellow spots between the veins along the hindwing margin. The abdomen is striped yellow and black longitudinally, with black inter-segmental rings, whilst the thorax bears white spots like a *Danaid*.

Status: Rare in Pakistan but may occur anywhere along the Himalayan foothill zone. Has been collected from Lahore and Murree, and seen by author in Dunga Gali at 2,500m (8,300ft) in early April.

Habits: The full-grown larva is velvety black or dark green with two dorso-lateral rows of carmine red spots and a cream-coloured dorsal band from segments 3 to 7, and again from segments 11 to 14. There are two lateral rows of sharp spines on segments 1 to 4, and a single row of spines along the other segments. The osmeterium is a watery indigo blue. The extraordinary pupa is attached by a silken girdle, upright to the tip of a slender twig, and is shaped and coloured exactly like a broken twig. The larval foodplants belong to the family Lauraceae, and include the Cinnamon, *Cinnamomum zeylanicum*, *Alseodaphne semicarpiefolia*, *Litsea sebifera*, and *Litsea deccanensis*. The butterfly is fond of sailing around, without fluttering its wings, from the tops of hills or around tall trees in open places, and the one I watched in Dunga Gali did this continuously, without settling, until I felt obliged to leave it!

Chapter 7
Pieridae—Whites and Yellows

Classification

The family Pieridae consists of medium to small sized butterflies, usually white or yellow in colour, with narrow black markings, and they are sun-loving, adapted to rather open country. Diagnostically they differ from the Swallowtails in that the forelegs in most species are imperfectly developed, and they have bi-fid pre-tarsal claws on their legs. The first pair of forelegs end in four claws. Also the hindwing vein 1b, lacking in the Papilionidae, is always present. Their wing scales contain pterine pigments (Haribal, 1992) which give rise to their predominant yellow and orange colours. In our region, and in fact the whole of south-east Asia, only two sub-families are represented. The Pierinae have the hindwing with a well-developed pre-costal vein, and the palpi are always hairy. The second sub-family, the Coliadinae, comprise those species in which the pre-costal vein is absent or hardly developed, and the palpi are not hairy. In nearly all the Pieridae, the sexes are different, the females being more heavily marked with broader dark borders. A few genera, like the *Pieris* spp. (Cabbage Whites), the *Apporia* spp. (Blackveins), and the *Delias* spp. (Jezebels), store unpleasant chemicals in their bodies, obtained from their larval foodplant, and so are avoided by some predators. This has given rise, over the course of evolution, to non-poisonous Pierids mimicing the above genera, such as the Gulls (*Huphina* spp.) and Albatrosses (*Appias* spp.), whilst the Wanderers (*Parenonia* spp.) mimic non-related Danaids. The males of some species have abdominal brushes to produce mate-attracting scents (*Appias* spp.), others have plume-like scales (*Delias* and *Pieris*), whilst the Clouded Yellows (*Colias* spp.) have brands.

In this family the eggs are very characteristic, being taller than wide and flask-shaped, often sculpted with longitudinal ridges, terminating at the top in a circle of small teeth (*see*, Fig. 2).

The larvae are generally smooth or pubescent with microscopic hairs, and usually green in colour, with paler longitudinal stripes. The pupae are of two types. One is suspended upright by a silken girdle like Papilionid pupa, with the anal claspers attached to a pad of silk on a stem or twig. These pupa have swollen wing cases and are smooth, and include *Leptosia, Catopsila, Ixias, Parenonia, Terias,* and *Colotis*. The second type of chrysalis is attached horizontally to a leaf and usually has spines down the mid-dorsal region of the thorax and abdomen, and the head produced into a point, often curving outwards. These include *Delias, Appias, Anaphaeis, Huphina, Colias,* and *Pieris*. Usually the pupae are pale yellow or green, some with black speckles, others unmarked. The foodplants of the larvae are varied. The Clouded Yellows (*Colias* spp.) feed on clovers and vetches, the Pierids on Brassicae, with some of the higher-elevation spp., or those adapted to arid zones, feeding on smaller cruciferi such as *Arabis*, and the *Colotis* spp. feeding on *Salvadora* spp. The Himalayan Blackvein feeds on *Berberis*, and most of the *Catopsila* and *Terias* spp. feed on *Cassia* spp. Pakistan, with much open arid country and also mountainous regions, is well endowed with representatives of this family, having a total of fifty-five species. There are ten different species of Clouded Yellows (*Colias* spp.) found in the high mountain areas, and twelve species of the genera *Pieris* and *Pontia* (Cabbage Whites and Bath Whites).

Sub-family Pierinae—Whites

These are mainly found in tropical regions and contain about 700 species worldwide. In our region they include the Cabbage Whites, Arabs, Jezebels, Puffins, Gulls, Wanderers, and Blackvein Whites. The larval foodplants are mainly Brassicaceae, Capparidaceae, Loranthaceae, and Santalaceae. They are mainly white in colour with the underside more cryptically marked, and often the adult butterflies contain unpleasant substances which repel predators.

Genus Leptosia

These are small, weak-flying butterflies with the forewing very rounded at the apex, and with vein 9 absent and vein 10 arising from the end of the cell.

The hindwing has veins 5 and 6 close together at their origin.

28. The Psyche

Leptosia nina
(Syn: *Leptosia xiphia*)
Wingspan 20–45mm, rarely up to 50mm

Description: This is a small butterfly, largely white on the upper side, with the upper forewing rather broadly rounded at the tip and along its outer margin. There is a large black discal spot and a black apex or wing tip. The base of the wings is dusted with black scales, especially the costa, which is speckled with black. The underside of the hindwing is finely streaked with greenish lines. When at rest, the butterfly characteristically folds the forewings inside the lower, giving it maximum camouflage by hiding the black apical spot. The sexes are alike.

Status: Collected banks of Ravi River near Lahore, also occurs sparingly in the foothills north of Sialkot. According to Bell (1912) it is not uncommon in Sindh, close to the coast around Karachi and Las Bela.

Habits: The egg is coloured blue, cylindrical, slightly swollen in the middle, and with thirteen longitudinal ribs, of which only six continue to the top, where they terminate in a ring of minute teeth. The larva is delicate in appearance, about 17mm long when fully grown. It is grey-green in colour with longitudinal lines, a dark one dorsally and another white one laterally. There are rows of short black bristles around the body at intervals and longer white hairs scattered mainly on segments 2, 12, and 13. The pupa is slight and delicate-looking, of a transparent green colour, often with a pink suffusion, and with dorsal and lateral brown lines, and black spots on each abdominal segment. The eggs are generally laid on the underside of a leaf close to the ground, and the larval foodplants are Capparidaceae. They include *Capparis spinosa*, *Capparis rheedii*, and *Crateva religiosa*, syn: *adansonii*, especially around Karachi. The butterfly, as stated above, is a very weak flyer, keeping close to undergrowth and settling frequently, when its forewing is covered by the hindwing closed over its back, so that the green lines on the under hindwing make it practically invisible amongst the grass stems.

29. Shaw's Dwarf

Baltia shawi
Wingspan 30–40mm

Description: This is a small butterfly with the base of the wings heavily dusted with black scales and protected by hairs. The forewing upper side bears four apical black spots, triangular in shape, and the end of the cell has a narrow black V-shaped bar followed by a wider black bar, with its apex downwards, narrowing to a point in the sub-apical region. The underside shows these black marks less strongly, whilst the underside hindwing is heavily irrorated with black scales and is hairy, with a row of greyish spots in the mid-costal region, followed inside beyond the end of the cell by a greyish curved line, its apex pointing outwards. Females average larger in size and have dark grey rather than black marks on the upper wing surface.

Status: Found above 3,950m (13,000ft) in high mountain regions of Chitral. Not rare.

Habits: Nothing seems to have been recorded about this species except the remark that the butterflies fly close to the ground.

30. Butler's Dwarf

Baltia butleri
Wingspan 35–45mm

Description: This is also a small Pierid which is strikingly different on the underwing pattern from *B. shawi*, lacking the black marginal spots on the underside of the forewing. The under hindwings have the veins white, margined narrowly on both sides with black, and no row of sub-marginal spots as in *B. shawi*. As in Shaw's Dwarf, the head and thorax are very hairy, and females average slightly larger in size.

Status: This species has been collected from the Karakoram range in Baltistan. It is on the wing in June, flying between 4,570 and 5,480m (15,000 and 18,000ft).

Habits: The butterflies have been described as having a very rapid wing beat and weak, fluttering flight, more reminiscent of a moth than a butterfly.

31. The Lemon White or Greenish Black Tip

Synchloe charlonia
(Syn: *Euchloe charlonia* or *Elphinstonia charlonia*)
Wingspan 35–45mm

Description: This little butterfly is bright yellow on the upper side, on both fore and hindwings, with the apex (tip) of the forewing slightly arched inwards. There is a broad black band at the end of the cell in the disco-cellular region, and the wing tip is black with a series of yellowish-white spots. The underside of the forewing is similarly marked, whilst the underwing is dark greyish-green with a few small white spots along the costa. The female is about the same size as the male but generally a much paler greenish-yellow, even almost greenish-white in a few individuals. Her forewing is also less prominently arched along the apex, and the black apical region of the forewing is generally more extensive.

Status: Widespread from Chagai to Zhob along western boundary of Balochistan, and also along the Mekran coast around Ormara.

Habits: According to Beazley (in Bell, 1912), the larval foodplants are *Mathiola tessala, Mathiola chenopodipholia,* and, in the Mekran, *Mathiola macranica,* (family Cruciferae), all of which are widespread in the dryer foothill zone inhabited by this butterfly. Beazley describes the butterflies as being swift on the wing, fluttering close to the ground, and often chasing conspecifics. When they settle on a plant they immediately close their wings, and their dark green underwing hides them sucessfully. They will, however, frequently bask on stones with spread wings.

32. Pale Lemon White

Euchloe lucilla
Wingspan 35–45mm

Description: This species is very similar in appearance and habits to *Euchloe charlonia*, but in the male the upper side is much paler lime green varying to greenish-white, without the primrose yellow tones of *charlonia*. It can also be distinguished from *E. charlonia* by the underside forewing apex lacking the grey-green patch, and by having larger costal white spots. The underside hindwing is also much paler grey-green, dusted with black scales. The majority of females are white on the upper forewing, with the upper hindwing only slightly more creamy. It was first recognized as a distinct species by Butler (1886) in describing specimens collected from Campbellpur (now Attock).

Status: This form was described as a distinct species by Butler in 1886, from a specimen collected near Campbellpur (now Attock) in Attock district. Subsequent authors place it as a sub-species of *E. charlonia* (Evans, 1932, and Talbot, 1939). It is now recognized as being confined to the northern foothill tracts of Pakistan in the NWFP and north-western corner of the Punjab, and it occurs in Waziristan, the Khyber Pass, in the hilly tracts of Peshawar district, spreading into Attock district (formerly Campbellpur district) in the Punjab. It is allopatric from the Lemon White, *E.charlonia*, which occurs from across the Mediterranean and Middle East to Balochistan.

Habits: According to Butler (1886) the larvae were found feeding on *Stachys parviflora*, which is common in the Punjab Salt range

33. The Striped White or Green Striped White of Europe

Synchloe belemia
(Syn: *Euchloe belemia*)
Wingspan 36–45mm

Description: Larger than *S. charlonia*, with a slightly less arched margin to the forewing. This medium-sized butterfly is white on the upper side with a broad black band in the disco-cellular region; the wing tip is also black, with a small white window in its tip and a dusting of white scales in the post-cellular black band. The hindwing shows darker greyish shadows of the underwing pattern. The under hindwing is strikingly patterned with zebra-like transverse, dark grey-green bars against a white background, and the forewing apex has four much narrower greenish bars. Females are often larger, with less densely black markings.

Status: Only found in extreme western Chagai and along Mekran coast near Ormara. The first record of its occurrence in Pakistan was from specimens collected near Robat on the border with Iranian Balochistan, by a Mr Cooper (Dover, 1923). It was found frequenting hilltops and sandy mounds. Later it was collected by Mr W.D. Cumming near Ormara during the laying of the telegraph. A typically North African species, widespread in the Middle East and Iran.

34. The Pearl White or Mountain Dappled White of Europe

Synchloe ausonia
(Syn: *Synchloe daphalis*)
Wingspan 40–49mm

Description: Upper surface white with a black narrow spot or band in the disco-cellular region of the forewing and black mottling in the wing tip. There is a grey shadowed pattern showing through the surface of the lower wings. The underside shows a more lunar shaped disco-cellular black spot, white centred, with the dark apical area mottled-greenish black and the underside of the hindwings bear scattered white spots against a background of green with minute yellow spots. The sexes are alike except that females have more rounded wing tips.

Status: Occurs Balochistan where collected Chiltan hills, Torghar, and upper Uruk valley. Collected TJR Saiful Muluk lake, Kaghan valley, at 2,700m (9,000ft) in June. Fairly common in Chitral, flying from March to April at 1,500–2,400m (5,000–8,000ft).

Habits: The egg is flask-shaped and brownish-yellow, later turning leaden grey. The larvae are green with three paler lateral stripes, and the spiracles along the side are ringed white. The larval foodplants, as in *S. belemia*, are Cruciferae. The pupa is strongly tapered at both ends and coloured brown with small black dots. Sometimes emergence of the pupa can be delayed for two years, depending upon local weather conditions. I found the adults numerous on alpine grassy slopes, with typical dancing up and down flight, and frequently settling on flowers. There is only one brood a year.

35. The Lesser or Small Bath White

Pontia chloridice
(Syn: *Parapieris chloridice*)
Wingspan 40–44mm

Description: Smaller than the Bath White (*Pieris daplidice*) with the black apical markings less extensive—the upper surface of the forewings with broad black disco-cellular spot, white-centred, and the apex bearing four or five black spots which are distally pointed. Inside this another row of sub-apical black spots. The hindwing shows a greenish-grey shadow of the underwing pattern. In the male, the hind underwing has yellow veins etched with olive-green in rather scattered pattern forming a narrow sub-marginal green band, and along the wing margin, green and white stripes. In the cell of the hindwing the white spot is more elongated than in the Bath White. The underside forewing apex is greenish-yellow rather than black as on the upper surface. Females have more extensive black apical markings, and on the upper hindwing there is a series of small black sub-marginal markings, whilst the underwing pattern shows through more darkly.

Status: Occurs in Balochistan where uncommon, but collected in the upper Uruk valley, and on the Shingar in Zhob. In Chitral it is common at low elevations, flying from March to May, and on the Shandur Plateau in July and August.

Habits: Larval foodplants Cruciferae.

36. The Lofty Bath White or Peak White

Pontia callidice
(Syn: *Parapieris callidice*)
Wingspan 42–52mm

Description: Like the Bath White, *P. daplidice*, in general pattern but having more extensive, dark greenish-yellow areas on the hindwing under-surface, which leave rather narrow, arrow-shaped white areas in between. The male has the upper wings white with black apical spots pointed inwardly, followed by a sub-apical black band, leaving white spots in between that suggest a spear-shaped pattern. There is also a rather restricted black band in the disco-cellular region which lacks any inner white spot as found in *P. chloridice*. Females have more extensive black apical markings as well as traces of black spots along the sub-marginal area of the hindwing, and much darker greyish shadows from the underwing pattern. Their underwing surface has yellowish-white areas in between the darker olive green, rather than white as in the male. In both sexes the under forewing apex has greenish bars rather than black.

Status: A high-mountain species flying in July and August from 2,700–4,570m (9,000–15,000ft), often on the edge of snowfields. Occurs Safed Koh, and in Chitral on the Shandur Plateau, and on the Baroghil Pass.

Habits: The eggs of all this genus are yellow, with longitudinal ribs and fine cross striations. Larval foodplants *Erysimum* spp. (Cruciferae) and *Reseda* spp. (Mignonette family).

37. The Bath White

Pontia daplidice
(Syn: *Pieris daplidice*)
Wingspan 42–50mm

Description: The male is white above with a rather broad black patch straddling the end cell and disco-cellular region, bisected by a thin white line on the vein. The apical region is black spotted with white, and the hindwing shows the darker grey shadows of the under surface pattern. The under surface of the forewing has the apical area greenish-yellow with white spots and shows a blackish spot in the outer half of the area between veins 1a and 1b. The underwing pattern shows largely greenish-yellow with a row of white spots along the margin, white irregular spots sub-marginally, and a rounded white spot in the mid cell. The female has more extensive black markings, and her hindwing upper surface bears quite prominent black spots in a curved band in the post-discal area and a large black spot near the costa, lacking in the male.

Status: Very common in Swat, and Chitral between 1,200–2,700m (4,000–9,000ft), flying from May to September in the hills, earlier in the foothills. Usually near irrigated cultivation. Collected TJR in Jabba valley, Swat, at 2,400m (8,000ft) in June. Occurs sparingly in the Murree Hills. In Balochistan it is common at higher elevations (above 1,500m or 5,000ft) including valley bottoms, flying from April to September. It has also been collected from Campbellpur (now Attock) in June and July, and from Islamabad gardens in July and August.

Habits: The eggs are flask-shaped and yellow with fine longitudinal ribs and are laid singly on the foodplant, either on the underside of the leaves or amongst the flower buds, where the larvae prefer to feed. The larva is bluish-grey with two pale yellow lateral lines and covered with black specks from which short hairs protrude only along the back. The head is yellow. The pupa varies from greyish-white to green or brownish-white, with dark-tipped horns on the front and rear of the thorax, and dotted with black specks. Like all the family, it is suspended upright to a stem attached by a silken body girdle and tail pad of silk. The foodplants are members of the Cruciferae, especially *Reseda* spp. and *Sysimbrium* spp. such as *S. altissimum* and *S. irio*. The butterfly is very numerous where it occurs and has a dashing fast flight, typically doubling back along the same route at the edge of a terraced hill field.

38. The Desert Bath White

Pontia glauconome
Wingspan 45–55mm

Description: Quite a large butterfly, with the black areas less extensive on the upper surface than in *P. daplidice*. The male, like *P. daplidice*, has black bars in the disco-cellular region and black wing tips with four teardrop-shaped white spots. The hindwing bears a series of black spots marginally and greyish shadows reflecting the underwing pattern. The under forewing is greenish-grey apically, the costa prominently darker greenish-grey also. The under hindwing has the greenish areas rather narrowly confined, leaving a row of teardrop white spots marginally and only a narrow greenish band in the disco-cellular region as well as the costal region anterior to the cell spot. Females have more extensive black markings on the apex forewing and distinct black lines on the margin of the hindwing enclosing circular white areas.

Status: Not common but occurs widely over Balochistan from Dalbandin in the Chagai, on the Mehtarzai Pass, and the Uruk valley, flying in April farther south and from May to July in central Balochistan. Collected Zhob (Fort Sandeman) in September. Also occurs quite commonly in lower Chitral in March and April, and again on the Shandur in July/August. Collected near Chitral town, 1,200m (4,000ft), by TJR.

39. The Green Banded White or Kreuper's Small White

Pieris kreuperi
(Syn: *Artogeia kreuperi*)
Wingspan 45-55mm

Description: Superficially like a rather small Cabbage White (*Pieris* sp.), it is distinguished by the forewing wing tip pattern, which includes a sub-apical triangular black costal spot with marginal black spots coalescing. There is also a black spot between vein 3 and 4, as in the Cabbage Whites. The costa is dusted with black and the hindwing bears a small sub-apical costal spot. In some specimens there are indistinct marginal black spots also on the hindwing. Females are much more heavily marked, with the dark green underwing pattern showing through, and marginal black spots always visible on the upper hindwings. The under hindwing bears two broad bands of dark grey-green, edged posteriorly with black, one band covering the base and the outer one, along the post-discal region, extending down from the costa, its outer margin extending to a triangular point between veins 2 and 3.

Status: In our area, confined to the extreme north-western part of the Pamirs and the Karakorams, extending eastwards into Ladakh. Collected Ladakh at 3,200m (10,500ft), and Nagar in Baltistan at 2,000m (6500ft). Elsewhere it extends through the Middle East from Greece through Iran and into Central Asia.

Habits: Like others of the genus, the eggs are yellow and flask-shaped with longitudinal ribs, and the larvae are grey-green with pale creamy longitudinal lines and some bristles along the dorsal region. Their larval foodplants in Europe are Mountain Alison, *Alyssum montanum*, and in Pakistan at high altitudes, *Alyssum canescens* is one of the probable foodplants. In July and August in Gilgit and Chitral, Sanders (1955) found adult butterflies descending mountain slopes to 2,700m (9,000ft) to visit flowers for a short time, between 11 a.m. and 1 p.m. and then retreating to higher, more barren slopes. He wrote that they were nowhere common and in behaviour very shy and wary.

40. Chitral Green-banded White

Pieris devta
Wingspan 50–55mm

Description: Very similar in appearance to Kreuper's Green-banded White, *Pieris kreuperi*, but differing in the underwing which is much more heavily dusted with yellow scales, especially on the under hindwing. The hindwing underneath and the marginal wing areas are suffused with yellow in specimens from Chitral, and the basal area is quite lightly dusted with yellow and green scales, whilst specimens from Balochistan are a darker grey-green but still suffused with yellow, and bearing larger, dark yellow-green marginal spots. Like *P. kreuperi*, the forewing bears the characteristic triangular black costal patch in the sub-apical area, well marked on both upper and undersides, but the upper forewing marginal black spots are more prominent and in female specimens tend to coalesce. Like *P. kreuperi*, the base of the forewing costa is dusted with black scales.

Status: Formerly treated as a sub-species of the more widely distributed *Pieris kreuperi*, this is now recognized as a full species, being allopatric with that species. Rare in Balochistan but collected Ziarat and upper Uruk valley, where it was considered very rare by Evans. It also extends through the NWFP and into southern Chitral, where it has been collected around Mirogram.

41. The Green-Veined White

Pieris napi
(Syn: *Pieris montana*)
Wingspan 40–55mm

Description: Similar to *Pieris canidia* in size and general appearance, but the veins on the upper side are tinged black, and the underside of the hindwing has the veins conspicuously veined black and dusted in between with pale greenish-yellow. The forewing tip is black with a discal black spot in between veins 3 and 4, and females are more heavily marked in the black areas, with the under hindwing tending to be yellower.

Status: Very common Murree Hills, and throughout lower Swat and lower Kaghan valley, favouring damper areas.

Habits: The eggs are laid singly, and they are greenish-yellow, narrow at the top and bearing longitudinal ribs which are slightly toothed. The larvae are dark green with a yellow spiracular stripe, and dotted all over with black and more prominent small white tubercles. The spiracles are also ringed with black. The pupa varies in colour according to the background to which it is attached, but usually it is bone coloured, and bears a row of black spots with angular bumps in the dorsal region, and a pointed beak at the head. The larval foodplants are variable, but all belong to the family Cruciferae, especially *Reseda*, and *Sisymbrium* spp. In the Murree Hills the butterflies can be seen early in April, having passed the winter in the chrysalis stage. They again become very numerous at the beginning of the monsoon.

42. The Murree Green-veined White

Pieris ajaka
Wingspan 40 -55mm

Description: This species differs from *Pieris napi*, in the male, mainly by having the black markings weak and, usually, on the upper forewing no trace of a black spot between veins 3 and 4; the female, by contrast, has more prominent black markings, with the veins heavily outlined in black and a prominent black spot between veins 3 and 4, and again between 1a and 1b, which extends into a black bar along the lower edge of the forewing towards the base. The underside is pale yellow on both upper and hindwings, with the margins dusted with grey and much dark grey dusting of scales. Compared with *P. napi*, the hindwing costal area is less bright yellow. There is considerable variation between individuals, with males generally being paler on the underside with less dark scaling.

Status: This species occurs in the Murree Hills, where it is less common than *P. napi* but still not uncommon after the monsoon. It also occurs further east around Gurais and in the northern valleys of Kashmir.

43. Alpine Large White or Kashmir White in Evans (1932)

Pieris deota
Wingspan 58–61mm

Description: Chiefly distinguished from the Large Cabbage White, *Pieris brassicae*, by the upper forewing margins bearing much heavier, broader black borders, and the upper hindwing distinguished by having a ring of black spots around the margin, whilst in *P. brassicae* the upper hindwing has no black at all along the margin. The under hindwing is also irrorated with black scales, showing very little trace of greenish-yellow as in *P. brassicae*. The female is patterned like the female Large Cabbage White, with prominent black spots on the upper forewing between veins 3 and 4 and a lower, larger black spot between veins 1b and 2. The black borders to both upper wings in both sexes are broader than in *P. brassicae* and are somewhat hastate along their inner margin, separated by white veins.

Status: It has been collected from the Shimshal valley and the northern ranges of the Karakoram, with specimens from the Saltoro Glacier valley in Baltistan, as well as further to the east in Ladakh, India. It also extends into Afghanistan in the Pamirs.

44. The Indian Cabbage White

Pieris canidia
Wingspan 50–60mm

Description: Very similar to *Pieris rapae* in wing pattern but with slightly bolder black markings, which in the forewing apical area are dentate inwardly from the margin. Also, on the hindwing underside the pre-costal area is bright yellow, the rest of the underwing being paler yellow. Males have only one black spot between veins 3 and 4, whilst females are more boldly marked and have a second black spot above vein 1b. The basal area of both wings is dusted with black scales. The underside of the forewing tips and the whole of the hindwing is pale yellow, dusted with grey.

Status: Very common in the hills from April to October from 1,500–3,660m (5,000–12,000ft) in

Chitral and Swat. In Murree Hills and foothills very common and can be seen on the wing almost every month of the year at lower elevations. Surprisingly, it occurs on the Shinghar Range in northern Balochistan, whilst *Pieris rapae*, the European Small Cabbage White, is the most widespread and common species over the rest of the Province. It has been collected from Campbellpur and a few individuals straggle as far south as Lahore.

Habits: Their eggs are often laid in batches as in *Pieris brassicae*, and in this respect it differs from *Pieris rapae*. Foodplants of the larvae are Cruciferae, particularly *Nasturtium* spp.; the larvae are green and difficult to see as they lie on the underside of leaves when not feeding. In lower Swat, Mohyuddin (1987) collected larvae in April from a Cauliflower crop. They usually travel away from the foodplant before pupating, attaching themselves to a more sheltered spot such as a rock face or tree trunk. The adults have a rather slower, more fluttering flight than many other Pieridae, and when they encounter a conspecific they indulge in a dancing up and down flight together, which is pretty to watch in the dappled sunlight of the Himalayan forest.

45. The Small Cabbage White

Pieris rapae
Wingspan 45–55mm

Description: Distinguished from *Pieris canidia*, which it closely resembles, by the black apical area on the forewing being less distinct and not dentate on the inner margin. Also, the hindwing underneath is evenly coloured pale yellow, dusted with grey, without the pre-costal area up to vein 8 being a darker yellow. The black spots above vein 1b and between veins 3 and 4 are generally smaller than in *P. canidia*.

Status: The commonest butterfly around Quetta valley and throughout Balochistan province above 4,000ft, to be encountered almost all year. In Chitral it is common from March to October between 1,200–1,800m (4,000–6,000ft). Collected on Takht-i-Suleiman by Rafiq Ahmad Rajput in June 1993.

Habits: The eggs are laid singly on both upper and under surface of leaves of the foodplant, which are Brassicas. They are yellow, shaped like a tall bottle, broader in the middle, and with ten to twelve longitudinal ribs and about thirty fine cross striations. The larva in its early stages is dark blackish-green, but later becomes leaf green with a broad dorsal yellow stripe and two yellow lateral stripes. Its head is brownish-green and the spiracles are ringed with black. The pupa is green, yellowish, or brown, with three paler creamy stripes. Generally it is speckled black all over. The head is pointed and there are pointed projections at each corner of the upper thorax. The butterfly frequently visits flowers and gardens and has a rather slow, fluttering flight, and it is as likely to be encountered in the gardens of Quetta city as high in the juniper-forested slopes.

46. The Large Cabbage White

Pieris brassicae
Wingspan 57–70mm

Description: This is the largest of the Pierids and more boldly marked with black in the apical area, the inner margin being smoothly rounded. The costa is also dusted with black, and in the male there are no visible black spots on the upper forewing but two conspicuous black spots on the underside above vein 1b and between veins 3 and 4. The underside of the forewing tip and the whole of the hindwing is pale yellow dusted with grey scales. Females have bolder black markings and two conspicuous black spots on the upper side of the forewing.

Status: Widespread in the northern Punjab, especially the Salt Range, and through Sialkot district down to Lahore where it is not uncommon, and the author has even encountred it in Pirawala Plantation, Khanewal district. It is extremely common in the Murree Hills, Swat, and Chitral, where it prefers lower elevations than the Indian Cabbage White, though these two species are sympatric. Further south, in Balochistan it is mainly found at higher elevations, and is much less common than the other two Cabbage Whites (*P. canidia* and *P. rapae*).

Habits: This species, like *Pieris canidia*, lays its eggs in batches, up to fifteen or twenty, on the foodplant, and the young larvae feed gregariously until they are well developed, when they split up and feed separately. The egg is yellow with longitudinal ribs and raised cross striations, but it is broader at the base and less swollen in the middle than the eggs of *Pieris canidia*. The adult larva is blue-green with an irregular broad yellow lateral line, and speckled all over with small black tubercles. The foodplants are members of the Cruciferae, often cultivated mustard crops, particularly in the Salt Range, but also Rock Cress (*Arabis* spp.), Mignonette, and other Crucifers in the Himalayas. Their larvae are reported to be heavily parasitized by tiny Ichneumon wasps. This species in Pakistan's northern areas is migratory, with broods hatching in the plains adjacent to the foothills during the cold-weather months and successive broods hatching in the Himalayas during the summer, adult butterflies during their brief lifetime travelling northwards into the Himalayan valleys from early spring onwards.

47. The Himalayan Blackvein

Aporia leucodice
(Syn: *Aporia soracte*)
Wingspan 40–70mm

Description: The male is white above with the veins clearly outlined in black. There is a black disco-cellular band usually curving around both ends of the cell, and the apical margin is narrowly black with a broader black discal band interrupted above vein 3. Underneath the apical area is very pale greenish-white and also the whole of the hindwing, with the pre-costal area deep chrome yellow, and all the veins outlined in black. The sexes are alike.

Status: Very common in the Murree Hills; on the wing in late April and May, with only one brood. Found sparingly in the better-forested side valleys of lower Chitral and in the Kaghan valley. Collected TJR in the Lutko valley, Chitral, at 2,100m (7,000ft) on 24 June 1966. Has been collected Lahore, where rare. It only occurs in Balochistan around Ziarat and around Zhargun, flying from May to June.

Habits: The eggs are yellow and milk-bottle shaped with a narrow neck, and are laid in batches. The larvae are brown in colour, covered with fine, soft white hairs. They live gregariously on the foodplant, spinning a tough silken web inside of which they can hide when small, and upon which they will bask when not feeding. Their foodplant is *Berberis lycium* and, no doubt, *Berberis baluchistanica* around Ziarat, where it occurs in Balochistan. The pupa is creamy white coloured but rather conspicuous, being covered with quite large black dots. Both the larva and butterfly are unpleasant smelling and tasting to potential predators. The adult butterflies in Pakistan have only one brood, in the early summer, and typically have a lazy, sailing flight, circling around tree tops. They can be very numerous during their short season as adults and have been reported as literally swarming together in thousands, both in Murree and near Dalhousie (Morshead, 1925).

48. The Dusky Blackvein

Aporia nabellica
Wingspan 50–65mm

Description: A very distinctive butterfly, with most of the upper wing surface yellow, heavily dusted with black, giving an overall blackish appearance, with white marginal bands separated by broad black veins. The under forewing is white with prominent black veins and a black post-discal band. The hindwing is dark yellow with the veins black and a post-discal band of spots shaped like arrow-heads. Females are paler, some showing more yellowish-white on the upper surface and a more conspicuous disco-cellular black band.

Status: Recorded from forested valleys of lower Chitral where considered very rare. It was collected by Evans (Leslie and Evans, 1903, and Evans, 1910) in the Utzen valley of lower Chitral in June and July, flying at 2,700m (9,000ft). It is common in the lower Kaghan valley between 2,400–3,000m (8,000–10,000ft), though only encountered by the author in Silver Fir forest on the slopes of Makrah and above Shogran.

Habits: Very little is known about this species, except that their eggs are laid in batches, and that the young larvae feed gregariously and shelter in a thick silken web strung between the twigs of their foodplant, which is *Berberis* spp. The adults are confined to forest and avoid open grassy slopes.

49. Hill Jezebel

Delias belladona
Wingspan 76–90mm

Description: This striking Jezebel is boldly marked with black on both upper wing surfaces, with hastate-shaped small white spots within the black areas along both wing margins, followed by larger white spots in the post-discal area. The cell has a restricted area of white along its middle, there are deep orange basal costal spots, and the tornal area of the hindwing has an extensive paler orange area extending up the dorsum. Females are slightly larger on average, with rather paler orange areas, but otherwise very similar to the males. The underside is very distinctive from all other *Delias* spp. because the cell area is largely orange, as well as the area between the costas and vein 7. The row of sub-marginal spots on the under hindwing are more rounded than on the upper surface, and instead of being white they are orange, as also the tornal area, extending well up the dorsum.

Status: It has been collected in the Murree Hills at the turn of the century and also from Poonch. It is probably a rare straggler into the south-eastern part of the Pakistan Himalayas, and is more plentiful further east in the Kulu valley, India.

Habits: The foodplant of the larvae are Mistletoes (*Loranthus* spp.) and the larvae are gregarious. They are unusual among the Pieridae in bearing tufts of long hair on each segment along their back. The adult butterflies will visit flowers, but spend a lot of time flying lazily around the tops of trees where the parasitic *Loranthus* grows, especially the females, which seldom descend to the ground. The adult butterflies contain toxic substances and so are avoided by predators, and their orange patches with conspicuous black and white markings are typical aposematic warning colours. Wynter-Blyth (1957) records that there are three broods during the summer in the north-western Himalayas and that they are on the wing from April to July.

50. The Pale Jezebel

Delias sanaca
Wingspan 69–94mm

Description: This Jezebel looks superficially like *D. belladona* but it can be distinguished, both when wings are closed and when basking, by the cell area being much more extensively white and the wing margins less extensively black when viewed from the upper side. In fact, the female has only a narrow black border to her hindwings, with no contrasting white sub-marginal spots. On the underside, the orange costal areas are much reduced compared with *D. belladona,* and in both sexes the underside of the forewing bears continuous white streaks from discal to sub-marginal region, which in *D. belladona* are broken up into two separate white bars. Like the Hill Jezebel, *D. sanaca* has orange tornal area and a series of orange marginal spots along the under hindwing, as well as the cell area and the inter costal being largely orange. Unlike the Hill Jezebel, the two sexes are quite different in appearance, the female being much whiter, with the veins only narrowly outlined in black.

Status: Because it closely resembles *Delias belladona* in appearance and the two species can be sympatric, its exact distribution in the Himalayas has not been well documented. This species has not definitely been recorded from within Pakistan, but it has been collected from the Banihal Pass in Kashmir and from various parts of Jammu, and probably strays into the Poonch area of Azad Kashmir. Its total distribution is tentatively given by Talbot (1939) as occurring from Kulu to Kumaon. Usually it favours lower elevations than *D. belladona*, flying typically around 1,500m (5,000ft).

51. The Common Jezebel

Delias eucharis
Wingspan 66–83mm

Description: This large and showy butterfly is white above with all the veins outlined in black and with a narrow sub-marginal black band, with the outer margin also black but enclosing seven white, teardrop-shaped spots, the upper ones narrower and more pointed, the lower ones almost square. The hindwing is also white with a black disco-cellular line and smaller black marginal triangular spots. The striking feature of this species is its underwings, which are heavily veined with black on the forewing and with the hindwing bright yellow with black veins, surrounded by a heavy black sub-marginal line and black margin enclosing scarlet spots, which are pointed at the margin. Usually the upper surface of the hindwings show pale pink in the post-discal region, reflecting the red underneath.

Status: Has been recorded around Lahore (Puri, 1931, and Rhe-Philipe, 1917) with adults feeding on the *Duranta plumieri* bush. Mohyuddin found larvae feeding on *Loranthus longiflorus* leaves in September in Kotli, Azad Kashmir. It is probably only a rare straggler in good monsoon years. This species, which is distasteful to predators, is very closely mimicked by the Spotted Sawtooth, *Prioneris thestylis*, and also by the Common Wanderer, *Valeria valeria*.

Habits: The eggs are yellow in colour and flask-shaped, with longitudinal ridges. They are laid in batches, sometimes as many as sixty on a single leaf (Gosh, in Talbot, 1939), and the larvae feed gregariously. Bell (1912) describes the larvae as having a black- and white-striped head, with the body a greasy, greenish-yellow brown covered with tufts of fine white hairs, and the body also covered with small white tubercles from which longer white hairs emerge. The larvae drop down by a silken thread when disturbed, and it has been recorded that they grow and develop unevenly, so that they do not pupate at the same time. The larval foodplants are members of the Loranthaceae (Mistletoes), such as *Loranthus eucharis, Loranthus elasticus, Scurrula* spp., and *Viscum* spp. The pupa is described as shiny surfaced, pale greenish-yellow, with small black spines along the dorsum, and some of the thoracic segments carinate with black blotches. Sevastopulo (1947) described a pupa as bright lemon yellow with black spines and black lines along the wing case, and having a short, forward-pointing snout. It is suspended at an angle of 45 degrees by a girdle and tail pad to a vertical or horizontal surface. Often the pupae are formed gregariously in batches on the underside of withered leaves. The butterflies visit flowers and have a fairly slow fluttering, flight in an irregular direction, whilst the females frequently circle lazily around the upper part of trees looking for the *Loranthus* parasite.

52. The Pioneer

Belenois aurota
(Sensu: *Belenois mesentina*, Evans, 1932)
Wingspan 40–55mm

Description: This is a medium-sized white butterfly with a broad black apical area on the forewing enclosing five white spots which are pointed at the outer margin, and having a broad black disco-cellular band. The hindwing has a narrower black marginal border enclosing rounder white spots. The underside has the veins broadly outlined with dark brown, though the disco-cellular band is jet black, the apical region as well as the whole of the hindwing being pale buff. The female has the apical area more broadly black with the white spots almost obsolete and the costa of the forewing black.

Status: This desert-adapted species occurs commonly throughout the four provinces of Sindh, Punjab, Balochistan, and the NWFP. It may be encountered in any month of the year but is most numerous during and immediately after the monsoon season. I have not encountered it in the higher parts of the Murree Hills, and in Swat and Chitral it is confined to the main valleys around 1,200m (4,000ft). Collected on slopes of Takht-i-Suleiman by Rafiq Ahmad Rajput in July 1993.

Habits: The eggs are white with longitudinal ribs, later turning orange. They are laid in batches, as many as 200 in a cluster, on the young leaves of the foodplant. The larvae are gregarious in their early stages, and even after separation they tend to herd together to pupate. The caterpillars are described (Bell, 1912) as greenish-ochraceous, with a broad purplish-brown supra-spiracular band mottled with white and brown. They are covered with small tubercles from which sprout hairs which are often beaded with a globule of liquid at their tips. Their skin looks oily. The larval foodplants are members of the family Capparidaceae, in Pakistan usually *Capparis aphylla, Capparis horrida*, and in lower Sindh *Capparis sepiaria*, and *Maerua arenaria*. Mohyuddin (1987) found the larvae on *Capparis decidua* near Peshawar. The pupae are green varying to greyish-blue and are carinated dorsally, with an upturned snout and the tubercles tipped black or yellow. They are often formed one close up against another on any vertical surface such as a stone or tree trunk. The adults are strong flyers compared with the Cabbage Whites, and are seen everywhere in the dryer, more open plains areas and foothills.

53. The Common Gull

Cepora nerissa
(Syn: *Huphina nerissa*)
Wingspan 40–65mm

Description: This butterfly is white on the upper surface with a broad black border to the forewing, widening near the apex and usually containing two small white spots in the apex. There is also a large black spot between veins 3 and 4, separated from the black area of the termen. The black border on the hindwing is narrower and dentate in shape, the points inward. The underside has both the apex of the forewing and the whole of the hindwing bright butter yellow, with the veins broadly dusted with greenish black. The upper surface also has some of the veins black. Females have the black areas more prominent and usually show small white spots on the hindwing in the black outer border. If there is a dry season brood it is less well marked, and the yellow underwing is paler and browner.

Status: This species is partly migratory, with individuals migrating to the Himalayan foothills in the monsoon season. Puri (1931) recorded it as very common in the Canal area of Lahore but did not state when it is on the wing. Might be overlooked because of its resemblance to the Pioneer (*A. mesentina*). It should be looked out for in Sialkot region.

Habits: Bell (1913) describes the eggs as white and bottle-shaped, and they are laid singly near the tip of a twig or thorn. Sevastopulo adds that the egg becomes orange later, and its top forms a small ring of teeth. The larvae change in appearance with each instar (moult), starting an oily yellowish-green with a black head and prominent tubercles, from which white hairs sprout, tipped with a globule of liquid. In later instars they become dark bluish-green covered all over with minute brown hairs and longer white hairs along the sides. They attain a maximum length of about 27mm. The pupae are coloured dark blue-green or grass green, with a slightly bi-fid head and contrasting white wing cases, and the dorsal surface toothed or carinated and speckled black sub-dorsally. The larval foodplants are Capparidaceae, especially *Capparis aphylla*, *Capparis sepiaria*, and *Capparis horrida*. The pupa takes about ten days to emerge. The butterflies are strong flyers, usually travelling on a continuous straight trajectory.

54. The Common Wanderer

Pareronia anais (Bouge, 1837)
(Syn: *Valeria valeria* in Talbot, 1939, and
Pareronia hyppia of Puri, 1931)
Wingspan 65–80mm

Description: This pretty butterfly is aquamarine in colour all over, with the veins prominently outlined in black. In wing shape the forewing costa is strongly arched, and there is a fairly narrow black border along both fore and hindwings, with a few small green spots in the apex region. The female is more heavily marked, with broader black venation and small green spots along both fore and hindwing borders. She is thus a passable mimic of the Blue Tiger (*Danais limniace*). The underside is whiter in both sexes, especially in the cell region, with the spots along the borders greenish and more prominent than on the upper surface.

Status: Rhe-Philipe (1917), writing of Lahore considered it should occur as it was common in south-east parts of Indian Punjab, and Puri (1931) collected three specimens during August from Bagh-i-Jinnah, Lahore. Recorded by Z.B. Mirza (1969) for Lahore. A rare straggler to the north-eastern border regions after good rains.

Habits: The eggs are described by Bell (1914) as white in colour, oval in shape with the top narrow and ending in a cicle of minute teeth, and the sides bearing seventeen longitudinal beaded ridges or ribs. They later show four bands of pale pink. They are laid in small batches of four to six, on the upper side of a leaf. The larva is dark green with brown spots and some white patches on the sides. There are diagonal brown lines across each spiracle in the middle part of its body. The larval foodplants are members of the Capparidaceae, most commonly *Capparis heyneana*. The pupae are distinctive in shape, having a swollen wing case and down-curving, pointed snout. Usually green in colour spotted with white. Adult females are rather slow and sailing in flight like their Danaid mimics, but the males are strong flyers, seldom settling. They prefer fairly well wooded country.

55. The Striped Albatross

Appias libythea
Wingspan 50–65mm

Description: The male has a rather curved costa and acutely pointed apex to the forewing. It is white on the upper surface, with the apex narrowly marked with black which extends very slightly along the veins. The hindwing is unmarked, but the underside hindwing is pale ochraceous, and both forewing and hindwing show a brownish-black band end cell. The female has the forewing with a broader black apical band and the costa broadly black from the base of the wing to the end cell.

Status: Collected Lahore after the end of the rains by Rhe-Philipe, who considered it rare. Also by Stockley from Haripur, Hazara district. Probably overlooked because of its resemblance in flight to *Catopsilia* spp. It has not definitely been recorded from Sindh.

Habits: Bell (1913) describes the egg as shaped like a short-necked bottle, with ten longitudinal ribs and a ring of five teeth at the top. It is coloured pearl white, turning orange later. They are laid singly on the young shoots of the foodplant, which belong to the family Capparidaceae. The larva has the body light yellowish-green, closely blotched with light purplish spots. It is covered with short bristles, with small-scattered white tubercles bearing longer hairs. Its length is up to 30mm. The pupa is generally light green or suffused brown-green, with a lateral row of black and below this a second row of yellow dots. The larval foodplants include *Crataeva religiosa* and *Capparis separia*.

Bell writes that the whole life cycle is rapid, lasting no more than twenty days from egg-laying till emergence of the imago. The butterflies are strong flyers with a fairly straight flight and seldom settling on flowers, preferring to rest fairly high up on a tree.

Sub-family COLIADINAE—Yellows

The predominant colour of this group is yellow or orange, and they include the Emigrants, Grass Yellows, Brimstones, and Clouded Yellows. According to Ford (1945) they possess no flavone pigments but, like all the Pieridae, their colours are derived from pterine-type pigments. They are mainly found in the tropics, and the males possess specialized scent scales or brands. The larval foodplants are Compositae (mainly Asteraceae), Leguminosae (mainly Fabaceae), Rhamnaceae, and Zygophyllaceae.

56. The Common Emigrant

Catopsila crocale
Wingspan 55–75mm

Description: This is a very variable species, but usually both sexes are pale greenish-white, while the male has a suffusion of yellow around the base of both fore and hindwings. Both sexes have narrow black borders to the wings but their undersides are unmarked. Females have broader black borders and the forewing costa is also black, terminating about the end cell. The male is hairy basally and bears a brand on the upper side of the hindwing near the base in the costal region, as do all the *Catopsila* spp. Both sexes have antennae which are black and rather stout.

Status: Widespread in all four provinces of Sindh, Punjab, Balochistan, and NWFP, though comparatively uncommon in dryer areas and usually on the wing only during and immediately after the monsoon season. Evans considered it only a migrant into Balochistan in August and September. It occurs in the main vale of Swat and lower Chitral, also as a migrant after the monsoon.

Habits: The eggs are white at first, turning yellow later. They are laid singly on the upper side of a leaf or on the leaf bud. The larvae are unusual in being able to defend themselves by exuding a green, strong-smelling liquid from the mouth when attacked, and they can also project their bodies in a leap of several inches. They also lie along the midrib of the leaf on which they are feeding, and consequently often escape detection. They are bright green in colour with a darker dorsal line and a broad white spiracular line, and the body dotted with bluish tubercles. The pupa is described by Bell (1913) as being smooth with only a slight bulge over the wings and the snout not sharply pointed. It is green, with yellow tips and yellow borders to the head and along the abdomen. Sevastopulo (1938) describes the pupa as being boat-shaped with pointed head and thorax keeled, and with two colour forms, one grey-green and the other apple green, and yellow or dark brown lines along the keel. The larval foodplants are *Cassia fistula*, *Bauhinia racemosa*, *Butea frondosa*, and *Cassia siamea* (common in Karachi gardens). This familiar butterfly is one of the swiftest Pierids on the wing, with a typical up and down bounding flight trajectory. They do, however, frequently settle on flowers. As their common name implies, they have strong migratory tendencies.

57. The Lemon Emigrant

Catopsila pomona,
now considered conspecific with *Catopsilsa crocale*
Wingspan 55–80mm

Description: Like *C. crocale*, this form is rather variable, but generally both sexes are less yellow than *C. crocale*, and whiter on the upper surface. The male has a very narrow black border in the apical region which does not extend down the termen, as in *C. crocale*. The base of his wings are often suffused yellow. Both sexes have red antennae, and on the underside of both fore and hindwings there are small silvery spots delineated with red at the end of the cell usually only one on the forewing and two on the hind. In the female, the narrow black apical border extends very faintly down the termen. Their undersides are more ochraceous yellow with darker striations.

Status: This appears to be a much less common form than both *Catopsila crocale* and *Catopsila pyranthe*, also occurring in Pakistan, and, like its congeners, occurs mainly towards the end of the monsoon season. It is an extremely swift and powerful flyer, subject to migratory movements even into the main valleys of the Himalayas, as well as into Balochistan.

Habits: Very similar to the former species, and the larvae have several protective devices against predators. They are capable of jumping up to eight inches (Wynter-Blyth, 1957) when disturbed, and also exude a strong-smelling green fluid from the mouth if attacked. The adult larvae are described (Sevastopulo, 1944) as being green with a rough skin and the head green with minute black spots. There is a white spiracular line along the body, bordered below by an olive green stripe and above by a dotted black line. Bell (1913) adds that the spiracles are oval and shiny white, and the anal flap is covered with small black tubercles. The pupa is apple green in colour and boat-shaped, with a yellow line along the thoracic keel and a faint blue dorsal line. It is suspended by a girdle and tail pad of silk. The foodplant is *Cassia fistula*. Like *C. crocale*, the adult is strong and swift on the wing, with migratory tendencies.

58. The Mottled Emigrant

Catopsila pyranthe
(Syn: African Emigrant, *Catopsila florella*)
Wingspan 50–70mm

Description: The male is greenish-white with fairly broad black apical borders on the forewing, and no markings on the hindwing. There is a clearly visible small dark spot end cell. In the dry season these borders are narrower and the discal spot may be absent. The female has broader black apical margins and traces of a black costal border on the forewing. Both sexes are distinguished from *C. crocale* and *C. pomona* in having fine dark-reddish striations on the under surface of both wings. Like *C. pomona* the underside of their wings are ochraceous with a reddish discal spot end cell. Their antennae are red.

Status: Common throughout all four Provinces, mainly during and immediately after the monsoon season, with migratory tendencies and the swift, strong flight typical of the genus. In August and September it extends into parts of Balochistan and the broader, lower valleys of Swat and Chitral.

Habits: Bell (1913) describes the egg as white at first, turning yellow, being spindle-shaped and broader in the middle, and with twelve longitudinal ribs. The larvae are like *C. crocale* in appearance but have tubercles on their venter. They are coloured dark green with the head paler green. The pupa is short and pyramidal in shape, with the head produced to a straight, not curved, pointed snout. It is dark green with a yellow dorsal line on the thorax, and a yellow line along the sides. The larval foodplants are *Cassia tora* and *Cassia fistula*, the former being common in both Sindh and Punjab. Kunte (1996) describes copulation with the female settling on the ground, with wings wide open and abdomen raised if receptive, when the male alights on her back, clasping the tip of her abdomen and immediately taking wing with the female suspended. He also describes watching near Pune, India, attempted copulation by several males of a dead female which had been imprisoned by a Crab Spider.

59. The Yellow Orange Tip

Ixias pyrene
Wingspan 50–70mm

Description: This is a smaller butterfly than the Emigrants. The male is brightly coloured, with lemon yellow wings broadly bordered with black around all margins, and with the apical area orange, bordered inwardly with black. The veins in the orange patch are black. Females are often white with the enclosed forewing apical patch also white instead of orange. Less commonly they are pale yellow with orange tips within the apical border. In both sexes the undersides are a paler yellow speckled with brown, and the black borders are not visible.

Status: Nowhere common but occurs sparingly in Lahore area, flying from July to September. Butler (1886) recorded it from Campbellpur (now Attock) in November. It is also found in the lower part of Chitral around Drosh, flying from June to September, where it is rare, and Evans recorded occasional specimens in the Bolan Pass on the south-eastern border of Balochistan. TJR encountered it at Fort Monro in late May. Curiously, it has not definitely been recorded from Sindh. Sakai (1981) also collected it from north-east Afghanistan adjacent to Chitral.

Habits: The eggs are laid singly on twigs or thorns of the foodplant. They are white and bottle-shaped, with twelve longitudinal ribs coalescing at the narrow crown to six teeth. The larvae are dark grass green, covered with small red spots, and with a white spiracular line bordered above and below by a reddish-brown line. They typically feed lying along the side or edge of the leaf. The foodplants are *Capparis sepiaria*, *Capparis divaricata*, and *Capparis aphylla*. They are described as wandering about a good part of the time before pupating, and not feeding but resting on branches and stems of the foodplant. The pupa is light green, with darker mottling of different-sized spots. The butterflies have a fairly straight, swift flight, but do visit flowers, and Haribal records them as feeding on *Salvadora* flowers and Compositae. They typically haunt fairly open, thorn-bush-dotted places, avoiding cultivated crops.

59b. The White Orange Tip

Ixias marianne

Status: Puri (1931) includes this species as occurring in Lahore, caught in the same area as the Yellow Orange Tip, but the specimen he figures (Plate iv), is a female of *Ixias pyrene*. Both sexes of the White Orange Tip have orange patches on the forewing, bordered with black. In the dry north-west of both India and Pakistan, the usual form of *Ixias pyrene* females is white, not yellow like the males, but these always have the area inside the forewing white not orange. I am not including it as occurring in Pakistan, as there are no other records.

60. The Common Grass Yellow

Eurema hecabe
(Syn: *Terias hecabe*)
Wingspan 40–50mm

Description: This small butterfly has the forewing costa rather rounded, with fairly straight margins. It is rich lemon yellow both on the upper and undersides, with terminal and apical blackish-brown border, which on its inner side is stepped, being oblique from the costa to the bottom of the apex and then cut square at vein 4, with the yellow area extending outwards to the margin through vein 4 down to 2. The underside is slightly paler yellow with red-brown markings which only cover part of the apex and include two small spots at the base of the cell, this being the main distinguishing feature between *E. hecabe* and *E. blanda*, the Three Spot Grass Yellow. There is also a larger kidney-shaped ring spot in the disco-cellular region. The under hindwing has a series of three small spots basally, with a prominent disco-cellular red-ringed blotch, and a series of curved lines in the post-discal region. The sexes are alike. In the monsoon season the black areas are wider.

Status: Very widespread in Punjab and Sindh Provinces, favouring canal banks and damper areas where grass grows. Can be found on the wing from the end of the monsoon up to February, and occurs in Las Bela, Balochistan, as well as the eastern cultivated parts of the NWFP, but it avoids dryer hilly areas, though it has been collected from Campbellpur. However, it spreads into lower Himalayan valleys where rice cultivation occurs, both in the main vale of Swat and Chitral, and the Murree foothills up to Bhagnotar at 1,800m (6,000ft).

Habits: Though the eggs are laid singly, the young larvae feed gregariously. The egg itself is white at first, turning yellow. It is spindle-shaped, thicker in the middle, bearing many fine longitudinal ribs and many fine transverse striations. The larvae are green and slender with a narrow white line along the sides which are also more blue-green. They are about 22mm long when fully developed. The pupae are usually green with a dark violet-grey dorsal line, and the wing cases bordered dark violet-grey. The larvae feed together until they have consumed the whole of a leaf, leaving only the stem or midrib. When they pupate, they do so in a close row hanging downwards from the leaf midrib and looking in shape and colour remarkably like dried leaflets of their foodplants (*see*, Fig. 4). The larval foodplants are Leguminosae, but very varied. The have been found on *Albizzia* trees, *Acacia* trees, *Sesbania* bushes, *Cassia* spp., and *Caesalpinia* spp. Mohyuddin (1987) found the larvae on *Cassia occidentalis* in October near Rawalpindi. The butterflies prefer shady places where grass is growing and they flutter weakly, close to the ground.

61. The Small Grass Yellow

Eurema brigitta
(Syn: *Terias libythea*)
Wingspan 30–40mm

Description: This is a very small butterfly with the same bright yellow colouring as *E. hecabe* but the black marginal area on the forewing is comparatively broader and along its inner edge rather smoothly rounded and not stepped as in *hecabe*. The hindwing upper surface also bears a fairly broad black margin which is slightly dentate inwardly. The underside forewing shows two small dark spots in the disco-cellular area, and the hindwing underside shows three irregularly placed red spots in the disco-cellular region, and the post-discal area bears a faint wavy red line.

Status: Occurs in some years only after the monsoon in the Lahore area and considered rare.

Habits: The adult likes swampy or wet places, and the eggs are laid singly on *Cassia kleinii*, which is a small annual growing in damp places (family Leguminosae). They are deposited on folded leaf buds or the upper side of young leaves. The larva is grass green, with a dark green dorsal line and a lateral yellowish-white line from segments 2 to 12. It is about 19mm in length when fully grown. The pupa has a pointed snout, which is pink and wrinkled, and the rest of it is green with a lateral white band and brown spots on each segment. Surprisingly the adults have been seen in the Himalayas up to about 3,660m (12,000ft). Like the rest of the genus, they flutter close to the ground, settling often.

62. The Spotless Grass Yellow

Eurema laeta
(Syn: *Terias venata*, wet season form)
Wingspan 30–45mm

Description: Like the two former Grass Yellows, this species is also bright sulphur yellow with the dry season form (illustrated) forewing tip more pointed and less rounded than in *E. brigitta*. The apical and termen area is black and is smoothly rounded in the wing tip region but towards the termen becomes irregular along the inner edge. The upper hindwing shows a very narrow black border along the termen. The underside is pale yellow densely shaded with pale reddish scales and with two curved red lines along the post-discal area of both wings. These bands are less distinct in the wet season form whilst the black border areas are broader.

Status: Occurs after the monsoon, flying up to October around Lahore where uncommon. Collected in the lower Murree Hills at Chattar on 9 October at 600m (2,000ft). Also in lower Sindh where common in some years, flying from May to June.

Habits: Apart from the fact that this species has two rather distinct monsoon and dry season forms, little else has been recorded about their life history since Col. Mosse (1929) succeeded in breeding them in captivity. The female lays her eggs on *Cassia pumila*. The eggs were laid singly and they hatched after four days. The larvae are like those of *E. hecabe*, but possess a dark dorsal line and a yellowish-white lateral line. The pupae are smaller than *E. hecabe*, green in colour and not shiny, with a white dorsal line, and their spiracles white. The snout is brownish and there are dark brown spots and markings along the abdomen and over the wing cases.

63. The Three Spot Grass Yellow

Eurema blanda
(Syn: *Terias silhetana*)
Wingspan 40–50mm

Description: This is one of the larger Grass Yellows like *E. hecabe,* and its underwing markings and dark border on the upper forewing closely resemble that of the Common Grass Yellow. The forewing black marginal area is sharply stepped inwards at vein 6 and again at vein 4, like *E. hecabe*. It can be separated from *E. hecabe* by the under forewing cell pattern, which has three red-brown spots as well as the larger ringed spot in the disco-cellular region. *E. hecabe* only has two spots inside the cell, besides the larger ringed disco-cellular spot. The underside hindwing also has a series of rusty red spots in the discal area, and the forewing underside usually has a more extensive red area at the apex than does *E. hecabe,* which always lacks the minute dark spot at the base of the cell on the underside forewing which character is constant for *E. blanda*.

Status: Probably the rarest of the Grass Yellows found in Pakistan. Only recorded from Lahore area (Puri, 1931) in October. It is the common Grass Yellow in the eastern part of the subcontinent.

Habits: Unlike the other members of this genus, the eggs are laid in clusters of twenty to thirty, usually on the upper surface of a leaf. They are white when freshly laid, turning yellow. The larvae feed gregariously, and are green in colour with the dorsal region dark bluish-green, and with a paler yellowish-green spiracular line, becoming more yellowish-green all over as they grow. Length 26mm. The pupae have a sharply pointed yellow snout, the rest of the body being brownish-green. They are invariably found hanging downwards, closely packed together on the underside of a leaf midrib, which they have eaten, or the underside of a twig. They are heavily parasitized by hymenopterous wasps. Their foodplants are *Delonix regia*, *Cassia* spp., and *Caesalpinia spicata*, all members of the Leguminosae. Female butterflies fly high up among their foodplant trees, but otherwise they are weak flyers, keeping close to the ground vegetation.

64. The Common Brimstone

Gonepteryx rhamni
Wingspan 60–70mm

Description: All species in this genus have a very characteristic forewing shape, with the costa near the apex very rounded and below the apex curving sharply inwards. In the male of this species the whole of the upper surface is lemon yellow, with the hindwing slightly more greenish-yellow. There is a dark orange spot in the end cell of both wings. The female is pale greenish-white, varying to creamy white. Tiny marginal reddish-brown dots are visible along the forewing tip and hindwing termen.

Status: This is a hill butterfly common throughout the forested areas and irrigated cultivation of the Murree Hills, Hazara district, Swat, and Chitral, to be seen on the wing from early spring (March) up to September, and between 1,200–2,400m (4,000–8,000ft). It occurs uncommonly in Balochistan in the juniper forest zone around Ziarat and on the Shingar, and extending into South Waziristan on the Takht-i-Suleiman. Also uncommonly in lower Gilgit around Chilas, and Astor. It has also been collected from Campbellpur (now Attock).

Habits: The eggs are laid singly, generally on the young leaf tips of the foodplant. They are yellow in colour, darkening later, and take about fifteen days to hatch. They are shaped like a narrow flask with the neck constricted and have ribbed sides, these being beaded. The larvae are a uniform dark green to start with but later develop a white spiracular line, and their dorsal surface is covered with minute black tubercles, each bearing a short hair. Because of their smooth shape and habit of lying along the midrib of the leaf, they are remarkably well camouflaged. The pupae also look like a folded or shrivelled leaf, being pale yellowish-green ventrally and darker green on the dorsum, with sharply pointed snout and anal extremity. The pupa is also dotted with black. The butterfly is a strong flyer but settles frequently on flowers, always with the wings closed. They are more commonly found in forested parts of the Himalayas but frequent open grassy glades. Their foodplants are shrubs belonging to the Rhamnaceae, and in Pakistan, *Rhamnus dahuricus*.

65. The Lesser Brimstone

Gonepteryx mahaguru
(Syn: *Gonepteryx zaneka/G. aspasia*)
Wingspan 50–55mm

Description: Despite its common name, this Brimstone is not much smaller than *G. rhamni* and often overlaps in wingspan size. The two easily distinguished features are the serrated margin to the tornal region of the hindwings, prominent in both sexes, and the bi-coloured pattern in males, with the forewing bright yellow and the hindwing pale greenish-white. There is a small dark orange spot in the end of the cell in both wings. The underside is slightly darker greenish-yellow on both fore and hindwing with faint darker green striations. Females have both wings creamy white on the upper side.

Status: Found only in the Murree Hills and lower Kaghan valley, where it is slightly less common than *G. rhamni*, flying from late June up to October.

Habits: Nothing has been recorded about the early stages of the Lesser Brimstone but it is known that the larval foodplants belong to the family Rhamnaceae. The adults are sympatric with *Gonepteryx rhamni* in the Murree Hills, and have much the same habits, being closely associated with Himalayan moist temperate forest at altitudes above 2,100m (7,000ft).

66. The Chitral Brimstone

Gonepteryx farinosa
(Syn: *Gonepteryx rhamni chitralensis*)
Wingspan 55–60mm

Description: This Brimstone is very similar to *G. rhamni* except that the male on its upper side has the cell and costal region of the forewing suffused bright yellow, whilst the rest of the forewing is pale greenish-white and the hindwing is wholly coloured creamy or greenish-white. Like the other Brimstones, there are small orange spots in the end cell region of both wings. Unlike *G. zaneka*, the Chitral Brimstone does not have a serrated margin to the hindwing. Females are almost indistiguishable from females of *G. rhamni* but their forewing disco-cellular orange spot is very small or absent and their colour is rather more chalky white than the greenish-white of *G. rhamni* females.

Status: Occurs in Chitral and collected from Drosh, Shishi Koh, Mastuj, and Utzen nullah, between 2,400–3,000m (8,000–10,000ft), where considered rare. Collected from Nuristan in eastern Afghanistan by Sakai (1981).

Habits: Larval foodplants are Rhamnaceae (Buckthorn Family).

Classification of Genus COLIAS

Talbot (1939), writes that about fifty-six species have been recognized worldwide, and in addition numerous sub-species. They are found throughout the Holarctic region, the Andes Mountains in Nearctica, and through Africa. The genus exhibits fairly marked sexual dimorphism, as well as polymorphism, hence the great variety of forms described. They are also found in south-west China, extending down into Taiwan, and in south India, besides the Sino-Himalayan region. According to Dr Harish Goankar, (pers. comm. 1996), ten species have now been collected within Pakistan territory.

67. The Dark Clouded Yellow

Colias fieldi
(Syn: *Colias croceus/Colias electo*)
Wingspan 54–65mm

Description: This is a very variable species but the male is orange on the upper surface with broad black outer borders lacking any small orange spots. There is a large black spot end cell on the forewing. The hindwing is dark basally being dusted with blackish-orange, and there is a prominent bright orange disco-cellular spot. Sometimes the veins on the forewing are outlined in yellow within the black margins. Females generally have broader black borders and these contain a series of small orange-yellow spots of uneven size. The underside is dark greenish-yellow in both sexes, with the basal part of the forewing more orange and bearing a large black spot, centred white, in the disco-cellular region of the forewing and a series of small black spots growing larger as they descend the post-discal area. On the hindwing, two silver spots ringed with red, followed by a faint ring of pinkish spots in the post-discal region. Both wings have their margins rimmed by pink cilia.

Status: Widespread from the Murree Hills, extending throughout Swat, Hazara district and Chitral from March to October and between 1,200–3,660m (4,000–12,000ft). Sanders (1995) collected a freshly emerged specimen at Shankargarh in southern Astor district at 3,350m (11,000ft) on 11 September. It extends down to the extreme north of Balochistan and South Waziristan on the Shingar and Takht-i-Suleiman Ranges. It is partly migratory and individuals will extend southwards along the base of the foothills to Sialkot, the Punjab Salt Range, where collected Campbellpur in May and June, and Lahore, where Rhe-Philipe considered it not uncommon in some years in February and March.

Habits: The female has been recorded as depositing her eggs on *Indigofera* spp. in the Himalayas, but doubtless other leguminous genera are suitable foodplants, including *Medicago*, *Trifolium*, and *Lotus* spp. The eggs are deposited singly on the leaves of the foodplant and they are flask-shaped, with narrow neck and longitudinal beaded ribs. They are yellow at first, darkening to orange. The larvae are green and pubescent, being covered with soft short hairs. They have paler yellow and orange-brown lateral stripes. The pupae are smooth, and are green with yellow lateral lines. They have pointed heads and tails, and are suspended on stems of the foodplant by a silken girdle and small tail pad. The butterflies are less adapted to dry areas than *Colias erate*, but have the same rather swift, direct flight, settling always with closed wings, whether feeding or resting. They are attracted to *Cnichus* thistle flowers, to *Taraxacum* spp., and *Gentiana* spp.

68. The Eastern Pale Clouded Yellow

Colias erate
sub-species *pallida* and *lativita* (Syn: *C. hyale*)
Wingspan 45–55mm

Description: This is also a very variable species, like *C. croceus*, but males are usually pale lemon yellow on the upper side with a similar broad black border as in *croceus*, without any yellow spots within the border. Females have their black borders containing a series of uneven-sized yellow spots, and there are black spots on the forewing in the disco-cellular region in both sexes. Another form *C. erate pallida*, found only in the females, is pale greenish-white with the black borders containing white spots and on the hindwing upper side a broad yellowish-orange spot showing through. The underside is variable, with some specimens showing a series of black spots along the border, or a continuous broad black outer border. The basal area of the hindwing is dusted with black and, as in *C. croceus* there is a black spot in the disco-cellular region of the forewing and two silvery spots outlined in red in the hindwing cell.

Status: Less common than *C. fieldi* in the Murree Hills, but more common in Swat Kohistan and all over Gilgit and Chitral, flying from March to October. Collected TJR in Chitral Bazaar at 1070m (3,500ft) on 22 June 1966, and in Balochistan from Zizri above Ziarat in late May. Collected by Rafiq Ahmad Rajput from Takht-i-Suleiman in July 1993. It occurs all over Balochistan above 900m (3,000ft) and is commoner there than *Colias fieldi*. This is a species which tends to be migratory and occasional specimens turn up in lower Sindh. Collected in Karachi by Swinhoe (1887) and by the Zoological Survey in the late 1960s. Butler (1886) records it from the Murree foothills at 1,800m (6,000ft) and from Campbellpur (now Attock).

Habits: The eggs are flask-shaped with narrow, rounded, not flat, top and bearing beaded longitudinal ridges. They are yellow at first, rapidly darkening to orange and then leaden. The female deposits these singly on leguminous field crops such as Lucerne (*Medicago sativa*), Lentils (*Lens esculenta*, and *Cajanus* spp.) and Clover fodder (*Trifolium alexandrinum*), and also many wild leguminous plants such as *Lotus* spp. and *Medicago* spp. The larvae are dark green, covered with silky short hairs, and with brown and orange lateral stripes or spots. The pupae are green with black spots on the abdomen and are attached to the foodplant by a silken girdle. The butterflies, which haunt open areas, are swift on the wing, generally keeping close to the ground and following a straight trajectory. They often settle on flowers to feed, such as *Oxalis* spp., and at higher altitudes *Geranium*, *Taraxacum* and *Gentiana* spp.

69. Marco Polo's Clouded Yellow

Colias marcopolo
Wingspan 40–45mm

Description: This is quite a distinctive species, the male being sulphur yellow with broad black borders, and on the upper forewing there is hardly any trace of an end cell spot, or it may be entirely absent. The female is paler yellow and has small yellow spots in the forewing margin with a very small black spot end cell. Her upper hindwing is darker, suffused with orange scales, and with the cell spot showing brighter orange. The underside of both sexes has the forewing clear pale yellow with a pale-sage green border, and the whole of the hindwing sage green with the disco-cellular spot white.

Status: Evans (1932) states that the true home of this species is in the Pamir range and the Hindu Kush, but he collected specimens from the Thui nullah near the Gilgit (Yasin)-Chitral border, where he considered it to be very rare.

Habits: It is believed that the larval foodplants belong to the family Ericaceae, and that these would be *Cassiope* and *Gaultheria* spp., which grow up to 4,260m (14,000ft) in the north-west Himalaya. Evans collected specimens on the wing in September.

70. The Greenish Clouded Yellow

Colias alpheraky
(Syn: *C. chitralensis* or *C. roschana*)
Wingspan 50–55mm

Description: A very distinctive *Colias* within this rather variable group. Both sexes are distinguished by their greenish-lemon colour, with a broad black border to the forewings containing arrow-shaped pale green spots in both sexes. The border along the hindwing margin is more interrupted, enclosing larger, round pale spots. The disco-cellular black spot is small and on the underwing surface not white centred, whilst the under hindwing bears only a single dull white spot in the cell and is darker blackish-green basally.

Status: Collected in Chitral just below the Shandur Pass, and by Shandur Lake, also Orghuty at 3,000m (9,800ft) in extreme north-east Chitral, and Yarkhun and near Darkot at 4,200m (14,000ft) in north western Chitral. Sanders collected it also at Ghizar in western Yasin district of Gilgit. Where it occurs it is fairly common.

Habits: Flying between 2,700–4,570m (9,000–15,000ft) in open alpine slopes, like all the genus, they frequently settle on flowers, but generally with wings closed. Adults are on the wing from June through July, but Sanders (1955) was able to collect a few in worn condition during mid-August at higher elevations.

71. The Broad-bordered Clouded Yellow

Colias wiskotti, form *leuca*
Wingspan 50–55mm

Description: This is a very variable species, as orange yellow, lemon yellow, and white forms occur in the same area. All three forms have noticeably broad dark borders. The specimens in the British Museum from Chitral, collected by Evans, are the white form *Colias wiskotti leuca,* illustrated in this book, but he also collected orange forms in August just below the Shandur Pass (Leslie and Evans, 1903). The species is characterized by having an unusually wide black wing border which on the upper forewing is irregular along the inner margin, and on the upper hindwing only extending down to vein 2. There is a black disco-cellular spot in both sexes, and the female has even broader black borders, with comparatively small white spots in the margin. Their underside is pale sage green on the hindwing with a white disco-cellular spot, and the base of the under forewing is also white. The fringe or cilia are white, tipped pink at the veins, with the costal margin of the forewing also dark pink—a useful character in separating *wiskotti* from *alpheraky,* which has wholly white cilia.

Status: Collected from northern border areas of Chitral, on the Garoghar Pass, and by Evans (1932), who considered it very rare, just below the Shandur Pass, but it was also earlier on collected by Col. Lorimer from the Baroghil Pass and Yarkhun in northern Chitral (Tytler, 1926). Sanders (1955) collected it from the Baroghil valley at 3,800m (12,500ft), and eastwards along the track to the Darkot Pass in late August, when they were in very worn condition. He also collected it from the Ghizar valley in Yasin district. It is commoner in Uzbekistan and Tajikistan to the north, where the yellow form occurs.

Habits: The larvae, according to Talbot (1939), feed on *Oxytropis* and *Astragalus* spp. (family Papilionidae), with many species of both leguminous genera growing up to 4,900m (16,000ft) in Chitral.

72. Orange Clouded Yellow

Colias stoliczkana
Wingspan 39–45mm

Description: Some specimens are quite small, especially females. There is marked sexual dimorphism in this species. The male has fairly broad, dark brown wing borders on the upper side, with rather dark orange-red spots on the hindwing end cell area. The female occurs in both white form and orange, but in the more usual orange colour morph the brown marginal areas are very extensive, with the upper hindwing wholly dark brown except for a restricted orange cell patch and a series of paler yellow sub-marginal elongated spots. The upper forewing is also a much lighter yellow tone than in the male.

Status: This is another very high-elevation species, and Talbot in the *Fauna* series (1939) gives its distribution as Kashmir and Ladakh, with another sub-species extending to Sikkim in the eastern Himalayas. This species has recently been collected from Gilgit, according to Dr Harish Gaonkar (pers. comm., 1996). It has been collected at 3,960m (13,000ft) in Ladakh.

73. The Pamir Clouded Yellow

Colias cocandica
Wingspan 35–45mm

Description: As can be seen from the above measurements, this is a small Clouded Yellow. It resembles *Colias ladakensis,* but *cocandica* has the upper surface more dusted with black scales, especially on the upper hindwing. Both sexes are pale lemon yellow with the black border bearing comparatively large yellow spots, though these are smaller in the male. There is a conspicuous black disco-cellular spot in both sexes, often bearing a tiny white-centred spot. The upper hindwing bears a comparatively narrow black border, the yellow spots coalescing from vein 2 downwards, bordered on the inner margin by a series of black spots. The underside hindwing is dark green heavily dusted with black scales and with the disco-cellular spot elongate and yellowish-white, never pure white. By contrast, the underside of *C. ladakensis* has a more conspicuous black border and continuous inner black margin, with the under hindwing darker green basally.

Status: Within Pakistan territory it has only definitely been collected from Misgar in northern Hunza (Tytler, 1926), but it is more plentiful in Rupshu in the south-eastern part of Ladakh, and there are specimens in the British Museum collection labelled Hindu Kush, without locality.

Habits: The larval foodplants are species of *Astragalus*.

74. The Ladakh Clouded Yellow

Colias ladakensis
(Syn: *Colias shipkee*)
Wingspan 45–50mm

Description: In this species the ground colour on the upper wing surface is a bright sulphur yellow with the black marginal border containing very large, prominent yellow spots of fairly even size. The forewing disco-cellular spot is small and black, and on the hindwing there is much dusting with black scales basally, with the cell spot rather orange, varying to green and not very dark. The under hindwing is very dark green basally with a small whitish cell spot, and in some specimens a brighter yellow streak along mid cell. The female is like the male, but with a slightly richer yellow colouring and broader black borders.

Status: This high-altitude species has recently been collected from the slopes of Tirich Mir in the Pamir Range of Chitral. It was previously only known from Ladakh, where it was collected in the Rupshu valley by Col. Stockley.

75. The Fiery Clouded Yellow

Colias eogene
Wingspan 40–50mm

Description: This handsome species is a rich orange-red on the upper wing surface, and in the male, the broad dark brown borders are unspotted and very even along their inner margins, whilst the dark spot end cell is prominent. In the female, the brown border bears a row of orange-yellow spots, and the red area is suffused with brown scales basally, so that only a small red spot shows in the outer half of the cell. The base of both upper and hindwings are suffused with bluish-white scales in both sexes. The under hindwing in both sexes is pale sage green with the small cell spot white, usually outlined with red. The under forewing is yellow basally with the same green border. In both sexes the base of the hindwings has bluish-violet hairs. A white form of this species has been collected from Skoro La in Baltistan.

Status: Collected in Chitral from Shishi Koh, the Bangol pass and on the Shandur plateau. From northern Hunza near the Baltoro glacier and the Khunjerab, up to the Mintaka Pass. From the Thui nullah in Yasin, and Rupal in Astor, both places in Gilgit. From Baltistan along the Shyok and at Kerulungma. Where it occurs, it is not uncommon and flying in July and August between 3,660–4,870m (12,000–16,000ft).

Habits: The larval foodplants are species of *Astragalus*.

76. The Glaucous Clouded Yellow

Colias leechi
Wingspan 45–50mm

Description: This is a striking species having pale greenish-yellow wings, heavily dusted with black scales, so that the overall appearance is greyish-green. The borders are also grey rather than black, and with large greenish-yellow spots inside. The forewing bears a larger than usual disco-cellular black spot, and the upper hindwing is glaucous blue-grey, with the disco-cellular spot very large and pale. The underside forewing is dull white basally with sub-marginal black spots, and the under hindwing is glaucous grey with a paler yellow-green border and a row of black spots dividing the discal and border areas.

Status: It has been collected from the Karakoram range between 15,000 and 18,000ft, where it was flying together with *Colias eogene*. Also from the Pamirs in northern Chitral, where Evans considered it rare, flying at 3,960m (13,000ft) in July.

77. The Small Salmon Arab

Colotis amata
(Syn: *Colotis calais*)
Wingspan 35–50mm

Description: This small butterfly, as its name implies, is salmon pink on the upper wing surface, patterned with black along the margin and forewing costa, the black extending down the end of the cell in the forewing and enclosing a series of small, squarish pink spots along the wing tip, with finer pink streaks in the apex. The hindwing has a similar dark outer border spotted with pink. The underwings are greenish-yellow, with a fine black marginal line, and the black markings of the upper surface showing through as grey bands, with the disco-cellular spot more prominently black and also a black spot between veins 2 and 1b.

Status: Very common in the dryer parts of Sindh and Punjab but absent from higher rainfall areas in the north of Punjab, and only entering Balochistan in Sibi district at lower altitudes. Avoids cultivation but is locally common wherever there are patches of scrub jungle with *Salvadora* and *Capparis aphylla*. In the Punjab, most frequently encountered from July to November, but a few may be seen throughout the cold weather. In Sindh can be seen all year round and most abundantly in what remains of the riverain forest, and usually in association with the Blue Spot Arab (*C. protractus*), and the White Arab (*C. vestalis*).

Habits: The egg is a truncated cone with longitudinal channels, and it is white at first, becoming yellow. The larvae have black heads and greenish-yellow bodies, varying to yellow-brown, with rings of tubercles with tiny hairs around the front margin of each segment. After the 5th instar they become grass green with a broad white dorsal band, and dotted with minute white tubercles. All the larvae of *Colotis* spp. have these bristles in the final instars, and these often terminate in minute transparent globules which reflect the light. Unlike the other *Colotis* spp., the eggs are laid in batches of up to fifty or sixty, and the young larvae feed gregariously. The larval foodplant is *Salvadora oleoides* in the Punjab, and *Salvadora persica* in Sindh. The pupae are green with the wing case edged yellow, and have a fairly short snout, and the larvae go off separately to pupate. The butterflies are weak flyers, but can occur in swarms. They flutter close to the ground, often in and out of thorn bushes, and are found in the hottest, dryest parts.

78. The Blue Spot Arab

Colotis protractus
(Syn: *Colotis phisadia*)
Wingspan 40–45mm

Description: Like the Small Salmon Arab, this species is salmon pink on the upper wing surface but having broader black margins on the hindwing, without any pink spots inside this margin. The forewing is also broadly bordered with black, but in this border are four bluish-grey spots, and the hindwing basally is covered with fine blue-grey hairs and scales. The cilia are pink. The underwings are greenish-yellow with the discal area tinged salmon pink on the forewing. There is a black spot or bar end cell visible on both upper and under forewings, and the under hindwing lacks any darker bars or spots. The female has the border areas dark brown rather than black.

Status: Less common than *Colotis amata* but occurring locally throughout Sindh and southern parts of Punjab up to Lahore. Also collected from Campbellpur. Particularly common in Multan district and extending westwards from northern Sindh into Sibi district of Balochistan. Often flying with *C. amata* and *C. vestalis*.

Habits: The eggs are laid singly, and hatch out on the third day. They are white, shaped like an ampule, and strongly ribbed. Fraser (1911) describes the larvae as grass green with crimson stippling, which fades with successive moults. After the 5th instar their bodies are covered with minute white tubercles, and these bear minute bristles. There is also a white dorsal line. The pupae are smooth and cream or flesh coloured, without black spots. The foodplants are *Salvadora persica* and *Salvadora oleoides*. Like the others of this genus, the butterflies are weak flyers, staying close to the ground or around bushes.

79. The White Arab

Colotis vestalis
Wingspan 40–50mm

Description: The upper wing surface is white in this species, with broad black borders similar in pattern to *Colotis protractus*, with no white spots inside the hindwing border, but the forewing having a large white spot in between veins 2 and 3, and with two or three smaller spots in the apical area. The costa and basal area of the forewing is dusted with blackish and grey scales, and the cilia is white. Females are similar except that the border is dark brown rather than black. The underside is greenish-yellow in the male with the discal area paler and the under hindwing is dull reddish, ochraceous in colour. There are three black spots in a curve along the sub-marginal area and a black disco-cellular spot on the forewing.

Status: Frequents the same habitat as other *Colotis* spp., and occurs more regularly during and after the monsoon season, being absent in the hottest months of May and June and the coldest months of December and January, but otherwise very numerous all over Sindh and Punjab, extending westwards into Sibi district of Balochistan. Often flying with *Colotis amata* and *C. vestalis*.

Habits: The egg is white and dome-shaped, bearing sixteen to eighteen longitudinal ribs. It later becomes yellowish-white with three transverse reddish bands. The eggs are laid singly, and the larvae are not gregarious. The larva in the first instar is grass green with crimson stippling which coalesces over segments 12 to 14 to form a crimson patch; later it is green, covered with minute light yellow and white spots, and is pubescent. The pupa is flesh coloured without markings. The larval foodplants are *Salvadora* spp. The butterflies can be extremely numerous where they occur, often in mixed company with the Small Salmon Arab. They have a weak fluttery flight, staying close to the ground.

80. The Large Salmon Arab

Colotis fausta
Wingspan 40–50mm

Description: The male is orange-pink or deep salmon colour on the upper side, with a narrow black border reduced to separate spots on the hindwing, and on the forewing separated by equally broad pink spots and an inner black band in the post-discal area. There is also a black disco-cellular ring spot. Females are generally paler salmon pink on the upper side, with broader black borders fully enclosing a series of pink spots marginally and inside this another band of pink spots in the post-discal region. Besides the disco-cellular black spot on the forewing there are traces of dark spots in an arc around the post-discal area of the hindwing. The underwing of both sexes is pinkish brown, paler and more yellow in the discal area of the forewing. The end cell ring spots are reddish-brown, not black, and there is a curved ring of dark ochraceous brown spots in the post-discal area of both fore and hindwings.

Status: Rare in the Punjab but has been collected from Campbellpur area from October to December, and also Lahore after the monsoon. Occurs in lower Sindh more commonly than further north but has been collected from Sukkur. The author collected it in Malir. In Balochistan it occurs uncommonly in the rocky gorge (Sheikh Wasil) just east of Nushki and in the Bolan Pass. It also occurs in Waziristan and has been collected in the Tochi valley. It is on the wing only from August up to October.

Habits: The eggs are laid singly on twigs of the foodplant. They are shaped like a short-necked bottle, with eighteen longitudinal ridges but no teeth at the top. Their colour is yellowish with three blood-red irregular transverse bands. The larva is grass green with an indistinct yellow dorsal band and on the sides, three larger pinkish-brown patches. The pupae often develop on the ground or close to it. They are bone-coloured with a rather blunt snout. The larval foodplant is *Maerua arenaria* (family Capparidaceae), a common climbing shrub in the desert tracts of Sindh. The butterflies are only on the wing for a short time during the monsoon, and they are stronger flyers than the rest of the genus. They come to gardens to feed off flowers.

PLATE 20: GRASS YELLOWS AND CLOUDED YELLOWS

A. *Catopsila pyranthe* - African or Mottled Emigrant, dry season form **B.** *Eurema hecabe* - Common Grass Yellow
C. *Eurema blanda* - Three Spot Grass Yellow **D.** *Eurema laeta* - Spotless Grass Yellow **E.** *Eurema brigitta* - Small
Grass Yellow **F.** *Colias fieldi* - Dark Clouded Yellow, female **G-1.** *Colias erate* - Pale Clouded Yellow, male, form *glicia*
G-2. Female **G-3.** Female, form *pallida*

PLATE 21: CLOUDED YELLOWS

A. *Colias marcopolo* - Marco Polo's Clouded Yellow **A-1.** Male **A-2.** Female **B.** *Colias alpheraky* - Greenish Clouded Yellow **B-1.** Male **B-2.** Female **C.** *Colias ladakensis* - Ladak Clouded Yellow **C-1.** Male **D.** *Colias eogene* - Fiery Clouded Yellow **D-1.** Male **D-2.** Female **E-1.** *Colias stoliczkana* - Orange Clouded Yellow, male **E-2.** Female **F.** *Colias wiskotti* - Broad Bordered Clouded Yellow **F-1.** Female **G.** *Colias cocandica* - Pamir Clouded Yellow **G-1.** Male **H.** *Colias leechi* - Glaucous Clouded Yellow **H-1.** Male

PLATE 22: BRIMSTONES AND ARABS

A. *Gonepteryx rhamni* - Common Brimstone **A-1.** Male **B.** *Gonepteryx mahaguru* - Lesser Brimstone **B-1.** Male
C. *Gonepteryx farinosa* - Chitral Brimstone **C-1.** Male **C-2.** Female **D.** *Colotis amata* - Small Salmon Arab **D-1.** Male **E.** *Colotis protractus* - Blue Spotted Arab **E-1.** Male **F-1.** *Colotis vestalis* - White Arab, male **F-2.** Female
G-1. *Colotis fausta* - Large Salmon Arab, male **G-2.** Females **H-1.** *Colotis etrida* - Little Orange Tip, male
H-2. Female **I.** *Colotis danae* - Crimson Tip, male

PLATE 23: NAWAB, COURTESAN, EMPEROR, WHITE ADMIRALS, MAP AND CASTOR

A. *Eriboea athamus* - Common Nawab **B.** *Sephisa dichroa* - Western Courtesan **C.** *Ariadne merione* - Common Castor **D.** *Apatura ambica* - Indian Purple Emperor **D-1.** Male **D-2.** Female **E.** *Limenitis trivena trivena* - Indian White Admiral **F.** *Limenitis trivena gilgitica* - Indian White Admiral **G.** *Calinaga buddha* - Freak **H.** *Cyrestes thyodamas* - Common Map

PLATE 24: LARGE FRITILLARIES

A. *Argynnis lathonia* - Queen of Spain Fritillary **B.** *Argynnis childreni* - Large Silverstripe **C.** *Argynnis pandora* - Western Silverstripe or Cardinal **C-1.** Male **C-2.** Female **D.** *Argynnis kamala* - Common Silverstripe **E.** *Argynnis jerdoni* - Jerdon's Silverspot **F.** *Fabriciana argyrospilata* - Afghan Fritillary **G.** *Argynnis aglaia* - Dark Green Silverspot **G-1.** Male **G-2.** Female **H.** *Argynnis adippe* - High Brown Silverspot

PLATE 25A: HIGH ALTITUDE FRITILLARIES AND REDBAND (MELITAEA) FRITILLARIES

A. *Clossiana hegemone* - Whitespot Fritillary **A-1.** Male **A-2.** Female **B.** *Bolaria pales* - Shepherd's Fritillary or Straight Wing Silverspot **B-1.** Male **B-2.** Female **C.** *Melitaea shandura* - Shandur Fritillary **C-1.** Male **C-2.** Female **D.** *Melitaea persea* - Desert or Spotted Fritillary **D-1.** Male **D-2.** Female **E.** *Melitaea lutko* - Balochi Fritillary **E-1.** Male **E-2.** Female

PLATE 25B: HIGH ALTITUDE FRITILLARIES AND REDBAND (MELITAEA) FRITILLARIES

F. *Melitaea trivia* - Lesser Spotted Fritillary **F-1.** Male **G.** *Melitaea minerva* - Pamir Fritillary **G-1.** Male **G-2.** Female
H. *Melitaea didyma* - Fiery Fritillary or Spotted Fritillary **I.** *Melitaea ala* **I-1.** Male **I-2.** Female **J.** *Melitaea arcesia* - Blackvein Fritillary **J-1.** Male **J-2.** Female

PLATE 26: COMMON LEOPARD, ADMIRALS AND COMMAS

A. *Phalanta phalanta* - Common Leopard **B.** *Argyreus huperbius* - Indian Fritillary **B-1.** Male **B-2.** Female **C.** *Vanessa atlanta* - European Red Admiral, sexes alike **D.** *Vanessa indica* - Indian Red Admiral, sexes alike **E.** *Cynthia cardui* - Painted Lady **F.** *Vanessa canace* - Blue Admiral, sexes alike **G.** *Polygonia egea* - Eastern Comma **H.** *Polygonia c-album* - European Comma **H.** sexes alike **H-1.** Early summer brood **H-2.** Post monsoon brood **I.** *Polygonia vau-album* - Comma Tortoiseshell

PLATE 27: PANSIES AND TORTOISESHELLS

A. *Junonia hierta* - Yellow Pansy **B.** *Junonia orytha* - Blue Pansy **C.** *Junonia lemonias* - Lemon Pansy **D.** *Junonia almana* - Peacock Pansy **E.** *Junonia iphita* - Chocolate Pansy **F.** *Aglais urticae* - Mountain or Western Tortoiseshell **G.** *Vanessa cashmiriensis* - Indian Tortoiseshell **H.** *Vanessa polychloros* - Black Leg tortoiseshell **I.** *Vanessa xanthomelas* - Yellow Legged or European Large Tortoiseshell

PLATE 28: EGGFLIES, SERGEANT, SAILERS AND BARON

A-1. *Hypolimnas missipus* - Danaid Eggfly, male **A-2.** Female, form *inaria* **B-1.** *Hypolimnas bolina* - Great Eggfly, male **B-2.** Female **C.** *Pantoporia opalina* - Himalayan Sergeant, sexes alike **D.** *Neptis mahendra* - Himalayan Sailer, sexes alike **E.** *Neptis hylas* - Common Sailer **F.** *Neptis zaida* - Pale Green Sailer **G.** *Neptis soma butleri* - Sullied Sailer, sexes alike **H.** *Euthalia garuda* - Common Baron

PLATE 29: TREE BROWNS, WALLS AND BUSH BROWN

A. *Lethe confusa* - Banded Tree Brown, sexes alike **B.** *Lethe verma* - Straight Banded Tree Brown **C.** *Lethe rorhia* - Common Tree Brown **D.** *Pararge menava* - Dark Wall, sexes alike **E.** *Pararge maerula* - Scarce Wall **F.** *Pararge schakra* - Common Wall **G.** *Pararge eversmanni* - Yellow Wall **H.** *Mycalesis perseus* - Common Bushbrown, sexes alike

PLATE 30: MEADOW BROWNS AND ARGUSES

A. *Maniola pulchella* - Tawny Meadow Brown **A-1.** Female **B.** *Maniola pulchra* - Dusky Meadow Brown, male **C.** *Paralasa kalinda* - Scarce Mountain Argus, sexes alike **D.** *Paralasa annada* - Ringed Argus, sexes alike **E.** *Paralasa mani* - Yellow Argus **F.** *Maniola hilaris* - Pamir Meadow Brown **F-1.** Male **F-2.** Female **G.** *Maniola wagneri* - Oval Spot Meadow Brown **G-1.** Male **G-2.** Female **H.** *Maniola lupinus* - Branded Meadow Brown, male

81. The Little Orange Tip

Colotis etrida
Wingspan 25–45mm

Description: The upper surface is white with the forewing apical area orange, bordered all around with black including the inner side, and with the veins outlined black as they cross the orange tip. There is a minute black spot in the end cell and the upper hindwing bears small black spots around the termen. Females have two black spots on the forewing below vein 1b and another below vein 3. The forewing orange area is slightly more restricted and more heavily outlined in black and the hindwing bears more prominent greyish-black spots in a curve around the hind margin termen. The underside of the forewing in the apical area is ochraceous yellow, as is the whole of the hindwing, and there is a ring of dark fuscous spots in a curve around the hindwing margin.

Status: Occurs mainly in southern Punjab and the Salt Range, where it is particularly abundant. Also throughout Sindh, frequenting uncultivated scrub jungle areas and any waste areas alongside cultivation, flying from August to December in Punjab, and almost throughout the year in Sindh. It also occurs in Sibi district of Balochistan and has been collected uncommonly in the Uruk valley and around Quetta.

Habits: The eggs are yellow with reddish blotching. They are flask-shaped with eight longitudinal ridges. They are laid singly on the foodplants, which are *Salvadora persica*, *S. oleoides*, and *Cadaba indica*, the False Caper. The larvae are dark green with a spiracular (lateral) white band, margined orange above and purplish-brown below, and they are pubescent. According to Bell (1914), the pupae are pinkish bone coloured with green bands, and when the pupa is formed amongst leaves these green areas are more pronounced, but when on the ground they are bone coloured.

82. The Crimson Tip

Colotis danae
Wingspan 40–50mm

Description: About the same size as the large Salmon Arab, with slightly arcuate tips to the forewing, this species is chalk white with a broad crimson-cerise tip to the forewing, enclosed marginally and on the inner side by black. The hindwing has a band of black spots along the termen. In the female, the crimson wing tip is paler and traversed within by a black line of spots. In both sexes there is a small black disco-cellular spot, and in the female traces of black spots below vein 1b and vein 2. The underside has the forewing apical area ochraceous, pink marked along its inner border with darker reddish-black spots. The basal wing area is yellowish-buff with the disco-cellular spot just visible. The under hindwing is generally ochraceous-buff.

Status: The least common of the 'Arabs' (*Colotis* spp.) in Punjab, occurring mainly in Bahawalpur and Multan districts. Common in Sindh, and extending into Sibi district of Balochistan, and only found flying from August to early October.

Habits: The larvae are glaucous green with many indistinct white spots and a lateral white line, bordered above by orange and below by purplish-brown. They are pubescent and when full grown about 20mm in length. The pupae are described as being stout, with a prominent wing bulge and thickly conical head process. They vary from bone coloured to green. The larval foodplants are members of the Capparidaceae, including *Cadaba farinosa*, *Capparis sepiara*, *Capparis divaricata*, and *Maerua arenaria*. The butterflies are, like *Colotis fausta*, rather stronger on the wing than the other *Colotis* spp.

Chapter 8
Nymphalidae

Classification

The family of Nymphalidae is very diverse in appearance, being exceeded in numbers of species within the Indian subcontinent only by the Lycaenidae. Wynter-Blyth in his classic *Butterflies of the Indian Region* (1957) lists a total of 307 spp. of Nymphalidae, nearly one-third of all butterfly species recorded from the subcontinent (his estimate, 1,443 spp.). Nevertheless, they can all be distinguished by shared characteristics. Besides being strong, swift flyers, with robust thorax and relatively short abdomen, they are usually very brightly coloured, often with the wing margins sharply indented or angled, sometimes with tails (Charaxinae), or toothed (*Vanessa, Nymphalis, Polygonia* spp.), and many predominantly patterned with black spots and streaks on a tawny background (Fritillaries). They all have the hindwing channelled to receive the relatively stout abdomen, fairly short antennae with prominent clubs, and the venation on the forewing with the cell closed, but on the hindwing the cell generally open (with the exception of the genus *Argynnis*). The males have the first pair of legs (thoracic) reduced to a brush-like structure, while the forelegs of the females possess four tarsal joints but are degenerate and useless for walking. The sexes can thus be easily separated by examination of the forelegs.

The eggs can be very varied in the sculpting of the shell (chorion) but are always ribbed longitudinally, with a flattish top, and are generally white. The larvae are also very varied, but the majority bear branched spines along the body, with a well-separated head and rounded anal end. In the Sailers (*Neptis* spp.) the spines are reduced to fleshy knobs, and the Barons (*Euthalia* spp.) have smooth bodies but bear long fern-like projections down each side of the body. The Emperors (*Apatura* spp.) are green and smooth-bodied but with fleshy horns on their head and tail. In colour they are generally green or brown. The pupae are suspended only by the tail claspers from a silken pad hanging head downwards. Many pupae bear short spines dorsally and are elongated, with angular projections over the thorax, and many of these bear metallic silver or gold spots. But there is much variation, and the chrysalises of *Apatura* spp. are smooth, without spines or metallic spots. There are examples within the family of mimicry of distasteful species, and of others with strong migratory tendencies. In contrast to groups such as the Satyrinae or Lycaenidae, the foodplants of the larvae are very varied. Bell (*JBNHS*, 1909–27) in his series of early life histories of common Indian butterflies, lists at least twenty different plant families as being selected by different Nymphalid species.

Pakistan does not have a rich variety of the more spectacular tropical genera, but does have one Nawab, one Baron, and one Emperor, with a total of forty-eight species within the family, so far recorded. Again, because of her extensive high mountain regions, the *Argynnis* and *Melitea* genera of Silver and Small Fritillaries are represented in great variety, presenting some difficulties for their identification and classification. Some of the characters of the sub-families represented in Pakistan, within the Nymphalidae, are for convenience detailed separately below.

Charaxinae

Charaxinae are tailed butterflies with a very powerful flight, associated with forest and often territorial in behaviour, patrolling the tree-top canopy and only descending to the ground to feed on fermenting fruit, carrion, or animal droppings. Their larvae are naked, with four horns on the head and two blunt tail points. The sub-family is largely confined to the tropics, with many representatives in Africa, and an estimated 3–400 spp. worldwide (Ackery, 1984).

83. Common Nawab

Eriboea athamas
(Syn: *Polyura athamas*)
Wingspan 60–75mm

Description: Above both wings with broad blackish-brown borders and a pale greenish-yellow, wide discal band. There are two tails on the hindwing, the only Nymphalid butterfly in Pakistan with this feature. The underside is more cryptically coloured, with the discal band greenish and the borders more greyish-brown. There is a sub-marginal row of small pale yellow spots on the upper hindwing, and these take the form of rusty lunar spots on the under hindwing.

Status: Occurs only in the Margallas and Lehtrar flying during and after the monsoon. Seen by author in October in the Margallas at 600m (2,000ft), but occurs up to 1,200m (4,000ft).

Habits: The eggs are laid singly on the upper side of young leaves of the foodplant. They are dome-shaped and smooth in texture with a shiny surface, and they are yellow in colour. The larva is green in colour with the head flat in front and hexagonal shaped, bearing four fleshy horns on its head; the tail is bifid, and the body segments bear a set of spines. It makes a bed of silk for itself on the upper side of a leaf, to which it returns between bouts of feeding. The larval foodplants are usually trees of the families Mimosoideae or Caesalpinoideae—of the former, *Albizia lebbek, Albizia pennata, Albizia molluccana, Adenanthera pavonia, Acacia catechu,* and of the latter, *Delonix regia, Caesalpinia sappan,* and *Caesalpinia bonduc* (syn: *bonducella*). The pupa is suspended head downwards by a silken pad under a leafstalk or twig, and it hangs rigidly. It is smooth and coloured a shiny yellow-green, with white streaks and spots. Though the butterfly is one of the most powerful Nymphalidae on the wing, I have seen *Pycnonotus leucogenys*, the Hill White Cheeked Bulbul, successfully catch one and eat it after severing the wings.

Apaturinae

Apaturinae are also strong flyers with robust bodies, some with beautiful purple reflected colours on the upper surface of their wings, and generally with ocelli on their undersides. Their larvae are naked, with two horns on the head and two projections on the tail.

Like the Nawabs, they don't come to flowers, but feed on carrion, droppings, and fermenting fruit.

84. Indian Purple Emperor

Apatura ambica
Wingspan 65–90mm

Description: Above dark brown with a broad white discal band across both wings, and a sub-marginal row of small white spots around the edge of both wings. Females are larger and the smaller male, as its name implies, reflects a beautiful iridescent purple at certain angles on both hindwings and the forewing post-discal region. The underside is pale aquamarine with a narrow ochraceous border on both wings, and on the under forewing a broader post-discal ochraceous band with black spots between veins 2 and 3, and 3 and 4. The Eurasian Purple Emperor (*Aptura iris*) differs markedly in the pattern of its underside, with buffy and white bands against a brown and black background. The larvae are also quite different (*see*, Plate 4).

Status: Nowhere common in Pakistan, but occurs in Chitral and Swat Kohistan from 1,200–2,400m (4,000–8,000ft) in predominantly Holly Oak (*Quercus ilex*) forest with scattered Himalayan Elms (*Ulmus wallichiana*). Further to the east, in Himachal Pradesh, they are not uncommon. It is on the wing in April and May, with a second brood after the monsoon from August up to early October. Evans (1903) caught specimens in worn condition at Ziarat, Chitral, in September.

Habits: The eggs are laid singly on the young leaves of the foodplant, and they are domed with a flat base. The larvae are apple green with two branched spines or horns on the top of the head. The tail ends in a triangular anal flap, and there are three pairs of short dorso-lateral fleshy spines on segments 4, 8, and 11. The larvae have been found feeding on *Ulmus wallichiana* in the north-west Himalayas (Mackinnon and Niceville, 1898), and the foodplants for other Asiatic species of 'Emperor' belong to trees of the genus *Celtis*. The pupa is greenish-white with two short horns on its head and is covered with a white powdery bloom. The male butterflies are more commonly seen, and are extremely fast flying. They like to bask with open wings on high branches of trees, seldom coming to the ground to sip from damp patches or feed on carrion.

85. Western Courtier

Sephisa dichroa
Wingspan 60–75mm

Description: This handsome butterfly is black above with tawny-orange spots and bands on the forewing and the discal area of the hindwing, followed by a narrow sub-marginal tawny band. The veins on the hindwing are outlined in black. The female has whitish apical spots on the forewing. The underside hindwing is frosted with white scales, and the underside forewing has greyish-white spots in the apical area.

Status: Occurs as far west as the side nullahs of lower Chitral from 1,500–2,740m (5,000–9,000ft), and in Swat Kohistan (where the author collected it in July and August), it is more common at 2,400m (8,000ft) in Holly Oak (*Quercus ilex*) forest. It is on the wing from July to September. It is very rare in the Murree Hills, where the author encountered it only once, flying around *Quercus dilatata*. Butler (1886) records worn specimens being collected around Murree in mid August, where he considered it rare.

Habits: Nothing seems to have been recorded about the early life history of the Western Courtier, except the foodplant of the larvae, which is *Celtis australis* (family Ulmaceae). The butterflies are strong flyers, fond of basking high up on trees and seldom coming down to the ground except around midday for an occasional visit to damp patches.

86. Indian White Admiral

Limenitis trivena
sub-spp. *trivena*, *hydaspes* and *gilgitica*
Wingspan 60–75mm

Description: This is a variable species with three recognizable forms or sub-species occurring in Pakistan, differing in the width of the white discal band which crosses both upper wing surfaces. Above it is dark brown, with a sub-marginal double row of dull red spots divided by black centres. The race *gilgitica* has the narrowest white discal bands, and the race *trivena* the broadest, with an intermediate race, *hydaspes*, occurring in Hazara district and Azad Kashmir. The underside is ochraceous yellow with the white bands outlined in black on the forewing. The antennae are conspicuously tipped yellow.

Status: Widespread in Himalayan forested areas, with three recognizable sub-species occurring. In Chitral, Swat Kohistan, and Gilgit, form *L. trivena gilgitica*, with an intermediate form *L. trivena hydaspes* in the Kaghan valley, and the nominate sub-sp. *L. trivena trivena*, with the broadest white discal band, in the Murree Hills. On the wing from June to September and from 2,130–2,700m (7,000–9,000ft).

Habits: The larval foodplants are species of *Lonicera* (Honeysuckle), family Caprifoliaceae. The eggs are laid singly on the foodplant, and in this family are spherical in shape, green in colour, with pitted reticulations, and bearing small white spines all over. The caterpillar is green with branched spines in two rows along the back. The pupae are green with metallic silver spots, and are attached hanging downwards by the tail. The dorsal ridge is sculpted into a snout-like curved projection, and the head bears two short horns. The butterflies are fond of visiting flowers and settling to bask on low bushes and plants, so are not difficult to catch.

Calinaginae

The sub-family Calinaginae is represented by only four species which inhabit the Sino-Himalayan region, of which most are confined to western China and only one occurs in the Himalayas. Aptly named the Freak, this butterfly, *Calinaga buddha*, has been placed within this mono-typic sub-family because it possesses characteristics which differ from any typical Nymphalinae. The butterfly has an almost hairless abdomen, like the Danaids, and a tough leathery body, and, like the Apollos, very light scale covering to the wings, giving them a papery texture. The thorax is partly covered with red or orange hairs and the antennae are very short for the size of the insect.

Because of its rarity the larval and pupal stages are only known from the Chinese species. Their larvae are smooth and thinly pilose, with the head armed with a pair of prominent horns, and the tail with two small projections (Ackery, 1984).

87. The Freak

Calinaga buddha
Wingspan 90–105mm

Description: This species has puzzled taxonomists, differing from all other Nymphalid genera in the papery texture of its wings, the abdomen also hairless and leathery, while the thorax is partly covered with orange red hairs. It is a large butterfly, with rounded apex to the forewings and the margin of the hindwing extending beyond the forewing costa. Above it is pale green with the veins and margins outlined greyish-black. There is also a sub-marginal row of pale green spots on both fore and hindwings. The underside is paler, almost white, with paler, smokey-grey in the corresponding dark areas. Races found in the eastern part of the Himalayas have much more restricted pale whitish areas on the wings.

Status: Occurs in Murree at 2,100m (7,000ft), where it is very rare. The British Nat. Hist. Mus. has three specimens collected from Murree. Both Wynter-Blyth (1957) and Haribal (1992) write that nothing is known about its habits.

Habits. The foodplant of the larvae belong to the family Moraceae (figs and mulberries), and the larvae are only known from related Chinese Calinaginae. They are covered with thin hairs (pilose), but otherwise smooth, with a pair of prominent horns on the head and the tail slightly bi-fid.

Nymphalinae

88. Common Castor

Ariadne merione
(Syn: *Ergolis merione*)
Wingspan 45–60mm

Description: The forewing termen is slightly concave in this species, and the apex square cut. The upper surface of both wings is rich chestnut brown with double wavy lines of black, interspersed in the cell and discal areas with paler, rusty brown. The underside is a more greyish brown with intervening dark brown narrow bands, and the male bears a triangular black patch of scent scales on the underside of its forewing, at the base of the cell.

Status: Collected (author) from Islamabad gardens, and the Margalla Hills, near Castor bushes (*Ricinus* sp.). Presumed rather local and uncommon.

Habits: The eggs are laid singly on the underside of a leaf, and look like miniature sea urchins (Sevastopulo, 1946). The larvae are green with longitudinal dark brown lines, and each segment is adorned with four rows of branched spines. On its head are two forward-projecting branched spines. The larval foodplants are *Ricinus communis*, *Tragia involucarta*, and *Tragia plukenetii*. The pupae are olive green, speckled all over with light green, and with a darker blackish-green patch on the abdomen. The butterflies are inclined to stay close to their foodplant and are not strong flyers, alternating rather jerky wingbeats with bouts of sailing with wings held horizontally. They invariably settle with wings outstretched. They are able to fly inside bushes when threatened. Haribal (1992) reports that the Common Garden Lizard (*Calotis versicolor*) catches the butterflies, but leaves the larvae alone, probably because of their protective armour of spines.

89. Common Leopard

Phalantha phalantha
(Syn: *Atella phalantha*)
Wingspan 50–60mm

Description: Resembling the Fritillaries, with upper wing surface orange-tawny covered with small black spots. The forewing apex is arcuate with the termen slightly concave. In this species there are four narrow wavy lines across the forewing cell, and the sub-marginal regions of both fore and hindwings bear looped black lines bordering the black spots along the margin. The underside is overlaid with violet-tinged scales making the black spots less distinct, except for a prominent black spot between veins 1a and 1b.

Status: Widespread and not uncommon in the Salt Range, Murree foothills, and down to Lahore, where Puri (1931) described it as abundant after the monsoon. Straggles in late summer to lower Chitral and Balochistan, where collected from Khuzdar. Has occasionally straggled to lower Sindh. Collected Karachi in July (Swinhoe, 1887). Occurs up to 2,280m (7,500ft) in the Murree Hills, and is not uncommon in Islamabad gardens.

Habits: The eggs are laid singly on young leaf shoots. The larvae are described (Bell, 1916) as being an oily brownish-purple, armed with black branched spines. There is a sub-spiracular waved white line, and no horns on the head. The ventrum is reddish-brown. The pupa is green with rose-crimson spines on the dorsal side and some spines gold tipped. It is formed on the underside of a leaf, attached horizontally. The larval foodplants are *Flacourtia ramontchi* (syn: *indica*) and *Flacourtia montana*, belonging to the family Flacourtiaceae or Bixaceae (related to Violaceae). Sevastopulo (1947) describes the pupa as having a brownish-yellow head, purple-brown body, minutely speckled with white with a double pale dorsal line and zigzag, cream-coloured sub-lateral line. The dorsal surface spined, and these metallic silver, tipped red, and with silver lines along the outer and inner margins of the wing case. Its life history is remarkable for its brevity, being completed in three weeks from egg-laying. Larvae bred in captivity pupated on 9 March and the butterfly emerged on 17 March (Sevastopulo, *op. cit.*). The butterflies come frequently to flowers and, though strong, swift flyers, settle frequently to bask or to sip nectar.

Fritillaries or Silverstripes

The tribe Argynnidi includes the genera *Argynnis, Fabriciana, Issoria, Bolaria, Clossiana* and *Melitaea*. All of them in the adult stage are tawny-orange in colour with a similar pattern of small black spots on the upper side. The larger *Argynnis* spp. have silver stripes or spots on the their under hindwing, and the smaller Fritillaries have alternating irregular bands of creamy yellow and rust brown, picked out in black, on their under hindwings. They have naked (hairless) eyes, and the wing margins are smooth without projections. Their larvae have branched spines, and many of the pupae bear gold or silver spots and are spiny.

90. Queen of Spain Fritillary

Argynnis lathonia
(Syn: *Issoria lathonia*)
Wingspan 50–60mm

Description: Above orange-tawny, paler in the forewing costal region. The upper wings are covered with small black spots, and the under hindwing bears more prominent silver spots than in any other *Argynnis* spp., the interspaces being ochraceous with tiny brown-ringed silver spots in the sub-marginal area, bordered outwardly by large silver marginal spots. The forewing apex is rather pointed, the termen being slightly concave.

Status: A hill butterfly, very common in the Murree Hills and extending westwards to Chitral, Dir, and Swat. On the wing from February to October between 1,200–3,000m (4,000–10,000ft). In Balochistan it is not uncommon around Ziarat in late summer and occurs around Quetta in the spring. It has been collected from Campbellpur (now Attock).

Habits: The early life history is only known from European populations. The caterpillar in its early stages, while still small, hibernates in or on the ground until the following spring, and the eggs are laid on wild spp. of *Viola*. The egg is flat-bottomed, yellow-green in colour, with dome-shaped sides bearing beaded longitudinal ribs. The larvae are black with bands of red-brown spots between the segments. The head is small and reddish-brown. The body is covered with short spines. The pupa is attached hanging head downwards from a leaf stalk. It is dark blackish-brown with a prominent white band across the upper part of the thorax, and the abdomen tip is humped. The butterflies are fast on the wing, and regularly visit flowers of many different families, especially thistles, and are found in open grassy glades in moist temperate Himalayan forest, as well as higher up on grassy alpine slopes. When at rest on a blade of grass and still numbed with the early morning cold at those altitudes, the silver spots on the hindwing resemble dewdrops glistening in the sun and the butterfly is not conspicuous, a helpful camouflage to those sight-hunting predators such as *Mabuya* Skinks, and birds.

91. Indian Fritillary

Argyreus hyperbius
(Syn: *Argynnis niphe*)
Wingspan 65–85mm

Description: This is a large Fritillary which exhibits sexual dimorphism, the female, with a white band accross the forewing apex and inky-blue bands below the white area, being a passable mimic of *Danais chrysippus*, the Plain Tiger. Both sexes are tawny-orange on the upper surface with scattered black spots, with a dull blue border to the hindwing margin outlined with double black looped lines, and a sub-marginal ring of black spots. The under hindwing is banded with buff and broken bands of olive green, separated by silver bands which are again margined in places with thin black lines.

Status: Nowhere plentiful but very widespread in northern Punjab, including the Salt Range (collected by author at 500m [1,900ft] near Chakwal), and Campbellpur in May. It extends into the lower hills from Murree, and westwards to Chitral. Collected author up to 2,400m (8,000ft) above Dunga Gali, after the monsoon.

Habits: The egg is straw yellow and dome-shaped or conical with honeycomb-like markings. The larvae have a black head and legs, with the body velvety black but with a broad orange-tawny dorsal stripe. There are spines, three on each segment, black in the forepart, turning rose pink or red towards the rear of its body. The pupa, which has a domed thorax and the head bi-fid, is sooty grey, varying to pale red-brown, with a fine network of black lines, and with spines on the first thoracic and first two abdominal segments which are metallic gold. It is suspended in the usual way from the anal segment, hanging head downwards. The various stages from egg-laying to emergence of the imago last from six weeks to two months (Fraser, 1916).

92. Large Silverstripe

Argynnis childreni
(Syn: *Childrena childreni*)
Wingspan 75–100mm

Description: This is an impressive butterfly on the wing, being the largest of the Fritillaries and of a rich golden-tawny colour. It exhibits slight sexual dimorphism, the male having the forewing more concave along the termen and with the apex more arcuate. Also, the male has veins 2 and 3 inflated and bearing scent scales. The female lacks the swollen veins, has a more rounded forewing and heavier black spotting, and is generally slightly darker chestnut-orange on the upper surface. In both sexes the tornal area of the hindwing is suffused inky blue. The under hindwing is dark green with five narrow silvery bands, including the terminal band, and the discal band is more or less straight.

Status: Fairly common in the Murree Hills and lower Kaghan valley, but much rarer further west in Swat and Chitral. On the wing from late June to September mainly from 1,200–2,700m (4,000–9,000ft).

Habits: All the silver-washed or silver-striped Fritillaries have eggs that are pale yellow in colour, dome-shaped with a flat bottom, and bearing longitudinal ribs. The female lays her eggs in bark crevices on tree trunks, not on the foodplants, which are *Viola* spp. The caterpillar upon hatching goes into hibernation in the bark crevice and has to crawl to its foodplant upon awakening the following spring. These produce the first brood, which are on the wing in Pakistan from early May. A second brood hatches out from late September into October. The larvae of all *Argynnis* spp. are very spiny; those on the front segment pointing forwards like two thin horns. The pupae are also very spiny, and decorated with metallic silver spots, and with a bi-fid head. They are attached by the anal claspers hanging head downwards from the foodplant stem, and are capable of wriggling violently if touched. This splendid Fritillary is easily seen as it often settles low down to bask and visit flowers.

93. Western Silverstripe or Cardinal Fritillary of Southern Europe

Argynnis pandora
(Syn: *Argynnis maia pasargades* of Evans, 1932)
Wingspan 65–75mm

Description: Slightly smaller than *A. childreni*, this is still an impressively large Fritillary. It is readily distinguished from the former by the under hindwing, which bears only two silver stripes, the discal one being broken. *A. childreni* has five silver stripes on the under hindwing. The females are typically much darker than the males, their upper surface being dusted with bluish-grey scales, and also with more prominent large black spots. In both sexes the under forewing is pale buffy-yellow at the apex with slightly darker ochraceous green tips, and the base of the wing pale pinkish-chestnut. The under hindwing is ochraceous green and the silver stripes may be quite narrow and faint. In both sexes the upper hindwing bears a continuous jagged narrow black line in the discal region, absent from *A. childreni*.

Status: Within our area, it has only been collected from western Chitral, from Shishi Koh Nullah, Drosh, and Sonogar. It flies from August to October at elevations between 1,500 and 2,400m (5,000–8,000ft). Evans considered it 'not rare'.

Habits: In southern Europe the larval foodplants are violets (*Viola* spp.). Adults are described as haunting flowery meadows, and having a strong fast flight.

94. Common Silverstripe

Fabriciana kamala
(Syn: *Argynnis kamala*)
Wingspan 65–75mm

Description: Smaller than *A. childreni,* this species is the same tawny-orange above but with heavier black spotting, and lacking the dark blue suffusion on the upper hindwing. The male also has veins 2 and 3 on the forewing swollen and bearing scent scales. The under hindwing in both sexes is dark green with five silver stripes, the discal one being angled outwards towards the termen.

Status: More common and widespread than the Large Silverstripe, and on the wing from late May to September, usually above 2,000m (6,500ft) and up to 3,000m (10,000ft). Fairly common from Chitral through Swat Kohistan and Hazara district to the Murree Hills.

Habits: Believed to be much the same as *Argynnis childreni*, with which they are sympatric in the Himalayas. They have the same very spiny larvae, and the pupae are decorated with metallic silver spots. There are two broods, the first butterflies emerging in May and the second brood on the wing in late September. The foodplants are Violaceae.

95. Jerdon's Silverspot

Argynnis jerdoni
Wingspan 35–45mm

Description: This is quite a small Fritillary, though some races in the eastern part of its range are larger. Tawny-orange above with black spotting, and the upper hindwing suffused with black scales basally. The under hindwing is dark olive green with silvery-white spots in a sub-marginal row, each ringed with dark green, and elongated white spots in the costal region, and another in the basal area of the wing being in the shape of a cross.

Status: Fairly widespread at higher altitudes from Chitral, through Swat Kohistan, Gilgit, and Baltistan. Collected from Chichele (Ochhili) Pass in Chitral, the Deosai, and Chogo Lungma in Baltistan. Collected TJR in the Kaghan valley from Battakundi and Burawai above 3,350m (11,000ft), where very common. On the wing from late June to end July, and found from 2,100–3,960m (7,000–13,000ft).

Habits: The author collected adults visiting the flowers of *Polygonum* (syn: *Bistorta*) *affinis*. They are presumed to have similar larval foodplants (*Viola* spp.) as other *Argynnis* spp. with those violets adapted to growing at higher altitudes.

96. Dark Green Silverspot or Fritillary

Argynnis aglaia
(Syn: *Mesoacidalia aglaja*)
Wingspan 55–60mm

Description: A medium-sized Fritillary with prominent silver spots on the under hindwing. Above tawny-orange with scattered black spots. Females are sometimes suffused with dark purple and this form was described by Evans as *aglaja purpura*. The under hindwing is dark olive green with three silver spots along the costa and a row of triangular sub-marginal spots outlined on the inner side with dark green. This species is very similar to *Argynnis adippe*, the chief distinguishing feature in *aglaja* being the absence of any tiny silver spots between the sub-marginal row and the larger post-discal row. In *adippe* there are five or six such spots, often circled with red-brown and an inner ring of black.

Status: Very widespread but also erratic in occurrence, in the northern areas, from Chitral, where collected from Garogar, Chichele Pass, the Utzen Nullah, and the Shandur Plateau; also Ghizar valley in Gilgit, Hunza, Rupal in Astor, and in Baltistan from Chogo Longma and Skoro La. Collected TJR from Swat Kohistan in the Utrot and Ushu valleys. Occurs between 2,700–4,570m (9,000–15,000ft), and on the wing from late June to August.

Habits: This species lays its eggs around the base of *Viola* spp. but the little caterpillars, after hatching and eating their eggshells, immediately go into hibernation low down on the plants. They emerge in the spring and then start feeding. The eggs are cone-shaped, but with a flat top and longitudinal ribs. The larvae are black with dark red spots along the sides, one between each segment, and paired spines on each segment. The larva before pupating draws *Viola* leaves around itself, held together with silk threads, and inside this shelter forms the chrysalis, which is attached hanging downwards by the cremaster or tail claspers. The pupa is dark brown in ground colour, decorated with pinkish-brown spots and lines, and with dorsal spines. The butterfly is only on the wing from late summer, and has a strong flight. The author encountered it in Swat Kohistan in August, in cultivated valleys around 2,700m (9,000ft), attracted to roadside flowers.

96.b Afghan Fritillary

Fabriciana argyrospilata
Wingspan 45–60mm

Description: Slightly smaller than *A. aglaja*, this Fritillary is generally also a paler tawny than *A. aglaja* or *A. adippe*, but the markings on the upper surface are very similar. The under hindwing is dark green all over, lacking the pale tawny margin of *aglaja*. It is profusely covered with silvery white spots, the mid-cell spot being narrow and bent outwards, in contrast to the rounded mid-cell spots in *adippe* and *aglaja*. Also the sub-marginal row of white spots are U-shaped, their base being outwards.

Status: This medium-sized Fritillary occurs in north-eastern Afghanistan in Badakhshan province (Sakai, 1981) in areas bordering Chitral, where it was collected between 3,200–3,500metres (3,400–11,500ft). Though there are no Brit. Nat. Hist. Museum specimens from Pakistan, it is most likely to occur when more collecting can be carried out in those areas. It is rare in Afghanistan.

97. High Brown Silverspot or Fritillary & Niobe Fritillary in Russia

Argynnis adippe
(Syn: *Fabriciana cydippe*)

Description: Very similar in appearance to *A. aglaja*, differing only in the presence of an additional row of five or six tiny silver spots between the sub marginal row and the post-discal row of spots. These tiny spots are encircled with red-brown sometimes with a trace of black encircling the silver between the russet surround. As in all the *Argynnis* fritillaries, the female is more heavily marked with black spots and, like *A. aglaja*, the male lacks swollen veins on the forewing as in *A. childreni*.

Status: A high-elevation species which is widespread but erratic in occurrence, from the Safed Koh, the Shishi Koh in Chitral, the slopes of Nanga Parbat, and around Skardu and Mushkin in Baltistan. Collected TJR from Jabba valley, Swat, at 2,400m (8,000ft), and in the Ushu valley at 2,700m (9,000ft), where locally common.

Habits: The early life history of this Fritillary is similar to *Argynnis aglaja*, except for the fact that the eggs, which are deposited in July and August, low down on *Viola* spp., do not hatch out until the following spring. The larvae are olive green in colour with narrrow bands of black between the segments. They are adorned with a dorsal and lateral row of branched or barbed spines coloured pink, and, perhaps so protected, do not need to hide themselves, as they are fond of basking in the sun on open leaves. Like *A. aglaja*, the caterpillar, when ready to pupate, draws leaves around itself with silken threads and then attaches itself by the cremaster, hanging head downwards to form the chrysalis. This is a rich brown colour adorned with brilliant silver-gilt cones. The author found this Silverspot much more plentiful in some localities than *A. aglaja*.

98. Whitespot Fritillary

Clossiana hegemone
(Syn: *Argynnis hegemone* of Evans, 1932)
Wingspan 38–42mm

Description: One of the smaller Fritillaries, rather pale chestnut in colour and bearing quite small black spots on its upper wing surface. The under hindwing is dull pale yellow, with a marginal row of white triangular spots only separated by the veins, and a sub-marginal row of very small spots encircled black, followed by a broad-pale yellow discal band edged black in a zigzag pattern.

Status: A high-elevation Fritillary only found in northern Chitral and Hunza, and considered very rare by Evans.

99. Straightwing Silverspot or Shepherd's Fritillary of Europe

Bolaria pales
(Syn: *Argynnis pales* of Evans, 1932)
Wingspan 32–40mm

Description: This is a small Fritillary with a pointed apex to the forewing and the hindwing margin also slightly pointed at vein 3. The black markings on the upper wing surface are quite small and the basal part on both fore and hindwings is dark blackish. The under hindwing lacks silver spots, these being white, interspersed with red brown bands, the basal red band edged black outwardly, and with a small white spot at the base of the cell.

Status: A high-elevation Fritillary, extending in range from the inner mountains of Chitral eastwards to Sikkim. It has been collected from the Kaghan valley, Misgar in Hunza, Kala Pani in Astor district, Chogo Lungma in Baltistan, and from the Deosai plateau. It is not uncommon where it occurs, and is on the wing from July to August.

Habits: This butterfly is rarely found below 3,000m (10,000ft), and has only one generation a year, the butterflies emerging in mid to late July. The larval foodplants are *Viola* spp.

Redband Fritillaries

100. Shandur Fritillary

Melitaea shandura
(Sensu: *Melitaea saxatilis shandura* of Evans, 1910)
Wingspan 45–50mm

Description: This small Fritillary is pale yellowish-tawny on the upper wing surface and the black spot end cell is very prominent. The female is often noticeably larger and lacks any post-discal row of black spots on the hindwing. The under hindwing lacks any red-brown bands, being dull ochre yellow with the veins outlined black, and two bands across the discal and post-discal areas slightly darker ochre and edged black.

Status: A high-elevation Fritillary found from Badakhshan in Afghanistan and eastwards into northern Chitral, where it has been collected from the Shandur plateau. It is on the wing from July to August, flying between 3,660–4,260m (12,000–14,000ft), and Evans (1932) considered it very rare.

Habits: As far as is known, all the *Melitaea* Fritillaries lay their eggs in batches of twenty to thirty on the foodplant leaf, and the tiny caterpillars, upon hatching, spin a thick communal web of silk, not necessarily on the foodplant, but often between the stems of nearby grasses or a bush. They use this for a shelter, emerging to feed gregariously. They do not reach full growth in the year that they hatch, but go into hibernation at different stages of growth, continuing to feed and grow the following spring. The caterpillars are armed with barbed spines, and are generally black on the body with paler inter-segmental bands or lateral spots. The pupae are generally formed inside a protective shelter made by the full-grown caterpillar drawing several leaves of the foodplant together with silken threads before developing into a chrysalis. The butterflies typically fly close to the ground surface, the alpine ones being adept at flying with the wind, and they settle frequently to bask or to feed from flowers.

101. Desert Fritillary

Melitaea persea
Wingspan 34–44mm

Description: This small Fritillary is one of a very complex and variable group which has given rise to many sub-species and synonyms. Sakai (1981) lists it as a distinct species from *M. lutko*, which is closely similar. The male is pale tawny with quite bold black spots on the upper wing surface, and the larger female usually has pale yellow discal bands. In the female there are no black spots on the upper hindwing post-discal area as in *M. didyma* and *M. ala*, but there is a series of black lunules around the sub-marginal area and no marginal black spots. The under hindwing is white and has the outer band of red spots inwardly convex and clearly separated from the inner line of black U-shaped lines and followed by a second inner line of variable-shaped black lines, both well separated from the basal red band.

Status: Collected from Quetta, Murgha Mehtarzai and the Khojak Pass, all in Balochistan. Flying between 1,500–2,400m (5,000–8,000ft). Collected TJR from Ziarat (Balochistan), and on the wing from May to July.

Habits: The larval foodplants are members of the family Plantaginaceae (Knapweeds), of which *Plantago ciliata* is the most likely, being very common in Balochistan. The eggs are pale yellow and ribbed, with a flat top. The females lay their eggs in batches, and the young larvae spin a thick silk web between grass stems or low down in a thorny bush as shelter, and live gregariously in their early stages. The larvae are black with pale grey bands separating each segment, and with a small red-brown head and dorsal row of short, branched spines. The pupa is grey, marbled with spots and streaks of black, yellow, and brown.

102. Balochi Fritillary

Melitaea lutko
(Syn: *Melitaea schoenis* or
M. robertsi lutko, in Evans, 1932)

Description: As stated above, this small Fritillary is very similar to *M. persea*, and, like the former species, the female is larger and has a pale yellow post-discal band on the upper forewing, and often pale yellow in the middle part of the cell. Both sexes lack any ring of black spots on the hindwing in the post-discal area, but both have small black marginal spots on the hindwing, and in the female only there is a sub-marginal ring of curved black lines above each spot. The under hindwing is white with three red bands, basally, in the discal area, and sub-marginally. This outer band is lined with looped black lines as in *M. persea*, but there is no second looped line as in that species.

Status: On the wing from late April through May in Chitral, and in Balochistan flying from late March to May. It has been collected from Drosh and the Lutko valley in Chitral, from Uruk valley, Quetta environs, Kuch, and the Khojak Pass in Balochistan, and also from the Murree foothills. It is fairly common in these areas on bare hillsides between 1,800–2,100m (6,000–7,000ft).

103. Lesser Spotted Fritillary

Melitaea trivia
(*Melitaea robertsi* in Evans, 1950)
Wingspan 30–44mm

Description: Similar to *Melitaea didyma*, but averaging smaller in size. The upper wing surface is reddish-chestnut with fairly bold black spots, and the under hindwing is creamy yellow with two red bands, the basal one being short but broad and the outer post-discal band less curved than in *M. persea* or *M. lutko*.

Status: This is one of the *Melitaea* spp. adapted to warmer latitudes and lower elevations. Collected from Campbellpur in the Salt Range, Attock bridge on the border between Punjab and NWFP, Miramshah in North Waziristan, Jamrud in the Khyber Agency, and from Drosh in Chitral. On the wing from April to June and quite common in suitable habitat.

Habits: The larval foodplant in the Mediterranean region is Great Mullein, *Verbascum thapsus*, which grows in the dryer regions of Chitral, Swat, and Gilgit.

104. Redband or Fiery Fritillary

Melitaea didyma
(Syn: *Melitaea saxatilis lunulata*, sensu Tytler, 1926 and Evans 1932)
Wingspan 30–44mm

Description: Like *M. persea*, this is a very variable species, but resembles *M. trivia* though averaging larger in size, and usually with more distinct black spots on the upper hindwing disco-cellular area. Some specimens are quite a dark reddish-chestnut on the upper wing surface and some have more dark brown scales at the base of the wings. The under hindwing pattern is creamy yellow with two red bands, the outer one being lined on its inner edge with fairly straight black lines, not V-shaped as in *M. lutko* and *M. persea*.

Status: This form is quite widespread in Central Asia but rather rare within Pakistan, and has only been collected from the border areas of Balochistan such as the Khojak Pass, and from Murgha Mehtarzai, and also at Misgar in northern Hunza. It is commoner further west in Afghanistan and Iran, and again in Tajikistan.

Habits: The larval foodplants in the eastern Mediterranean region are Plantains, family Plantaginaceae, and Toadflax (*Linaria* sp.), family Scrophulariaceae. In Chitral, Swat, and Gilgit, *Linaria odora*, and *Linaria simplex* grow in the valleys where this species has been collected.

105. Pamir Fritillary

Melitaea minerva
Wingspan 35–40mm

Description: This is a heavily-marked Fritillary, with the female showing distinctive pale yellow bands on the upper forewing. The base of the upper hindwing is quite black, and there is a continuous line of black in the disco-cellular region rather than a series of separate spots as in the closely similar *M. ala*. The under hindwing bears the two red-brown bands common to the genus, the basal one being more or less continuous and the outer one bordered on its inner side by a fairly straight black line. The female is handsome, with yellow bands end cell bordered above and below with broad black bands and a second disco-cellular yellow band.

Status: Only recorded from Chitral, where fairly widespread between 2,280–3,660m (7,500–12,000ft). Collected from the Shandur plateau (Tytler, 1926, syn: *Melitaea balba*), Madaglash valley, and Sonoghar. On the wing from July to early August.

106. Blackvein Fritillary

Melitaea arcesia
Wingspan 34–40mm

Description: This species shows fairly marked sexual dimorphism, the male being lightly marked with black spots on the upper wing surface, while the female, besides being heavily marked, has yellow bars on the upper forewing and an extensive black area on the base of the hindwing covering the whole of the cell. The upper hindwing in the female also bears a marginal series of pale yellow-orange lunules in a broad black border. In the male the upper hindwing has smaller, tawny lunules framed inwardly by a narrower, black-looped border.

Status: Occurs in the eastern fringe of the Deosai plateau in Baltistan and has been collected on the Karakoram Highway in Chilas at lower altitudes. It is rare in Chitral, and there are no specimens from Chitral in the Brit. Nat. Hist. Mus. collection. It flies between 2,700–4,260m (9,000–14,000ft), and is on the wing in July and August.

107. *Melitaea ala*

(Syn: *Melitaea pseudoala*)
Wingspan 35–40mm

Description: Similar in appearance to *M. arcesia*, this Fritillary shows sexual dimorphism, with the female being much more heavily marked and having more extensive, paler yellow areas on the upper forewing than *M. arcesia*. These pale areas extending into mid-cell. The male is a dark tawny-red with a series of bold black spots across the disco-cellular region and a ring of slightly smaller spots on the hindwing disco-cellular region. In *M. arcesia* males there are only traces of dark spots in the disco-cellular region. The female has much less extensive black areas at the base of the upper hindwing compared to *M. arcesia*, but shows a paler yellow area tornally. The under hindwing in both sexes shows a broken band of red basally, whilst the outer red band is lined with fairly straight black lines, and in the intervening creamy-yellow band there are a series of black spots, entirely lacking in *M. arcesia*.

Status: Only recorded from Chitral where collected from Madaglash and the Utzen nullah flying between 2,100 and 3,000m (7,000–10,000ft). On the wing in July and early August.

Pansies

The Pansies, *Junonia* genus, are smallish butterflies, usually confined to the plains and open country, and are prettily patterned on the upper wings with bright colours and prominent ocelli. They are often very abundant, and as they regularly visit flowers and have the habit of flying low to the ground and gliding between bouts of fluttering, they are familiar to many people not otherwise concious of butterflies. Their larval foodplants belong to the family Acanthaceae, and the larvae all have branched spines, tubercles or hairs along their body, with contrastingly-coloured longitudinal stripes.

108. Yellow Pansy

Junonia (Syn: *Precis*) *hierta*
Wingspan 45–60mm

Description: Most of the *Junonia* genus are brightly coloured and the Yellow Pansy, with chrome-yellow panels on both fore and hindwings, is one of the prettiest. The male is marked differently from the female, with no ocelli or black band across the end cell forewing. Both sexes have a narrow black marginal border on fore and hindwings, with the forewing apex black and bearing small yellow spots. On the hindwing is an iridescent blue oval panel at the base of the wing. The female, a duller yellow, has two small ocelli on the hindwing post-discal area, and a larger blue-pupilled ocellus on the forewing, with a broad black bar across the disco-cellular region of the forewing. The under hindwing is greyish buff with brown band across the disco-cellular area, and narrow brown marginal and sub-marginal lines.

Status: Very widespread and common in the plains from Sindh through Punjab and the NWFP. It occurs as a straggler only in Balochistan after the monsoon season in the plains. It can be seen on the wing throughout the year in Sindh, but mainly from February to December in the more northern parts of its range.

Habits: The larvae are described by Haribal (1992) as cylindrical and light greenish-brown in colour, with the dorsum velvety black on segments 3 and 4, and a double yellow line behind each segment. All the *Junonia* spp. have their bodies armed with nine longitudinal rows of many-branched spines, and the head is covered with bristles. The larvae of *Precis hierta* are described by Davidson and Aitken (1890) as dark brown or grey, with a broad dorsal stripe formed by minute white and blue spots, and covered with black spines, the head covered with minute conical yellow tubercles. The pupae are greyish brown with the wing cases whitish. It has been found (Mohyuddin, 1987) feeding on *Barleria cristata* buds in November at Rawalpindi, and elsewhere in India it has been recorded on other *Barleria* spp. and *Hygrophila auriculata*. The butterflies are fond of basking on the ground with wings outstretched, and usually fly close to the ground surface, with frequent gliding on wings held horizontally. They are less inclined to enter gardens, preferring more open, dry areas.

109. Blue Pansy

Junonia orithya
(Syn: *Precis orithya*)
Wingspan 40–60mm

Description: The male has the upper forewing black except for the apex, which is a buffy white with narrow brown bands and a small panel of blue in the tornal area. The upper hindwing in the male is iridescent blue, with two red-brown centred, black ringed ocelli. In the female these ocelli are more prominent and the blue areas are duller. The under hindwing is pale greyish-buff as in *P. hierta*, but the brown bands are almost obsolete, and the under forewing has narrow black paired lines across the cell.

Status: Possibly more common than the Yellow Pansy, as better adapted to irrigated regions. Widespread throughout the Indus plains and occurring in lower Balochistan also, and it has been collected from Quetta. In Sindh can be seen on the wing throughout the year. Collected Takht-i-Suleiman by Rafiq Ahmad Rajput in July 1993.

Habits: The larva is described (Haribal, 1992) as plumbeous black in colour with the first segment orange, and a jet black dorsal line, spotted finely with white, and these spots ringed with orange, and having tubercles which are black, bordered with yellowish-white. The larva lies along leaf stems by day and feeds at night. The pupa is slaty dull grey. The larval foodplants are Acanthaceae family such as *Justicia procumbens, Justicia migrantha*, and *Lepidagathis prostrata*; also Plantaginaceae family, such as *Plantago ovata*, all of which occur in Pakistan. This species readily enters gardens and is fond of settling on flowers. Like the Yellow Pansy it has a low swift flight, keeping close to the ground, and often basking with wings outstretched on the ground.

110. Peacock Pansy

Junonia (Syn: *Precis*) *almana*
Wingspan 60–65mm

Description: Unlike the two previous species, the forewing of the Peacock Pansy has the apex square cut, not pointed, and the termen concave. Also, it is slightly larger than the previous two species, and both sexes are uniformly orange-tawny on the upper surface, with two conspicuous ocelli on both fore and hindwings. The upper ocellus on the hindwing being very large, black rimmed with inner yellow ring and dark chestnut and black centre. There are three narrow black lines along the border of both upper fore and hindwings. The under hind and forewings are pale tawny-buff with three black marginal border lines and the same row of ocelli, the hindwing upper one being bi-pupilled, and the tornal ocellus much larger than on the upper wing surface.

Status: Though common in the Indus plains, this Pansy prefers more mesic conditions and is especially abundant in the rice-growing tracts of Sindh, in Thatta and Larkana districts, and Punjab, such as Gujranwala, Sialkot, and Sheikhupura. Consequently it is more often seen on the wing during and after the monsoon.

Habits: The larvae are described as smoky black, with the first segment orange, and the surface of the body covered with minute hairs. They bear spines on each segment which are light orange with black tips. The pupae are greyish-green with black and cream markings. They are attached firmly by the cremaster and hang downwards. The larval foodplants are *Barleria* spp., *Gloxinia* spp., *Osbekia* spp., and *Hygrophylla auriculata*, all of the family Acanthaceae. Mohyuddin (1987) found a pupa attached to a grass stem. The butterfly, like its congeners, alternates swift low flight with glides, and is fond of basking low down on grass stems or the bare earth.

111. Lemon Pansy

Junonia (Syn: *Precis*) *lemonias*
Wingspan 45–60mm

Description: This is a more sombrely-clad Pansy, with both fore and hindwings brown, a darker brown on the forewing, and with a series of four paired pale lemon-yellow spots across the forewing from end cell to apex. There are two ocelli on the upper forewing, the apical one small and inconspicuous but the tornal one large, with a black pupil and red surrounding ring. This ocellus is also ringed by a circle of scattered yellow spots. On the upper hindwing a large bi-pupilled ocellus stretches from vein 4 to 6. The underside is more cryptically marked with reddish-fawn and pale brown bands and streaks.

Status: Rather rare and local in Sindh, but common in Punjab and the plains of the NWFP. It has not been recorded from Balochistan. Can be seen on the wing from February to early December.

Habits: The larvae resemble those of the Chocolate Pansy, *P. iphita*, being grey with a black dorsal line, and covered with branched red-yellow spines. Each segment bears eight of these spines. The body is also covered with minute whitish tubercles. It differs from the other *Junonia* spp. in having the head bi-lobed. The larval foodplants are *Hygrophila auriculata*, *Nelsonia canescens*, *Cannabis sativa*, *Sita rhombifolia*, *Lepidagathis* spp., and *Barleria prionitis*. Like the Blue Pansy this species readily enters gardens and is fond of settling on flowers, but otherwise is similar to all the 'Pansies' in flight pattern and habits.

112. Chocolate Pansy

Junonia (Syn: *Precis*) *iphita*
Wingspan 55–80mm

Description: Averaging larger than the other Pansies, the forewing, like that of *P. lemonias* and *P. almana*, is square cut at the apex with the termen concave. Both sexes are dark brown all over the upper wings with rather indistinct, darker brown vertical bands and wavy lines, and on the upper hindwing a series of very small, paler brown ocelli around the disco-cellular region. This pattern is repeated in reverse on the underside, with the paler areas being darker and the corresponding dark brown areas on the upper side being silvery brown.

Status: The rarest of the Pansies in Pakistan as this species prefers forested areas. Has been collected from around Lahore and Sialkot, but is more likely to be encountered in the foothills, as in the rest of the Himalayas.

Habits: The larvae are black in the early stages, becoming red-brown. Their bodies covered with dirty-yellow pedicles from which brown spines, shorter than in the other 'Pansies', emerge. The young larvae make a shelter for themselves by binding leaves together with silken threads and hide in this during the day, feeding at night. The pupae are grey or brownish-black, with a pale spot on the head and side of the thorax, and are formed low to the ground on the underside of a leaf, branch, or rock. The larval foodplants are *Carvia callosa*, *Justicia micrantha*, and *Hygrophila auriculata*.

Admirals and Tortoiseshells

These butterflies belong to the tribe Vanessidi. They have their wing margins angled and slightly toothed, to varying degree, with contrastingly coloured, narrow wing borders and red pigments often predominating. Their eyes are covered with hairs, and their antennae have pear-shaped clubs. On the underside their wings are always very cryptically coloured. The adults regularly visit flowers. The larvae are always spined and often feed gregariously in their early stages. Their principal foodplants belong to the family Urticaceae. The adult butterflies usually hibernate over winter.

113. Indian Red Admiral

Vanessa indica
Wingspan 55–65mm

Description: This and the next species are closely similar in appearance, with the forewing square-cut at the apex and the termen concave below. The general impression is of a dark blackish-brown butterfly with a prominent scarlet diagonal band across the forewing. The hindwing margin is also scarlet, containing a series of small black spots. The forewing apex is jet black with two rows of small white spots, whilst the upper hindwings are actually dark brown, not black. The broad scarlet band across the forewing appears to be interrupted along its lower edge by three large black spots.

Status: A hill butterfly, flying between 1,200–2,700m (4,000–9,000ft) and especially common in forested regions of the Murree Hills, and Kaghan valley, and can be encountered from March to October, being double-brooded, the second brood appearing from late August to September, and many of these individuals hibernating through the winter. It also occurs across lower Chitral, Swat Kohistan, and in the Neelum valley, Azad Kashmir.

Habits: According to Hannyngton (1911), the eggs are laid singly on the upper leaves of the Himalayan Nettle, *Girardinia palmata*, but in the Murree Hills they have been found on the smaller nettle, *Urtica dioica*. Like its European congener, the small caterpillar makes a shelter for itself by binding together the margins of the leaf upon which it has hatched, and then commences feeding inside, until the leaf becomes withered and unpalatable. As it grows bigger it selects a larger leaf, or two or more leaves, to bind together with silken threads, until it finally makes a tent of several leaves and, hanging head downwards by its tail claspers, pupates. According to Wynter-Blyth (1957), the eggs are sometimes laid in batches and the larvae feed together and finally make a tent of leaves 'the size of an orange' in which they pupate. The larvae are described as sluggish, rolling themselves into a ball if touched and falling to the ground. The larvae at first are like those of *Vanessa atlanta*, and are dark reddish-black, with a pair of black branched spines on each segment and a thin, sub-spiracular pale yellow lateral line. In the fifth or final instar, however, all their spines are pale lemon yellow. The pupae are darker than those of *V. atlanta*, but with the same rows of dorsal spines which are tipped bronzy-gold, and with the wing cases smoky green. The pupa takes about ten to twelve days to develop and emerge, and in the Western Himalayas there are two broods. The butterfly is strong on the wing and very fond of visiting flowers, especially *Buddleia*, often sipping with wings outstretched.

114. European Red Admiral

Vanessa atlanta
Wingspan 55–65mm

Description: This Palearctic species, which ranges across western Europe, is remarkably consistent in wing pattern, and the single known specimen from Pakistan is identical to those from Britain, which is to be expected from a migratory species. It differs from the Indian Red Admiral in the shape of the scarlet band across the upper forewing, which is narrower and straighter than in *V. indica*, with the base of the forewing an even black triangle and the upper hindwings being darker than in *V. indica*, though often more a shade of brown than black. There is a small tornal blue-centred ocellus on the hindwing. The underwing pattern is closely similar to the Indian Red Admiral, the under hindwing being cryptically patterned with dark browns, white speckling, and some black spots.

Status: Col. Stockley collected a specimen from the Shingai Gah at 2,400m (8,000ft) in *Pinus gerardiana* forest, on 27 August. This is the only record for the subcontinent. I have examined this specimen, which cannot be differentiated from western European specimens. It probably occurs on the Takht-i-Suleiman range also.

Habits: This has been well documented in the European population. They differ from other *Vanessa* in that the female lays her eggs singly. They are always laid on *Urtica* spp. of stinging nettles. The eggs are pale greenish-white with widely spaced prominent longitudinal ribs, and horizontal fine striations. The little larva binds together the sides of the leaf upon which it hatches and feeds inside the rolled leaf until it withers and becomes unpalatable. It then moves to a larger leaf, and as it grows bigger it will bind together several leaves and feed within these. Finally, in a tent of leaves held together with silken threads, it hangs head downwards by the tail claspers and pupates. The larvae are reddish-black, with a thin cream or buff lateral line, and on each segment a pair of branched black spines. The pupae are olive green or brown, depending upon their background, and have spines along their dorsal surface that are metallic gold tipped. They are also covered with a waxy sort of bloom. The butterfly has strong migratory tendencies and is a powerful flyer, as would be expected.

115. Painted Lady

Cynthia cardui
(Syn: *Vanessa cardui*)
Wingspan 55–70mm

Description: Newly emerged specimens are a beautiful pinkish-orange colour on the upper wing surface but this soon fades in older specimens to a dull tawny. Both wings are heavily spotted with black and with white spots in the black area of the forewing apex, which is rounded at the tip, not square cut as in the Red Admirals. The base of the hindwing is covered with silky brown hairs (cilia). The sexes are alike, as in all the *Vanessa*. The under hindwing is cryptically marked with fawn and grey bands, spots, and lines, with a row of four dark-centred and buff-ringed ocelli around the disco-cellular region.

Status: This most cosmopolitan of butterflies occurs throughout the plains though it is uncommon in Sindh (and I have seen it in Karachi) and occurs in all the northern hill areas. I have encountered it in all parts of Balochistan, and in the Himalayas up to 4,570m (15,000ft), and right up to northern Hunza. It can be found in the plains mainly in the winter months, and from April to November in the Northern Areas, and there is evidence that it is partly migratory to the north in the spring.

Habits: The eggs are barrel-shaped and taller than those of the Indian Red Admiral, but with the same prominent, widely spaced longitudinal ribs which are white, the spaces in between being pale green with horizontal striations. They are laid singly on a great variety of foodplants, as might be expected from a species which is so cosmopolitan in latitudinal and altitudinal distribution worldwide. Bell (1910) lists Compositae, Leguminosae, Urticaceae, Boraginaceae, and Malvaceae among the families used as foodplants. Found on *Malva parviflora* in Rawalpindi by Mohyuddin (1987). In the Murree Hills, *Cnicus wallichii* and *Cnicus arvensis* thistles are favoured. In Swat, Mohyuddin (1987) found larvae on *Carduus nutans*, in the northern regions of Pakistan *Artemisia,* and in the forested regions of the Himalayas *Debregeasia salicifolia*. The larvae are covered with white hairs and even the head is bristly. Its body ground colour is yellow, but this is smudged with varying amounts of black, some specimens being almost wholly black. There are double yellow dorsal lines and a broader marginal yellow band, and branched spines on the front of its body and in the middle segments 5 to 12. The caterpillar initially, when small, makes a shelter for itself by spinning the edges of a leaf together and feeding inside this. When large, it spins an untidy web of silk and hides inside this, feeding at the same time on the leaf to which this web is attached. If alarmed it drops to the ground. The pupa is suspended in the usual manner from a horizontal or vertical surface by the tail clapsers, and the chrysalis is similar in shape to that of the *Precis* spp. being constricted at the junction of the thorax and abdomen with a series of short spines along the abdomen. It varies from green to brown to pinkish-grey in colour with the spines brilliantly decorated with metallic gold or coppery hues. The butterfly has strong migratory tendencies, and is swift and direct in flight, but readily settles on flowers and in the early morning is fond of basking on stones with wings spread. The author has caught it in irrigated farm fields in Khanewal district, Punjab, in alpine meadows above Battakundi in Kaghan, and seen it in gardens in Gilgit town in November.

116. Blue Admiral

Vanessa canace
(Syn: *Kaniska canace*)
Wingspan 55–70mm

Description: This very striking butterfly is deep blue-black on the upper wings with a broad silvery-blue discal band across both wings. This band contains small black spots between the veins. The forewing is square cut across the apex and strongly concave below along the termen and the hindwing bears a small tail at vein 4. The underwings are dark, mottled with brown and black, resembling the pattern on the underside of Tortoiseshell butterflies.

Status: Much less common than the Indian Red Admiral, this handsome species occurs from lower Chitral, where it is rare, eastwards through Swat; it is less rare in Hazara district and the Murree Hills. It is a forest loving species, commoner at lower altitudes (for example the Manga valley) than higher up, and on the wing from March to May in the foothills, and from July to September above 2,100m (7,000ft).

Habits: Little has been recorded about the early life history of this species, but it is much more confined to forested areas than the other *Vanessa* spp. The male butterflies are inclined to be territorial, keeping to the same flight boundaries and chasing off intruders. The larval foodplants are monocotyledons belonging to the families Dioscoreaceae, in Pakistan *Dioscorea deltoidea*, and Liliaceae, in Pakistan probably *Smilax vaginata*, which grows commonly from 1,670–3,350m (5,500 to 11,000ft). The author found the adults particularly attracted to tree sap in early spring, with several feeding together on the wounds in the trunk of *Mallotus philippensis*.

117. Eastern or Southern Comma

Polygonia egea
(Syn: *Vanessa egea*)
Wingspan 40–65mm

Description: Specimens from the dryer regions such as Balochistan are smaller. This species is notable for the toothed borders to both fore and hindwings, with the forewing tip square cut and strongly concave along the termen, and the hindwing bearing a short tail from vein 4. Above it is tawny-orange with a brown marginal border along both wings, that on the hindwing bearing small yellow spots between the veins. There are small-scattered black spots on the forewing and three on the hindwing. The underside hindwing is brown with darker striations, and bears a faint silvery comma-shaped mark end cell.

Status: Nowhere common but widespread in the juniper forest areas of Balochistan (collected TJR in Ziarat in early June), the dryer areas of Chitral and in Baltistan. It has been collected from Nathia Gali in the Murree Hills, where it must be considered very rare. In Balochistan there are two forms, as in the case of the European Comma (*Polygonia c-album*). The darker form, particularly in the shade of the underside, appears at the end of June, with some adults hibernating over winter, and a paler form hatches from eggs laid the previous summer and is on the wing from May to June.

Habits: The female lays her eggs singly on the upper surface of the leaves of the foodplant. The eggs are quite dark green with prominent longitudinal white ribs, flat topped, and with curved sides like a barrel. The larvae, when they hatch, crawl to the underside of the leaf and feed from underneath. Unlike the Tortoiseshells, they are not gregarious. The larvae in Europe are described as bearing branched spines on each segment, and being dark bluish in colour, with the second segment striped transversely with yellow and black. The larval foodplants belong to the family Urticaceae, and in Europe they feed on *Parietaria* spp. In Pakistan the species *Parietaria debilis* and *Parietaria judaica* are common in Balochistan and Chitral. The pupae are distinctive in having two short horns on the head and a prominent hump just behind the head, and no spines on the abdomen. They are brownish-buff in colour, decorated with six metallic silver discs. The butterfly is swift on the wing, but typical of the family in settling frequently to feed on flowers, or low down to bask. It is usual to encounter this species in the hotter, dryer low valleys of the outer Himalayas, or in the more sheltered 'Tangis' in Balochistan.

118. European Comma

Polygonia c-album
(Syn: *Vanessa c-album* in Marshall and de Niceville, 1886)
Wingspan 44–48mm

Description: This species in Pakistan has two broods, both very distinctive in the underwing colouration. The spring brood, which hatches from eggs laid the previous winter, has the underwings very pale ochre brown with darker striations, whilst the late summer brood has the underside a much darker umber brown varying to a slatey-brown shade. The upper wing surface of both broods is a reddish-chestnut colour, with darker brown narrow wing borders and scattered black spots of varying intensity, but always more prominent than in *P. egea*. Some specimens from the same locality appear much paler in general colouration, but generally the darker, more richly coloured forms are hatched in late summer. The wing margins are strongly indented, with the hind margin of the forewing concave, and on the hindwing a more pronounced tooth between veins 3 and 4. The underside is cryptically marked with darker striations and irregular, darker brown bands. The under hindwing bears a small silver-white comma shaped mark more conspicuous than in *P. egea*.

Status: In the Himalayas this species has a very wide distribution from Pakistan across to Sikkim, but it is rather uncommon everywhere, and erratic in occurrence. It has been collected throughout the Murree Hills, from Thandiani to Murree, and in Dunga Gali on the summit of Mukhshpuri at 2,740m (9,100ft). The first brood is on the wing in June and July and the second brood after the monsoon up to the end of October. Evans (1903) recorded it as very common between 1,200–1,500m (4,000–5,000ft) from February to April in the main valley of Chitral, with a few from the second brood flying in August in side valleys up to 2,700m (9,000ft).

Habits: The eggs are laid singly on the top of leaves, but the little larva upon hatching immediately crawls to the underside before starting to feed. The eggs are green with white longitudinal ribs. The foodplants of the larvae are nettles (family Urticacea) and elms (*Ulmus* spp.), in the Murree Hills probably *Ulmus wallichii*. The full-grown caterpillars are curiously like a bird dropping in appearance, having a broad white dorsal band on the rear half of the body, between segments 7 to 13, and being blackish-brown with narrow red-brown lines framing the black spiracles. Their heads are black with two horn-like projections, and the first segment is also black but covered with bristly warts. Each segment bears a tuft of branched spines, yellow on the forepart and white over the rear. The pupa hangs head downwards from the claspers, being freely suspended, and, like that of *P. egea*, it bears two short horns on the head and a prominent hump in the middle of its thorax.

119. Comma Tortoiseshell or False Comma of Europe

Nymphalis v-album
(Syn: *Vanessa l-album* in Evans, 1932, and *Vanessa vau-album* in Marshall and de Niceville, 1886)
Wingspan 65–75mm

Description: Larger than the Tortoiseshells, this is a handsome butterfly with the upper surface tawny chestnut, bearing white bars between a black area on the forewing apex, and the hindwing costal region. The forewing bears scattered black spots, with two between veins 2 and 3, whereas there is only one black spot in this area in all the other Tortoiseshells. The wing margins are less prominently toothed than in the true Commas, but with the forewing termen, square cut at the tip and strongly concave below. The wing margins are pale brownish-buff without any trace of blue lunules. The underside is brown varying to ochraceous, with darker irregular bands and striations, and usually traces of a small white comma mark end cell.

Status: It is apparently nowhere common in the Himalayas though its range extends from Eastern Europe to Central Asia. In Pakistan it has been collected from Chitral from the Shishi Koh valley, and around Ziarat. From the Murree Hills around Miranjani and Mukhshpuri. In Kashmir it has been collected from Gulmarg and the Sind valley. It flies between 2,700–4,260m (9,000–14,000ft), and is on the wing from July to early September.

Habits: In Eastern Europe the larvae are described as bearing branched spines on each segment, with their body brownish-red. There is a darker mid dorsal line and yellowish-white lateral stripes. The young larvae feed gregariously on Poplars and Willows, *Populus* and *Salix* spp. (Marshall and de Niceville, 1886) whilst Whalley (1996) lists Beech (*Fagus*) and Elm (*Ulmus*) as foodplants. The adults haunt forest and alpine meadows, at the edge of forest.

120. Large Tortoiseshell or Yellow Leg Tortoiseshell

Vanessa xanthomelas
(Syn: *Nymphalis xanthomelas*)
Wingspan 60–70mm

Description: Confusingly like *Nymphalis polychloros*, and only reliably separable from the latter by the yellow-brown hairs covering middle and hind legs, which are black in *N. polychloros*. If a series can be compared with *N. polychloros*, the black spots on the upper forewing are rather wider and square edged, those on the latter being smaller and rounder. The sexes are alike; a rich reddish-orange above with a dark marginal border to both wings, speckled yellow outwardly and inwardly margined blue and then black. The forewing costa is noteable in bearing stout black cilia along its margin. The base of the hindwings are covered with silky brown cilia. The underside is dark brown finely striated with yellow-brown speckling between darker bands. It is generally less dark brown than that of *N. polychloros*.

Status: Much less common than the Small Indian Tortoiseshell but widespread from Chitral eastwards to the Murree Hills. On the wing early in the spring from March to end of May. It flies between 600–2,700m (2,000ft–9,000ft) (Cumming, 1916).

Habits: The females lay their eggs in batches and the young larvae live and feed gregariously. The egg is dome-shaped and orange, with widely spaced white longitudinal ridges. The young larvae spin a thick web in which to shelter when not feeding. They are blackish-brown in colour, varying to ferruginous, thickly sprinkled with ochraceous spots, with thin paler lateral lines, and each segment armoured with branched spines, the dorsal pair of spines being the longest. Their heads are small, black, and covered with small tubercles. The pupae are attached to any vertical or horizontal surface, such as a rock face, as they fall down from the foodplant when ready to pupate and crawl some distance in search of a sheltered niche. They are grey-brown in colour, with the head bi-fid, and short spines along the abdomen. The larval foodplants in the Himalayas are the shrubby willow, *Salix denticulata*, which is common in forest clearings throughout Hazara district and the Murree Hills. Mohyuddin (1987) found larvae on the leaves of *Salix acmophylla* near both Peshawar and Rawalpindi. The butterflies may also lay their eggs on *Ulmus wallichiana* as the author has watched females frequently fluttering continuously around such trees in the Murree Hills and *Ulmus* spp. are used by *Vanessa polychloros*. The adults are on the wing early in the spring and do not usually come to flowers, but are attracted to wood sap and the smell of resin, particularly if any tree or log has been freshly chopped.

121. The Black Leg Tortoiseshell

Vanessa polychloros
Wingspan 50–65mm

Description: Closely similar to *V. xanthomelas*, this Tortoiseshell is the same species which spreads into western Europe, but its dark brown legs are covered with black, not yellow, hairs. It also averages slightly smaller in size than *N. xanthomelas*, and often appears rather darker orange-red on the upper wings with a narrower black marginal border and more rounded black spots on the forewing. Also, along the forewing costa there is more yellow between the black patches. Compared with *V. cashmiriensis*, the upper hindwing lacks the black basal patches, being uniformly tawny. The underwing is the same as in *V. xanthomelas*, with three dark brown bands across the forewing and a broad discal paler speckled band across the hindwing.

Status: Found only at high elevations on the western border regions of Balochistan, Waziristan, the Safed Koh, and Chitral. The author caught specimens during the monsoon season on the summit of Mukhshpuri, in Dunga Gali where it is very rare.

Habits: The early life history has been well documented in the European population. The female lays her eggs in a batch, which is glued together and deposited, in a neat ring around the twig or branch of an Elm or *Sorbus* tree. The eggs are deep orange coloured, darkening to the colour of the tree bark with age. The caterpillars on emerging weave a communal web of stout silk upon which they sometimes bask in a group, and under which they can shelter. They are armoured with branched yellow spines and are black with two thin dorsal lines. When ready to pupate they roll themselves into a ball and fall to the ground unharmed, even from high up in a tall tree. They then crawl to a suitable site for pupation, fixing themselves by the tail claspers in a hanging position. The pupa is greyish-brown, with sharp points along the abdomen and a slightly humped thorax. The butterfly does not visit flowers but is fond of basking on vertical tree trunks and is rarely seen away from forest. The larval foodplant is probably *Ulmus wallichiana* or *Celtis australis*.

122. The Indian Tortoiseshell

Vanessa cashmiriensis
(Syn: *Aglais cachmirensis*)
Wingspan 50–65mm

Description: As can be seen from the above dimensions, the Indian Tortoiseshell is often as large as *V. xanthomelas* but is very variable in size, the majority of specimens being much smaller. Patterned like the Large Tortoiseshell, the upper forewing shows more prominent yellow patches along the costa with the apical spot being almost white. The upper hindwing is black basally, with a fairly narrow orange-red band around the post-discal region. The hindwing margin is broadly black with blue lunules along the inner border and in the discal tawny band a wavy brown border. This is the only character separating this species from the much rarer Mountain Tortoiseshell, *Aglais urticae*. The underwings are darker than in the Large Tortoiseshell, lacking the paler speckled band on the hindwing.

Status: This is perhaps the commonest Nymphalid butterfly in the Northern Areas, found from 1,200–3,000m (4,000–10,000ft), and on the wing almost throughout the year at lower elevations. It is found from Chitral eastwards through Swat, Gilgit, the Murree Hills, and the Kaghan valley. It has not been recorded from Balochistan.

Habits: Hannyngton (1911) collected larvae of this species at 2,100m (7,000ft) in early May, on the Common Nettle, *Urtica dioica*, which is found throughout the Himalayas in waste places. The female lays her eggs in batches on the underside of the leaves of nettles. The eggs are green and flat topped, with longitudinal ribs of the same colour and fine horizontal striations. The young larvae spin a thick web across the top of several nettle stems and feed gregariously, using the web for shelter at night and to bask upon in the day, between bouts of feeding. They are yellow, thickly speckled with black, and with branched yellow spines along their dorsum. When they have consumed their first plant, the whole brood moves to a fresh patch of nettles and spins a fresh web. They disperse at the final instar, and then each caterpillar conceals itself by folding a leaf, held together with silken threads. Normally they crawl away to pupate on nearby twigs, not on the nettles. The pupae are usually dark brown suffused with black, but may have gold burnishing on their spines. They have a bi-fid head. Eggs collected by Hannyngton (*op. cit.*) in early April hatched on 20 April, the larvae were ready to pupate by 7 May, and butterflies hatched out between 20 and 30 May. The butterfly can be encountered in open country or forested regions; it visits flowers frequently and likes to bask with open wings.

123. Mountain or European Tortoiseshell

Aglais urticae
(Syn: *Vanessa urticae*)
Wingspan 45–55mm

Description: Exactly like the Indian Tortoiseshell, *A. cachmiriensis*, except for the absence of any brown wavy border inside the dark marginal border on the upper hindwing. There are blue lunules in the hindwing border like the former, but on the underside forewing there is a noticeably paler buffy band across the discal region compared with the Indian Tortoiseshell. If a large series can be compared, the Mountain Tortoiseshell averages smaller than the Indian Tortoiseshell

Status: This is the same species as occurs in western Europe, and is found in the inner Himalayan regions from Chitral, and in northern regions of Gilgit and Baltistan. The author collected it in July at 3,000m (10,000ft) in the Lutko nullah in Chitral. It is rather rare everywhere. It flies between 2,400–3,000m (8,000–10,000ft) from May to August, but will normally only be encountered well above the tree line.

Habits: The life history of this species is very similar to that of the Indian Tortoiseshell, *Aglais cachmiriensis*, the female laying her eggs in untidy batches on the underside of nettle leaves, and the young larvae spinning a thick communal web over the foodplant under which they hide by night and on top of which they often bask in a group by day, when not feeding. The larvae are black with a thin yellow lateral line and double yellow dorsal lines. The back bears black spines. After the last instar they separate and feed alone. The pupae are brownish-pink or brown suffused with black, and have a bi-fid head. Sometimes the chrysalis is decorated with metallic gold patches. The only butterfly seen by the author (*see above*) was flying over a steep stony scree, and settling on the flowers of *Aster* spp., (Compositae) when caught.

(123b. The Mongol

Araschnia prorsoides
Wingspan 50–55mm

Description: This is a small butterfly, tawny-orange above with the base of the wings, and two rows of discal spots, dark brown. There are yellow bands across the forewing cell and an oblique row of three yellow spots across the apex. The sexes are alike. The underside is beautifully marked with variegated bands and lines of chestnut purple, dark brown, and creamy yellow.

Status: This Eastern Himalayan species was recorded by D.R. Puri [1931] from Lahore, where he considered it rare. Based upon known distribution, which is largely confined to Manipur in Assam, where it is rather rare and local, and to Myanmar [Burma], it is excluded from the present list of Pakistan butterflies.)

Genus HYPOLIMNAS

Though only represented by two species in the subcontinent, Eggflies, are however widely represented, in Far Eastern countries. The genus *Hypolimnas* is interesting in that the sexes are strikingly different and the females in some forms mimic Danaid butterflies, even to the point of females being slower in flight with more gliding, like their Danaid models, in contrast to the fast-flying males.

124. Danaid Eggfly

Hypolimnas misippus
Wingspan 70–85mm

Description: The common name of this species indicates two of the striking features about this largely Indo-Malayan genus. The females are quite different in appearance from the males, and mimic Danaid butterflies, whilst the males have prominent white oval patches on fore and hindwings, resembling eggs in shape. The male is black above, with the four large white patches on the wings mentioned above, and these are surrounded by iridescent purple. The underside of the male bears a wide white band across the hindwing, bordered at the base and post-discally with chestnut, and there is a prominent black spot on the hindwing near the costa above vein 7. The female resembles remarkably closely the Plain Tiger, *Danais chrysippus*, with black and white apical bands on the forewing and the wing margins with a narrow black and white speckled band. There is a second form of female, which has the same colouring as *D. chrysippus* but lacking the white on the forewing. This form (illustrated) is called *inaria*. The main distinguishing features between this mimic and the Plain Tiger is that the hindwing margin of *Hypolimnas missipus* is waved, and on the upper and underside of the hindwing there is a black spot above the middle of vein 7, whereas *Danais chrysippus* has three or four small costal black spots.

Status: Occurs rather uncommonly throughout the better-wooded portions of Sindh and southern Punjab, extending throughout Balochistan, but rare around Lahore and north-east Punjab, though it has been collected from Campbellpur. It has not been recorded from Islamabad. Collected TJR from Khanewal and Rahimyar Khan. On the wing mainly from July till November.

Habits: The larvae are described by Bell (1910) as velvety black with smoky greenish tinges on the sides, and each segment armed with spines which are glassy white in colour. The anal end is covered with a triangular flap. Its head is orange with black lines and bears two small horns, and the pseudo legs and ventrum are orange. The pupae are exactly similar to those of *Hypolimnas bolina*, described below. The female lays her eggs either singly or in batches of six or seven on the underside of leaves of the foodplant, but near the ground. The larval foodplant is *Portulaca oleracea*, which is a small succulent herb with yellow flowers, common in all four provinces of Pakistan. The butterflies are strong flyers with a rather up and down trajectory. They like to visit flowers and are inclined to chase other butterflies, often returning to the same leaf.

125. Great Eggfly

Hypolimnas bolina
Wingspan 70–110mm

Description: This species also exhibits sexual dimorphism, with the female besides being larger, mimicking the Crows, *Euploea* spp. The male is very dark brown above, with iridescent pale mauve patches on fore and hindwings, surrounded by darker purple. The upper hindwing has a row of small white spots in the post-discal region, and the hindwing margins are scalloped. The female mimics the Common Crow, *Euploea core*, having a double row of creamy marginal spots on both fore and hindwings, but it can easily be distinguished from the latter by its broader wing shape, with the forewing termen concave and the hindwing margin scalloped. The forewing has two violet-blue spots in the end cell region, but these are often obscure.

Status: Occurs only during the monsoon around lower Sindh (collected TJR early July in Malir), and north-eastern Punjab in Sialkot and Lahore areas, where it is slightly more plentiful than the Danaid Eggfly.

Habits: The female lays her eggs singly or in small groups of five to seven on the underside of young leaves of the foodplant. The eggs are spherical and green in colour. The larva is cylindrical-shaped with a black, velvety body with grey marbling and bearing spines along the dorsum which are a dirty yellow-red. There is an orange band behind the head, and an anal flap, which is brown. The head is heart-shaped and yellow with two long black horns. When full-grown it is 57mm in length. The pupa is suspended by a thick pad of silk from a twig or rock face, and hangs downwards quite loosely, not rigid. It is thick and stout in shape, the head being square and nearly as broad as the thorax. It is coloured dark grey-brown blotched with lighter grey, and its surface is rough or carinated, with a dorsal row of sharp-pointed conical tubercles along the abdomen. The larval foodplants are *Sida rhombifolia*, *Portulaca oleracea*, *Elatostemma cuneatum*, and *Fleurya interrupta*. The butterfly is very aggressive, frequently chasing other butterflies and returning to the same perch high up on a tree. They are long-lived as adults and keep to the same territory.

126. Common Map

Cyrestis thyodamas
Wingspan 50–60mm

Description: This rather delicate butterfly, with semi-transparent wings, has a bi-lobed tornal area, a short tail at vein 4 on the hindwing, and the forewing tornus is indented. It is a pale creamy-white above, with fine vertical dark lines crossing both wings and some pale green bands in the forewing apex and hindwing post-discal area. The sexes are alike and the underside is marked like the upper.

Status: Rare and only known so far from the Murree foothills and Margallas. Collected by TJR at 600m (2,000ft) in October above Nurpur Shahan.

Habits: The egg is shiny yellow and dome-shaped with longitudinal ridges. At the top there is an aperture covered by a cogwheel-shaped lid, the teeth of the cog fitting into the indentures corresponding to the ribs. The female lays the eggs singly on the underside of young leaves of the foodplant. The larvae emerge by lifting the egg flap. They are spindle-shaped being narrower at both ends and unusual in appearance. There are two long fleshy horns curving out and backwards from the head, another stouter, sword-shaped pair on segment 6, and another on segment 12, also curved forward. Its body is dark brown varying to reddish-brown, with a broad yellow or green spiracular band along the side. The larval foodplants are tree figs (family Moraceae), including, in the higher rainfall parts of India, *Ficus religiosa*, *Ficus glomerata*, and *Ficus bengalensis*. However, in Pakistan, where it only occurs in the foothills, it is more likely to feed on *Ficus auriculata*, or *Ficus virgata*. The pupa, which is attached by the tail claspers, hanging freely, is shiny and olivaceous brown, but is a peculiar shape, resembling somewhat a twisted dead leaf, having two foliaceous curved projections over the thorax. The butterfly itself has a very weak flight, sailing slowly on outstretched wings without flapping for prolonged periods, but when alarmed can fly fast in a jerky manner, and typically dives into a leafy branch, settling with wings held out flat on the underside of a leaf. It visits flowers and is fond of basking quite high up on a tree leaf.

Sergeants and Sailers

The tribe Neptini comprises the Sailers, which are all medium-sized butterflies, black in colour with broad white diagonal bands or, outside our area, yellowish-chestnut bands. Their habit of sailing lazily around on outstretched wings, seldom flapping, has given rise to their common name. They can be distinguished from the similarly patterned Sergeants (*Pantoporia* spp.) by the absence of any white band at the top of the abdomen, and by the fact that the upper forewing bears only two white marks, a streak and an end spot, inside the cell. The Sergeants have the abdomen with a white narrow band and often with three white marks inside the forewing cell. Both groups are very similar in appearance between species and their identification presents some challenge.

127. Himalayan Sergeant or Hill Sergeant

Pantoporia opalina
(Syn: *Parathyma opalina*)
Wingspan 55–70mm

Description: The Sergeants, *Pantoporia* spp., are similar to the Sailers in colouration, being black above with banded white markings, but distinguished by the white upper band on the hindwing being continuous across the top of the abdomen, whilst in *Neptis* (Sailers) this is discontinuous. In this species, the diagonal white line through the forewing cell is well separated from the arrow-shaped white spot end cell, and it is narrowly divided in two places across the cell. The lower band of white spots on the forewing slopes sharply out towards the apex. The underside of both wings have the same white bands but the spaces in between are bright chestnut, the outer brown band is unmarked, whilst in *Pantoporia perius*, the Common Sergeant, this band bears a row of black spots. However, in Pakistan only *P. opalina* has been recorded, *P. perius* being confined to the eastern Himalayas.

Status: Not as common as the Himalayan Sailer but sharing the same habitat of Himalayan moist temperate forest. Occurs rarely in the lower Kaghan valley, more commonly in the Murree Hills and Neelum valley of Azad Kashmir. On the wing from June to September, between 1,800–2,700m (6,000–9,000ft).

Habits: The early life history is similar to the entire *Pantoporia* genus, with the egg being spherical and prickly like a sea urchin. The larvae are cylindrical, slightly thicker in the middle segments and with tubercles bearing spines along the sides of the forepart of the body and another set on segment 12. The head is also spiny. The body colour is dull chestnut, with white and grey patches. The larval foodplants are species of *Berberis*, particularly *Berberis lycium*, and, in the eastern part of its range, *Mahonia nepalensis*. The young larva makes a pad of silk alongside its body and throws up a rampart of its droppings along the edge of this web, which serve to camouflage its outline. The pupae are dull chestnut marked with grey, and sometimes with a golden metallic sheen on the wing cases. They bear two foliaceous projections on the back of the thorax and resemble a dead or twisted leaf in shape. They generally hang from the underside of a leaf very close to the ground. The butterflies have a strong flight, often sailing with wings open flat and keeping close to the ground, but they generally settle on low bushes or trees, not on the ground. They are always found within forest clearings or on the edge of forested areas.

128. Common Sailer

Neptis hylas
Wingspan 50–60mm

Description: Difficult to distinguish from the Himalayan Sailer, *Neptis mahendra*, though generally averaging slightly smaller, and on the upper hindwing the white discal band is of even width, whereas in *N. mahendra* this band widens towards the costa. The upper forewing has the minute upper costal spot shorter than the rest of the apical band of white spots; in *N. mahendra* this costal spot is as long as the other two. On the underside the chestnut bands are quite dark in colour and edged with black. Compared with the Sergeants of the genus *Pantoporia*, their bodies are more slender and delicate.

Status: Fairly common in a restricted zone around the Murree foothills, from the Margallas to Lehtrar, and between 450–1,200m (1,500–4,000ft). May be seen on the wing from April to October.

Habits: The eggs are laid singly on the upper side of leaves, near the tip, and are like a sea urchin, with spines all over and conical in shape. The larva is pale varying to dark grey-green, covered with yellow tubercles on segments 3, 4, 6, and 12. The head bears two small conical points on top, and this feature is common to all *Neptis* spp. It makes a bed for itself with silk, on the tip of a leaf midrib, gradually eating the surrounding leaf but leaving the midrib, which hangs down but remains attached by the larval silk. It actually feeds on the withered leaves and may entirely conceal itself inside bits of hanging leaf, drawn together with its silk. The pupa is a pearly greenish-yellow, with olive brown lines on the wing cases, and a golden suffusion on the abdomen. The larval foodplants are *Nothapodytes nimmoniana* (family Olacaeae), *Grewia tiliaefolia*, *Helicteres isora*, and *Macuna purpurea*. The butterflies, like the entire *Neptis* genus, are easy to recognize by their rather continuous, sailing, circling flight, interspersed with a few rapid shallow flaps of their wings.

129. Himalayan Sailer

Neptis mahendra
Wingspan 55–65mm

Description: This species, as mentioned above, is closely similar to the Common Sailer, which occurs at lower elevations. It can be separated from *N. hylas* by the upper hindwing white discal band being wider at the costa, whereas it is of even width in the latter. Also, on the upper forewing, the small costal white spot is equal in width to the other two spots below. The white spot end cell is well separated from the white cell streak and sharply pointed. Like *N. hylas*, the underside hindwing has the chestnut bands quite dark and black-edged, particularly the outward side of the discal band.

Status: Very widespread in the Himalayas from Chitral, through Dir and Swat Kohistan (collected TJR, August, above Miandam), extending to the Murree Hills and one of the commonest butterflies in Himalayan moist temperate forest in Pakistan. On the wing from May to end of August.

Habits: The larval foodplants are *Bombax ceiba* and *Helicteres isora* (family Sterculiaceae), and the larva, like others of the genus, makes a shelter of broken, half-eaten leaves for itself, held together with its silk, and prefers to feed on withered leaves inside this shelter. The caterpillar, like all the *Neptis* spp., bears tubercles on its sides but not spines. The pupae are usually formed close to the ground and hanging from a horizontal stem or twig, but attached rigidly, not swinging freely. Mohyuddin (1987) found its pupa attached to *Berberis ceratophylla*. When touched it is capable of wriggling. The larvae are green or grey green with lateral tubercles bearing spines, and the pupae are red-brown or yellow-buff according to their background, and often burnished with metallic colours on the abdomen.

130. Pale Green Sailer

Neptis zaida
Wingspan 65–75mm

Description: This is one of the larger Sailers, and rather distinctive within this very homogenous genus. The white bands on the upper wings are tinged with greenish-buff, and the cell streak is continuous through to its pointed end, extending between veins 3 and 4. Also, there are no white costal spots on the upper forewing apex, as in the two previous spp. The under forewing is a paler chestnut than the two previous species, and the band of apical white spots is dark-rimmed on its inner edge, while the large spots in veins 2 and 3 are bordered inwardly by a broad black patch. The under hindwing lacks any black borders to the two white bands.

Status: Has been collected from Murree, where very rare, and it is more likely to be encountered in eastern parts of Kashmir, and is only common around Mussoorie. It prefers lower elevations from 900–2,400m (3,000–8,000ft).

Habits: The larval foodplant has been recorded as *Celtis australis*. The butterfly prefers the vicinity of streams in forested areas.

131. Sullied Sailer

Neptis soma butleri
Wingspan 50–60mm

Description: This species has the upper forewing cell streak only narrowly separated from the end cell spot. Though Eliot (1969) assigns the Pakistan specimen to *N. soma*, which typically has the upper hindwing discal white band relatively narrow, the band is about as broad as in *N. yerburyi*. The under forewing has rather obscure white spots between the broad white discal spots and the outer apical band of small white spots. The under hindwing has quite dark chestnut bands and these are not black edged, and there is a double line of dull white along the margin.

Status: A specimen was collected by Butler on 12 October 1886, from the Dhum Tank, Abbotabad, at about 1,370m (4,500ft). He wrongly identified it at the time as *Neptis yerburyi* and this distributional record has been cited by later authors (Hasan, 1994). In a recent review of the *Neptini*, Eliot (1969) examined Butler's original specimen and found that it was *N. soma*, not *N. yerburyi*, and named it as the type specimen for *Neptis soma butleri* (Yerburyi Auctt.). This is a low-elevation 'Sailer', frequenting foothill regions of the western Himalayas from Chitral through Hazara district. Due to paucity of collected material and its similarity to other *Neptis* spp., its status is undetermined, but it is thought to be rather uncommon within Pakistan.

Habits: The butterflies inhabit foothill regions and lower elevations, and the larval foodplants are thought to be Leguminous trees such as *Dalbergia* spp.

(131b. Yerbury's Sailer

Neptis yerburyi

See above account of *Neptis soma*. The single specimen from Pakistan was incorrectly identified by Butler [1886] and is in fact *Neptis soma*.)

Genus EUTHALIA

Though only represented by one species in Pakistan, the Barons (genus *Euthalia*), are strong-flying, robustly-bodied butterflies associated with forested areas of higher rainfall, and some spp. in the eastern part of the subcontinent are brightly coloured. They don't visit flowers but are attracted to rotting fruit and tree sap. Their larvae are unique in possessing long feather-like processes extending horizontally out from the body. They instinctively lie along the axis of the leaf midrib upon which they are resting, and these feather-like processes actually resemble the leaf venation and serve well to camouflage the caterpillar.

132. Common Baron

Euthalia aconthea
(Syn: *Euthalia garuda*)
Wingspan 55–75mm

Description: Females are larger, and both sexes are dark brown above, with two slightly darker bands within the cell of both fore and hindwings which are black-edged. There are traces of darker brown spots along the sub-marginal area of the forewing, and white spots across the disco-cellular area of the forewing. The upper hindwing also has a series of small black spots around the sub-marginal region. The female is a paler brown and her forewing white spots are much larger. The underside of both sexes is ochraceous, with pale fawn spots on the forewing.

Status: Occurs not uncommonly around Islamabad and the Margalla ravines, straggling down through Sialkot and Lahore, mainly associated with mango (*Mangifera indica*) gardens, and on the wing from April to October, but more plentiful during the monsoon season.

Habits: The larvae are unique amongst the Nymphalidae in having ten pairs of feathery branched spines or 'scoli', each about 9mm long and extending horizontally outwards from the sides of its body except for the front pair, which extends forward over the head. The body is leaf green and there is a paler yellow or yellow-green mid dorsal stripe, and a thin mauve band between each segment. The larval foodplants are Anacardiaceae, and are the mango, *Mangifera indica*, and the cashew, *Anarcardium occidentale*. The pupae are attached on the underside of a leaf, and are green in colour, burnished with gold metallic spots. Their shape is rather squat with the abdomen conical and, when viewed from the side, like a pyramid in outline. The head bears two lobes, and there is a brown or white line encircling the highest point on the thorax. The larva instinctively takes up a position along the upper side of a leaf on its midrib and thus blends in with the leaf venation in a perfect camouflage.

Chapter 9
Satyrinae—Browns, Satyrs and Walls

Classification

The Satyrinae are, according to recent thinking, placed in a separate sub-family within the Nymphalidae, not being considered sufficiently distinct to warrant raising to full family rank. They are a distinctive group, however, characterized by being mostly brown in wing colour, usually with ocelli or eye spots, and in having a rather weak, jerky flight. Diagnostically, they share the same feature with Nymphalidae in that the adults' forelegs (prothoracic legs) are degenerate, being brush-like in the males and though having several joints in the females, are not used in walking. The males usually possess scent scales in long brands on the forewing, and another distinguishing feature of the sub-family is the construction of the veins on the forewing, some of which are always dilated at their base. On the hindwing the pre-costal vein is absent (*see*, Fig. 1), except in the Palmflies (genus *Elymnias*), which are extra-limital to Pakistan. The eggs are characteristic also, being melon-shaped, with the surface beautifully sculpted with vertical ridges and often with cellular or hexagonal indentations in between (Garcia-Barros and Martin, 1995). The larvae are covered with small hairs and are pubescent, generally slender and fusiform in shape, i.e., fattest in the middle, with the head bearing two horns (except the *Hipparchia* and *Ypthima* genera), and the anal plate also having two horn-like projections. They are usually coloured green, brown, pink, or yellow, with paler longitudinal lines, and often they are quite shy, feeding at night and staying on the underside of leaves. The pupa is rather fat and squat-like that of the Danainae, and is always suspended head downwards from the tail claspers (cremaster) only. Sometimes the pupae are green, and others dull buff colour, ornamented with black specks, or sometimes with bright gold. The larvae feed mostly on grasses (Poaceae), but some tropical families feed on Bamboos, Arecaceae, and Cyperaceae (sedges).

There are two rather distinct ecological types of Satyrinae: those that are shade-loving, weak flyers, and confined to forest and heavy bush country, like the Bush Browns, Tree Browns, Evening Browns, and the Rings (*Ypthima* spp.), mostly of Oriental affinity, and those that inhabit open rocky areas, mostly of Palearctic affinity, and living in mountainous areas. Not surprisingly, Pakistan is well endowed with the latter mountain-dwelling species, with a total of only ten of the former, but twenty-nine inhabiting the open mountainous places, including Rockbrowns, Meadow Browns, and Satyrs. Some mountain-adapted genera have speciated so exhuberantly, with their restrictive territorial lifestyle and little chance for mutations to be 'bred out' through normal genetic exchange, that even within Pakistan there is a bewildering variation, for example within the Arguses (*Erebia* spp.) and the Meadow Browns (*Maniola* spp.), presenting considerable identification and taxonomic difficulties.

133. The Straight-Banded Tree Brown

Lethe verma
Wingspan 55–60mm

Description: This is a sombre brown butterfly, with the whole of the upper surface dark brown except for a prominent white diagonal bar across the forewing, and no white spots above the bar in the apex as in the other two *Lethe* spp. inhabiting Pakistan. The hindwing upper side bears two small, inconspicuous, white-centred ocelli. The termen of the hindwing is very slightly crenulated and is narrowly bordered by paler buff and black lines. The underside forewing shows two wing-tip ocelli, and the hindwing underside bears a row of six white-centred ocelli in an irregular arc, with two silvery-brown zigzag lines across the basal area.

Status: Common in the Murree Hills, flying from April to late September. Occurs from 1,500–2,700m (5,000–9,000ft). Collected TJR May and June in Dunga Gali at 2,100–2,700m (7,000ft–8,900ft).

Habits: According to Sevastopulo (1946), the female lays her eggs singly, on grasses. The eggs are spherical with a flat bottom, blue coloured, with the surface unsculptured. The larvae when full-grown have a green head with two pink horns tipped black close together on the head. The body is apple green, with a blue-green dorsal stripe edged with yellow, and a zigzag darker green sub-dorsal stripe, edged below with yellow. Its spiracles are dark red, and the anal segment is produced into two horns of pale blue-green. Its whole body is minutely shagreened with yellow points. The larval foodplants are grasses. The pupa is suspended hanging downwards attached by the cremaster. It is yellow-green in colour with the head bearing two frontal points, and with two fuscous stripes along the wing case. The butterfly, like the entire genus, is decidedly crepuscular in habits, spending the whole day until evening, and again all night till early morning, hanging down with wings closed from the underside of a plant leaf. Its flight is very weak and jerky.

134. The Common Tree Brown

Lethe rohria
Wingspan 58–70mm

Description: This is a larger butterfly than the Banded Tree Brown, with the forewing arcuate at the tip and the hindwing markedly dentate along its border. The upper surface is a paler brown than in *L. verma*, darker brown in the apex, and with two small creamy white spots at the apex of the forewing, followed behind along the costa by three creamy elongated spots close together. The hindwing upper surface shows the ocelli of the underside as dark spots, and two narrow sub-marginal black lines. The underside is elaborately patterned with pale creamy-brown bands and a row of small black-centred ocelli in the distal area of the forewing. The hindwing bears one much larger, white-centred ocellus in the apical region, followed by three disintegrated smaller ocelli and then a black-centred ocellus. There are marginal and sub-marginal paler creamy-brown lines, as on the forewing.

Status: Common in the Murree Hills, Margallas, and Lehtrar, from 600–2,600m (2,000–8,500ft). At lower elevations it is on the wing from March to August, emerging from mid-May at higher elevations.

Habits: Though in south India the larval foodplants are bamboos, in the Himalayas the eggs are laid singly on grasses including *Apluda* spp., *Capillipedium* spp., and *Microstegium* spp., which form the foodplants. The larvae have a heart-shaped head coloured green, with the apex pink and slightly bi-fid. Its body is blue-green, with a bluish dorsal stripe, edged whitish, and a yellow sub-dorsal line, with prominent yellow spots along the sides. The anal plate is produced into a long triangular process and is bi-fid at the tip. The whole body is clothed with short, colourless hairs. The pupa is smooth, with the head bearing two short points and the anal segment curved, so that the pupa hangs down at an angle of 45 degrees. According to Sevastopulo (1946), it is usually grey-green in colour, finely streaked fuscous. The butterfly may be encountered down to 600m (2,000ft) and up to 2,400m (8,000ft), always in well wooded or forest areas, and it is sluggish in habits, being most active in evening and early morning. It rarely comes to flowers, but is attracted to sap. In this species, the male has no scent brands.

135. The Banded Tree Brown

Lethe confusa
Wingspan 55–60mm

Description: Very similar to *Lethe verma* in appearance, but separable by two main features: on the underside there is a whitish discal band across both wings, and on the forewing upper side there are two small white spots in the apical area above the broad diagonal white bar, both features lacking in *L. verma*. The general colour is dark brown with the basal area of both wings slightly paler brown. The underside forewing bears two ocelli, the top one larger in the apical area, and besides the white bar across the discal area, there is a narrower white line across the basal region, which is duplicated across the under hindwing, with a zigzag white line in the discal area and a series of six ocelli across the sub-marginal area, of which the top one is largest and the lower ones show small white centres. The margin of the hindwing is slightly dentate, and on its upper side the ocelli underneath show through as black spots, except for the bottom tornal one which is ringed with a small eye.

Status: Only recorded from Murree area where very rare. Commoner further east, especially around Simla, and preferring lower elevations between 600–1,200m (2,000–4,000ft).

Habits: As far as is known, the female lays her eggs singly on grasses, including *Miscanthus* spp. and *Capillipedium* spp. The early stages have not been described. The butterfly, according to Talbot (1947), has three broods, the first from April to early June, the second from late June to July, and the third from August to October. The male is reported to emit 'a delicious scent like vanillah'. According to Haribal (1992), this species will visit flowers, and is quite aggressive, chasing other butterflies of any species, returning to a favourite tree perch.

136. The Common Wall

Pararge schakra
(Syn: *Lasiommata schakra*)
Wingspan 55–60mm

Description: There are only very slight differences between the next three species. The upper side is brown with the marginal area having an orange band, wider at the top and separated by the veins outlined dark brown. There is a prominent eyed ocellus in the apex, and on the hindwing a series of four smaller, lunar-shaped orange spots containing three black ocelli from the tornus upwards, the top orange spot smallest and unmarked. The Dark Wall, *P. menava*, is largely distinguished from this species by having only two black ocelli on the upper hindwing. The under hindwing is more prettily marked, with a buffy-orange marginal band and the discal area deeper orange-tawny, with the apical ocellus prominently ringed with ochre. The under hindwing bears six to seven ocelli, each white-centred and double-ringed with ochre. Inside the disco-cellular region is a thin dark brown line, which in this species is broken at the junction of vein 4, distinguishing it from *P. menava*. There is also a narrow, darker brown band across the upper side of the forewing between vein 4 and the costa which in the male is a sex brand. This is lacking in *P. maerula*.

Status: Widespread from extreme northern Balochistan on the Shingarh, the hills above Parachinar, and from Chitral through to Gilgit. It has also been collected from Campbellpur. Favouring open, dryer slopes, it occurs from 1,830–2,700m (6,000–9,000ft), and is on the wing from April to October.

Habits: The eggs are greenish-white and spherical with a flat bottom, delicately sculpted with longitudinal lines and horizontal cross striae. They are laid singly on grasses. The larvae do not have any horns or projections on their head but the anal segment is bi-fid. In the first instar they are dark and hairy, but as they grow and moult, they become apple green with pale yellow lateral stripes, and are then hairless. The middle of their body is slightly thicker than either end, so they are spindle-shaped. They are believed to take some months to develop, hiding at the base of grass tussocks and not feeding for prolonged periods. The larvae are also formed low down amongst the grass tussocks. They are smooth and hang head downwards, suspended by their tail claspers. They can be grass green, brown, or blackish-green. The butterflies are quite rapid flyers, readily coming to flowers to feed and fond of basking on warm stone or rock faces.

137. The Scarce Wall

Pararge maerula
(Syn: *Lasiommata maerula* or *P. schakra maerula*)
Wingspan 55–60mm

Description: Like the Common Wall (*P. schakra*) in colour and wing pattern, with the same orange-tawny margins and large apical ocellus on the forewing, which is ringed with ochre yellow on the underside. The underside hindwing bears the same narrow dark brown line across the discal region, but this is interrupted between veins 6 and 7, whereas it is continuous in *P. schakra*. The underside ocelli on the hindwing are slightly smaller and more compact than in *schakra*, and the forewing apical area has more extensive pale yellow around the apical ocellus. The main difference, only discernible under magnification, between this and *P. schakra* is the absence of any brand on the forewing of the male.

Status: Now considered as a sub-species of *schakra*, the Common Wall. It has only been recorded from the lower valleys of Chitral and at Ziarat (Chitral), below the Lowari Pass, flying between 1,500–2,400m (5,000–8,000ft) and usually near a stream.

138. The Dark Wall

Pararge menava
(Syn: *Lasiommata menava*)
Wingspan 50–60mm

Description: Again very similar to *P. schakra* in appearance, though when a series of specimens can be compared they are a darker brown on the upper wings, and, in the male, there is a more extended brand between the costa and vein 4. Like two of the other 'Walls' in Pakistan, there is a prominent white-eyed ocellus on the forewing tip, and usually a second tiny black ocellus in the tip of the apex, which never occurs in *P. schakra*. The hindwing bears only two ocelli in the posterior region. This is the main distinguishing feature between *P. menava* and *P. schakra*, the latter always having three ocelli on the hindwing. On the underside there are dark brown lines across the sub-marginal and discal areas, and the latter line is not interrupted at vein 4, as is the case in *P. schakra*, the line being smoothly continuous. The orange and yellow area on the forewing underside is rather brighter and more extensive than in typical *P. schakra* specimens, extending in *menava* inside the cell. The sexes are alike.

Status: Occurs uncommonly, from the Khojak Pass to the juniper forest zone in central Balochistan above 1,800m (6,000ft), also collected Khawas valley at 2,400m (8,000ft). In Chitral it has been collected between 1,800–2,700m (6,000–9,000ft) near Ziarat and around Drosh, where it is not uncommon. It is on the wing from June to September, and prefers a dryer habitat than *Pararge schakra*.

139. The Yellow Wall

Pararge eversmanni
(Syn: *Kirinia eversmanni* and *Amecera cashmirensis*)
Wingspan 55–60mm

Description: This is one of the prettier species among this rather sombrely-coloured family. The forewing in both sexes is a golden tawny-yellow with a very dark brown border which extends inwardly between veins 3 and 4, and there is a broad, dark brown band in the disco-cellular region. The hindwing is darker orange-brown with four brighter orange ocelli along the sub-marginal area bearing solid black eye spots. The underwing is pale orange, with three small dark lines crossing the cell and a broader disco-cellular bar. The under hindwing is dark brown with two wavy dark blackish lines across the discal and basal areas, and six uneven-sized ocelli along the margin, each ringed by dull orange. The post-discal area is also dusted with whitish scales.

Status: Collected TJR, at 2,200m (7,200ft) on open grassy hillsides below Bunyan in Swat Kohistan where not uncommon, and occurs Chitral between 1,800–2,700m (6,000–9,000ft), in the lower side valleys. Considered rare in the north-west Himalayas by Talbot (1947), but has been collected from Gulmarg, Kashmir, at 2,700m (9,000ft). Collected by Sakai (1981) from Nuristan, Afghanistan.

Habits: The author found this species quite plentiful in one valley of Swat, around 2,100m (7,000ft). It frequently came to flowers (Compositae spp.) to sip nectar and, like its congeners, was fond of basking on warm rock faces. It was not seen near forest but on open terraced fields.

140. The Dusky Meadow Brown

Maniola pulchra, var. *neoza* (syn: *Hyponephele pulchra*)
Wingspan 41–44mm

Description: This species belongs to a rather variable and complex group of butterflies which, because of their sedentary habits, have evolved into a number of slightly different forms. The Dusky Meadow Brown is distinguished by having the orange areas on the forewing rather dark, being completely overlaid with darker brown scales. Like the other *Maniola* spp., the forewing tip bears an unpupilled black ocellus. In the female the upper side forewing is brighter in the orange area, being less heavily overlaid with dark brown scales, and the apical ocellus surrounded by paler buffy-orange. There is also a second black unpupilled ocellus in the forewing tornal area. The underside hindwing bears a dark zigzag line in the discal area as well as traces of a second dark line in the sub-apical area. The underside forewing apical ocellus is pupilled with white and ringed with paler buff. It can be distinguished from *Maniola pulchella,* which has a broad, paler buffy-brown post-discal band on the under hindwing and no thin zigzag post-discal dark line.

Status: Normally found at high elevations, up to 4,260m (14,000ft), from northern Chitral through

Swat Kohistan and Gilgit around Astor, including Rama lake at 3,660m (12,000ft). Collected TJR 22 August from above 3,350m (11,000ft) near Bunyan. Very common around the shores of Shandur Lake.

Habits: The eggs are laid singly on grasses and are flat-topped with curved sides, like a barrel in shape, and blotched with darker spots. The larvae are secretive, feeding at night and sheltering by day at the bottom of grass tussocks. They are spindle-shaped and green with a velvety pubescence, and have two short points on the anal segment. It is believed that all *Maniola* spp. have a fairly prolonged development, the caterpillars spending some time during the early part of their life hibernating at the bottom of grass clumps, and after pupating, take about one month before the butterfly emerges. Before pupating, the caterpillar spins a pad of silk on a grass stem or leaf sheath to which it attaches itself by the cremaster. The pupae are attached hanging downwards from the stem and close to the ground, and are green in colour, with black specks along the abdomen, and the wing cases picked out in black. The butterflies haunt alpine meadows and are often sympatric with *Maniola pulchella*. They visit flowers, and have a rather up and down fluttering flight. Where they occur they can be very abundant.

141. The Tawny Meadow Brown

Maniola pulchella (syn: *Hyponephele pulchella*)
Wingspan 38–45mm

Description: Very similar to *M. pulchra*, except that the forewing upper side is a brighter tawny-orange and the underside hindwing barely shows the dark zigzag discal line of *M. pulchra*. Instead there is a paler buffy-brown wide band curving around the post-discal area. The forewing has a black eye in the apical area which is only pupilled white on the underside. The hindwings are paler brown than in *M. pulchra*, and show fine black speckling.

Status: Collected TJR at above 3,000m (10,000ft), in Swat Kohistan at Banyan, in August, but also occurs down to 1,800m (6,000ft), and is on the wing in June and July. Occurs Chitral, even at high elevations, up to 3,960m (13,000ft), where Evans found it the commonest butterfly around the shores of Shandur Lake. Collected by J. Biddulph from Astor, Gilgit, at 2,340m (7,700ft) in September.

142. Tawny Branded Meadow Brown

Maniola narica (syn: *Hyponephele narica*)
Wingspan 48–50mm

Description: This is very like *Maniola davendra*, but averages smaller in size and with rather paler yellow-tawny areas on the forewing. The male bears the same black, unpupilled ocellus on the wing tip, with the brown border clearly defined and regular along its inner border, and with a prominent dark brown brand curving up from the dorsum to the end cell. The underside forewing is pale tawny-yellow with the apical ocellus pupilled white and ringed with paler yellow. The hindwing underside lacks any marginal ocelli, which distinguishes this species from *M. davendra*, whilst it is also thickly freckled with white and reddish brown, lacking any clearly defined darker discal and marginal lines as in *M. davendra*. It differs from *Maniola lupinus*, which also lacks ocelli on the under hindwing, in the white dusting on the underwing, and from *M. wagneri*, which does have a prominent dark discal line. Females have two black ocelli on the forewing, the lower one very small and the wingtip one with a white pupil.

Status: Only occurs in Balochistan within Pakistan limits, and collected from Khojak, Bogra and Murgha Mehtarzai Passes.

143. The Branded Meadow Brown

Maniola lupinus
(Syn: *Hyponephele lupinus* and *Epinephele interposita*)
Wingspan 45–50mm

Description: This Meadow Brown is dull brown on both upper wings without any orange discal panels. The male does show a darker brownish-black brand on the forewing and has an apical black ocellus. The hindwing is scalloped along its margin and on the underside lacks any marginal ocelli, or any discal darker lines, white bordered as in *M. davendra*. The under forewing is tawny-orange in the discal and cell areas, the apical ocellus being ringed paler orange. The female has a second small black ocellus between veins 2 and 3, and in some specimens shows a trace of tawny-orange around the discal region.

Status: This Meadow Brown is adapted to warmer latitudes and lower elevations than the previous species. It is common in the Khojak and Bogra Passes of Balochistan in June and July, and in lower Chitral between 1,800–2,400m (6,000–8,000ft) in the Blue Pine (*Pinus wallichiana*) forest, from June to August.

Habits: The same as for the other *Maniola* spp., with the eggs being laid singly on grasses, and the larvae spindle-shaped, green in colour, covered with short fine hairs, and with a bi-fid tail. They spin a pad of silk to the base of a leaf sheath or stalk and pupate hanging head downwards by the cremaster. The butterflies fly in the usual rather jerky, up and down manner, frequently settling on warm rocks or stones to bask.

144. White-Ringed Meadow Brown

Maniola davendra latistigma (syn: *Hypenophele latistigma*)
Wingspan 52–55mm

Description: This is a more distinctive Meadow Brown due to the forewing pattern being brighter orange-tawny in the male, with narrower, more clearly defined dark brown borders, and a prominent dark brown brand curving upwards from the dorsum almost to vein 4. There is an apical black ocellus, and on the forewing underside this is pupilled white and ringed by buffy-orange with another outer dark brown narrow ring. The hindwing has a narrow dark brown line in the discal area, outwardly margined with white, and another broken white line along the margin. In between are three to four white-ringed black ocelli. The hindwing margin is strongly scalloped. The female has the forewing upper surface a paler, yellower orange with a second black ocellus in between veins 2 and 3. Very like both *Maniola narica* and *Maniola tenuistigma* in appearance, but with the hindwing margin more strongly dentate and averaging larger in size than *M. narica*, and its under hindwing has ocelli which *narica* lacks. *M. tenuistigma* also averages slightly smaller in size, and with a much narrower brand on the forewing of the male, as well as clearer orange-tawny forewings without the darker brown irregular discal band on the forewing of *M. davendra*.

Status: This is a very common species at lower elevations, from above 900m (3,000ft) in Balochistan, and between 1,200–2,400m (4,000–8,000ft) in the NWFP, extending up to Chitral and Swat. It is on the wing from May to September, usually associated with pine forest in the north and open, scrub-dotted hillsides in Balochistan. Collected TJR on grassy hillside behind Mingora city in Swat at 1,200m (4,000ft).

Habits: The author found this species in lower Swat, sharing the same habitat as *Eumenis parisatis*, and in Chitral sharing the same habitat with *Maniola lupinus*, and occurring up to 2,400m (8,000ft).

145. Lesser Whitering Meadow Brown

Maniola tenuistigma (syn: *Hypenophele tenuistigma*)
Wingspan 48–50mm

Description: A smaller version of *Maniola davendra*, as can be seen from the dimensions above. It bears the same orange-tawny band on the forewing, but the male has a narrower brand which extends slightly above vein 3. The upper hindwing is plain brown and its margin slightly dentate as in *M. davendra*. It differs slightly from *M. davendra* in the underside forewing, with the post-discal line strongly waved below vein 4, and anteriorly edged white on its outer side. The genitalia in the male are different from *M. davendra*, and described (Evans, 1932, and Talbot, 1947) as having the uncus short and stout, whereas in *M. davendra* the uncus is long and thin.

Status: Not really known, because of its similarity to *M. davendra*. Evans (1933) considered it not uncommon above 2,100m (7,000ft) in Balochistan around Ziarat, the upper parts of the Uruk valley, and on the Khojak Pass. In Chitral he considered it rare. Sanders (1955) collected specimens from Astor at Doyan and the Dashkin forest, between 2,100–2,400m (7,000–8,000ft). Sakai (1981) collected it from north-eastern Afghanistan.

146. Oval Spot Meadow Brown

Maniola wagneri, (syn: *Hyponephele wagneri*)
(Syn: *Epinephele mandane*)
Wingspan 50–52mm

Description: This species has a large black ocellus in the forewing tip which is elliptical or oval in shape, and which is surrounded by a narrow pale ochre ring. In the female, inside this in the post-discal area is a wider ochraceous band extending down to the dorsum. In the male this ochraceous band is absent or very obscure, but there is a dark blackish-brown brand on the forewing which is rather wide distally, being triangular in shape, the apex extending up to vein 3. The hindwing is brown, darker basally, and the lower margin is scalloped or slightly dentate. The under forewing is pale orange with brown costa and margin, and the apical ocellus is without any white pupil. The under hindwing is light greyish-brown speckled with darker brown, and having a thin dark discal line which is angled inwards at the end cell and edged outwardly with white. There are two small black ocelli without white pupils in the tornal region.

Status: It was collected from Balochistan in the vicinity of Quetta early this century and there is a Brit. Museum specimen, labelled Bogra, from Balochistan. Evans (1933) never came across it, and it must be considered very rare in Pakistan. It is widespread in Iran and Kurdistan.

147. Pamir Meadow Brown

Maniola hilaris (syn: *Hypenophele hilaris*)
Wingspan 34–36mm

Description: As can be seen from the above measurements, this is a small Meadow Brown. Like the others, it has a prominent black ocellus in the forewing apex which is unpupilled, and with orange tawny area on the forewing only. The female has a larger unpupilled black ocellus on the forewing and an irregular brown post-discal band across the forewing. The under hindwing lacks any ocelli, being marked with wavy dark brown lines in the basal and post-discal areas, against a buff background, speckled and vermiculated darker brown.

Status: This species was collected by Evans (1932) from the Baroghil Pass, and before descending on to the Wakhan plateau, still inside Chitral territory. For some reason Talbot, in the *Fauna of British India* series (Vol. 2, 1947) treats this species as extra-limital. Sanders (1955) also collected this species from the Baroghil valley in extreme north Chitral, and also north of the Darkot Pass, and found many *Maniola hilaris* at both places between 21 and 23rd August, at which time they appeared to have freshly emerged. These passes are south of the river which forms the boundary between Afghanistan and Chitral.

The butterflies were collected between 3,960–4,100m (13,000–13,500ft).

148. The Balochi Heath

Coenonympha myops
(Syn: *Lyela myops*)
Wingspan 34–40mm

Description: Though placed in the genus *Coenonympha*, this species does not share many of the characters typical of that genus, lacking the usual number of ocelli, but does share the small size and wing shape of other 'Heaths'. This rather small species is plain dark brown on both fore and hindwings, without any distinguishing feature except for an apical black ocellus which is ringed with dull ochre on both upper and undersides. There are no ocelli on the hindwing, nor any scalloping along the hindwing margin. The upper apical ocellus is without any white pupil, and usually the under one also. The sexes are alike.

Status: Recorded from the western borders of central Balochistan around the Khojak Pass and Gwal Forest, where it is common. Also collected from the Uruk valley and near Hannah Lake. It is on the wing from late March to early May and found from above 1,800m (6,000ft).

Habits: Presumed similar to the *Maniola* genus, with the larval foodplants being grasses and the caterpillar in its early stages hibernating and developing very slowly.

149. The White Edged Rockbrown

Hipparchia parisatis
(Syn: *Eumenis parisatis*)
Wingspan 65–70mm

Description: This is often a very large butterfly, with a striking broad white outer border on both wings which is much wider on the hindwing, and which is toothed along its lower margin. The upper surface of both wings is dark blackish-brown, within certain lights a bluish velvet sheen. On the forewing there are two to three small white dots between veins 3 and 4, and 5 and 6, with a barely visible black ocellus in the apical area. On the hindwing there is also a small black pupilled ocellus in the tornal area. The

PLATE 31: BRANDED MEADOW BROWNS, WHITE EDGED ROCKBROWN AND COMMON ARGUS

A. *Maniola davendra* - White Ringed Meadow Brown **A-1.** Male **A-2.** Female **B.** *Maniola tenuistigma* - Lesser White Ringed Meadow Brown **B-1.** Male **B-2.** Female **C.** *Maniola narica* - Tawny Branded Meadow Brown **C-1.** Male **C-2.** Female **D.** *Coenonympha myops* - Balochi Heath, sexes alike **E.** *Hipparchia parisatis* - White Edged Rockbrown, sexes alike **F.** *Erebia nirmala* - Common Argus, sexes alike **F-1.** Form *materta* **F-2.** Form *kala* **G.** *Erebia nirmala* - Common Argus, form *daksha*

PLATE 32: ROCKBROWNS AND SATYRS

A. *Eumenis mniszechii* - Tawny Rockbrown **A-1.** Male **A-2.** Female **B.** *Eumenis thelephassa* - Balochi Rockbrown **B-1.** Female **C.** *Hipparchia persephone* - Dark Rockbrown **C-1.** Male **C-2.** Female **D.** *Hipparchia heydenreichi* - Shandur Rockbrown **D-1.** Male **D-2.** Female **E.** *Karanasa digna* - Chitrali Satyr **E-1.** Male **E-2.** Female **F.** *Karanasa actaea pimpla* - Dark Satyr, sexes alike **G.** *Karanasa heubneri* - Tawny Satyr, sexes alike **H.** *Karanasa moorei* - Turkestan Satyr, sexes alike

PLATE 33: SATYRS, YPTHIMA RINGS AND EVENING BROWN

A. *Aulocera padma* - Great Satyr, sexes alike **B.** *Aulocera swaha* - Common Satyr, sexes alike **C.** *Aulocera saraswati* - Striated Satyr, sexes alike **D.** *Ypthima asterope* - Common Threering, sexes alike **E.** *Ypthima nareda* - Large Threering **F.** *Ypthima inica* - Lesser Threering **G.** *Ypthima sakra* - Himalayan Fivering **H.** *Ypthima avanta* - Jewel Fourring **I.** *Ypthima bolanica* - Desert Fourring **J.** *Melanitis leda* - Common Evening Brown

PLATE 34: DANAIDS, 'BEAKS' AND 'PUNCHES'

A. *Danaus genutia* - Common Tiger **B.** *Danaus chrysippus* - Plain Tiger **C.** *Tirumala limniace* - Blue Tiger, male **D.** *Euploea core* - Common Crow, female **E.** *Acraea violae* - Tawny Coster **F.** *Libythia lepita* - Common Beak **G.** *Libythia celtis* - European Beak or Nettle Tree Butterfly **H.** *Dodona dipoea* - Lesser Punch **I.** *Dodona durga* - Common Punch **J.** *Dodona eugenes* - Tailed Punch

PLATE 35A: PIERROTS AND BABUL BLUES

A. *Castalius rosimon* - Common Pierrot, male **B.** *Tarucus nigra* - Black Spotted Pierrot, male **C.** *Tarucus extricatus* - Rounded Pierrot, male **D.** *Tarucus theophrastus* - Pointed Pierrot **D1.** Male **D2.** Female **E.** *Tarucus rosacea* - Balochistan or Mediterranean Pierrot **E-1.** Male **E-2.** Female **F.** *Tarucus callinara* - Spotted Pierrot, male

PLATE 35B: PIERROTS AND BABUL BLUES

G. *Tarucus venosus* - Himalayan Pierrot **G-1.** Male **G-2.** Female **H.** *Syntarucus plinius* - Zebra Blue **I.** *Azanus uranus* - Dull Babul Blue **I-1.** Male **I-2.** Female **J.** *Azanus ubaldus* - Bright Babul Blue **J-1.** Male **J-2.** Female **K.** *Azanus jesous* - African Babul Blue, female

PLATE 36A: EVERES CUPIDS, HEDGE BLUES AND JEWEL BLUES

A. *Everes argiades* - Chapman's Cupid **A-1.** Male **A-2.** Female **B.** *Everes buddhista* - Shandur Cupid **B-1.** Male **B-2.** Female **C.** *Everes hugelii* - Tailed Cupid **C-1.** Male **C-2.** Female **D.** *Arletta vardhana* - Dusky Hedge Blue, male **E.** *Celastrina puspa* - Common Hedge Blue **E-1.** Male **E-2.** Female

PLATE 36B: EVERES CUPIDS, HEDGE BLUES AND JEWEL BLUES

F. *Celastrina ladonides* - Silvery Hedge Blue **F-2.** Female **F-1.** Male **G.** *Celastrina argiolus* - Hill Hedge Blue or Holly Blue of Europe **G-1.** Male **G-2.** Female **H.** *Philotes vicrama* - Chequered Blue **H-1.** Male **H-2.** Female **I.** *Plebejus christophi* - Small Jewel Blue **I-1.** Male **I-2.** Female **J.** *Plebejus sephyrus* - Balochi Jewel Blue or Zephyr Blue **J-1.** Male **J-2.** Female

PLATE 37A: ARGUSES, VACCINIA JEWELS, AND GREEN UNDERWINGS

A. *Aricia agestis* - Orange Bordered Argus, sexes alike **B.** *Aricia chiron* - Streaked Argus **B-1.** Male **B-2.** Female
C. *Aricia astorica* - Astor Argus **C-1.** Male **C-2.** Female **D.** *Turanana cytis* - Spotted Argus **D-1.** Male **D-2.** Female
E. *Vaccinia khwaja* - Dark Jewel Blue **E-1.** Male **E-2.** Female

PLATE 37B: ARGUSES, VACCINIA JEWELS, AND GREEN UNDERWINGS

F. *Vaccinia iris* - Jewel Argus, sexes alike - Alpine Argus **G.** *Agriades jaloka* **G-1.** Male **G-2.** Female **H.** *Agriades pheretiades* - Greenish Mountain Blue **H-1.** Male **H-2.** Female **I.** *Albulina pheretes* - Mountain Blue **I-1.** Male **I-2.** Female **J.** *Albulina omphisa* - Dusky Green Underwing **J-1.** Male **J-2.** Female

PLATE 38A: GREEN UNDERWINGS AND POLYOMMATUS MEADOW BLUES

A. *Albulina galathea* - Large Green Underwing **A-1.** Male **A-2.** Female **B.** *Albulina metallica* - Small Green Underwing **B-1.** Male **B-2.** Female **C.** *Glaucopsyche alexis* - Western Green Underwing **C-1.** Male **C-2.** Female **D.** *Iolana gigantea* - Gilgit Meadow Blue **D-1.** Male **D-2.** Female **E.** *Polyommatus sieversi* - Pale Jewel Blue **E-1.** Female **E-2.** Male

PLATE 38B: GREEN UNDERWINGS AND POLYOMMATUS MEADOW BLUES

F. *Polyommatus loewii* - Large Jewel Blue **F-1.** Male **F-2.** Female **G.** *Polyommatus sarta* - Brilliant Meadow Blu **G-1.** Male **G-2.** Female **H.** *Polyommatus gracilis* - Dusky Meadow Blue **H-1.** Male **H-2.** Female **I.** *Polyommatus bogr* - Balochi Meadow Blue **I-1.** Male **I-2.** Female **J.** *Polyommatus icarus* - Violet Meadow Blue **J-1.** Male **J-2.** Femal underwings show much larger ocelli, which are pupilled and ringed with yellow and dark brown. The rest of the underwings are greyish-brown striated with fine white lines and with a dark brown line across the discal area and a second sub-marginal line, the discal line sharply angled outwards at vein 3 and broadly margined with white on the outer edge. The female resembles the male, often with a broader white marginal border.

Status: A very widespread butterfly in dry rocky areas throughout Balochistan, the NWFP, and the inner Himalayas from Chitral through Swat, Gilgit, and Hunza to Baltistan. Collected TJR from the hills behind Mingora, and from Takht-i-Suleiman by Rafiq Ahmad Rajput in June 1993. It also occurs on the outer, more barren hills of the Murree hill range, and has been collected from Campbellpur (now Attock). It is on the wing from May to October.

Habits: The full-grown larvae are pubescent, with their tail bifurcate, and the body marked by darker lateral stripes. It is known that the foodplants of the larvae of Rockbrowns are grasses, and that the larvae take a long time to develop. They are nocturnal in feeding, hiding at the base of grass tussocks during the day. The pupa is formed in an unusual situation for a butterfly: the fully-grown larva burrows under the soil and, by wriggling its body, smooths out a chamber for itself which it lines with its body silk and then pupates. The butterflies are strong flyers, swift on the wing and fond of settling on bare cliffs or rocks, and they are typical inhabitants of open dry hills with no trees. Wynter-Blyth (1957) reports that in hot areas they shelter during midday in rock hollows or overhangs, flying only in early morning and late afternoon.

150. The Tawny Rockbrown

Eumenis mniszechii gilgitica
(Syn: *Hipparchia mniszechii,* Syn: *Eumenis baldiva*)
Wingspan 58–65mm

Description: Not as large as *Eumenis parisatis,* this species is similar in general appearance to *Eumenis thelephassa.* It is dark brown on the upper forewing, with a broad yellowish-orange discal band which is relatively straight along its inner edge, and a narrow dark brown marginal border. In *Hipparchia thelephassa,* this forewing orange band is sharply indented at vein 4 where the basal brown area extends outwards. Within the orange bands, there are two black ocelli, the upper apical one large and the second, between veins 2 and 3, being smaller. In the male these are unpupilled, whilst in the female they are white-pupilled. The hindwing lacks any ocelli and is paler brown basally. The under forewing is whitish-yellow, margined brown along the costa and outer margin, and darker orange basally, with the two black ocelli unpupilled. The under hindwing is dull white, finely speckled all over with dark brown, but more densely in the basal area, and with a whiter band in the discal area.

Status: This is a widely-distributed butterfly in the higher ranges of central Balochistan, through the NWFP, and northwards through Chitral, Gilgit, and Baltistan. In Yasin and the Ghizar valley (Gilgit), Tytler (1928) recorded it as not uncommon. It can be found on the wing from June to August, and from as low as 1,200m (4,000ft) up to 3,900m (13,000ft).

Habits: This Rockbrown flies at higher elevations than *H. parasitis,* and has been collected as high as 3,960m (13,000ft) on the Shandur plateau. Evans (1932) reported it as common in Balochistan during June and August, but flying earlier, during May in Ziarat. The larval foodplants are grasses, and as far as is known the early stages are like *H. parasitis.*

151. The Balochi Rockbrown

Eumenis thelephassa
(Syn: *Pseudochazara lehana* and
Hipparchia thelephassa)
Wingspan 55–65mm

Description: Very similar to the Tawny Rockbrown, *Eumenis mniszechii*, being dark brown above with both wings crossed by a broad orange-tawny discal band, but, unlike *mniszechii,* the inner border of this band is not straight but sharply indented where the brown basal area intrudes at the junction of vein 4 and the cell. There are two black, white-pupilled ocelli on the forewing. The tawny bands are outlined with a narrow black line along the outer edge and are somewhat darker, being dusted with brown scales, when compared with *mniszechii*. There is a small black ocellus with a white pupil on the hindwing, and the inner edge of the tawny discal band is more toothed than in *mniszechii.* The under hindwing is white, thickly speckled and striated with dark brown, and bearing two dark blackish-brown zigzag lines in the post-discal and sub-marginal areas. The male has a black brand, quite prominent along the middle of the cell.

Status: Very common in Balochistan and the western border regions of the NWFP, including the Khyber Agency. Collected Takht-i-Suleiman by Rafiq Ahmad Rajput in July 1993. On the wing from May to September and found from 1,200–2,700m (4,000–9,000ft). Collected TJR Swat Kohistan, near Kalam, at 2,100m (7,000ft) on 19th August.

Habits: The butterflies are strong on the wing and favour hot stony slopes, where they frequently settle on the bare rocks; with forewings concealed under the closed hindwings, they become invisible.

152. The Dark Rockbrown

Hipparchia persephone
(*Chazara persephone* in Sakai, 1981, and *Eumenis persephone* in Wynter-Blyth, 1957)
Wingspan 60–70mm

Description: In Talbot (1947) the Dark Rockbrown is described as occurring in two colour morphs, following Evans (1932). The usual one is dark brown with creamy-white discal bands, and another form, named *analoga* by Evans, has the bands a rich orange-chestnut instead of white. The author, examining all the specimens in the British Nat. Hist. Mus. collection, noted that all the white-banded specimens were males with a black brand, extending up to the base of the cell, clearly visible on the forewing, whilst all the orange specimens were without a brand and appeared to be females. This separation of the sexes is also followed by Sakai (1981). This species is rather similar to *Hipparchia heydenreichi*, being dark brown on the upper surface of both wings, but with an interrupted orange-tawny discal band in the female and a similar creamy-white band in the male, which is broken between veins 5 and 6, and with two black unpupilled ocelli inside the orange band, one at the wing tip and the second between veins 1b and 2, the lower ocellus being bigger. The hindwing has a curved white band in the male, or orange band in the female, in the post-discal area, and its margin is scalloped. Unlike *H. heydenreichi*, there is no white patch inside the cell on the forewing. The underside forewing is orange with dark brown lines crossing the cell and the costal areas, and two black ocelli in the same position as the upper side. The under hindwing is greyish-white, densely freckled and streaked with fine dark lines and two irregular dark blackish lines, the outer one down the post-discal area, bordered outwardly by a paler whiter band, and a more jagged line in the sub-basal area also bordered with a whiter area on its outer side.

Status: Not uncommon in Balochistan on bare hillsides from 1,200–1,800m (4,000–6,000ft), and collected on the Khojak Pass and Uruk valley. Also in the western border regions of the NWFP and Chitral, in which latter region it is common, flying from May to October. Sanders (1955) collected specimens from Astor district in Gilgit, between 1,800–2,400m (6,000–8,000ft) between 20 and 22 July.

153. The Shandur Rockbrown

Eumenis heydenreichi
(Syn: *Chazara heydenreichi*, and *Hipparchia shandura*)
Wingspan 50–65mm

Description: Similar to *Eumenis persephone* in general appearance, being the same colour in both sexes, but differing mainly in the presence of a creamy-white triangular patch in the basal half of the cell, and the broad creamy white borders, not orange as in the female of *E. persephone*. Two black ocelli on the forewing are placed in the middle of the white band between veins 5 and 6 and a much larger one between veins 1b and 2. The hindwing bears the same broad curved creamy-white band which is speckled lightly with dark brown, and the wing margin is scalloped. The under forewing is creamy on the cell and post-discal areas, with no trace of the orange which is visible on the under forewing of both sexes in *H. persephone*, and a broad dark band at the end of the cell and into the disco-cellular area. The under hindwing is brown and white in bands, with fine dark striations and the veins outlined white. Females generally average larger than males, with more prominent black ocelli, and the males show a black brand on the forewing curving up to the base of the cell apex.

Status: Only found at high elevations in the extreme northern areas, from Chitral to Hunza. Collected in Chitral on the Baroghil Pass and the Shandur plateau at 3,660m (12,000ft) in August. From the Braldo Glacier, Skoro La, and Koch Boozla, all in Baltistan, and Gilgit (without locality) by Biddulph.

Habits: Evans considered it uncommon in the localities where he encountered it. No doubt the ability of the larvae to pupate underground enables them to survive at the high altitudes which they frequent.

154. The Dark or Black Satyr

Satyrus pimpla
(Syn: *Hipparchia actaea* and *Karanasa actaea*)
Wingspan 45–65mm

Description: This is a rather large butterfly, wholly dark brown on the upper side, except for traces of paler hazel-brown flecks on the hindwing and a small black-pupilled ocellus in the forewing apical area. The underside forewing is rich orange-tawny in the discal area, with the apical ocellus larger and surrounded by dull ochraceous-tawny. The underside hindwing is overall dark brown, being strongly freckled with brown against a paler brown background. The veins are outlined in white and there are two zigzag dark lines encircling the post-discal area and the outer sub-marginal area. The sub-marginal line is sharply dentate, the post-discal line less so, and both are bordered outwardly with paler brown. Females tend to be paler brown on the upper side, with larger apical ocellus showing a trace of an ochraceous outer ring.

Status: Widespread in Balochistan over 2,100m (7,000ft) from Ziarat and Zhargun and Kawas, and in Chitral from Drosh northwards, but nowhere common. Usually on the wing from June to September. Collected TJR in the Beori nullah, Chitral, at 1,980m (6,500ft) on 24 June. Occurring occasionally at higher elevations up to 3,000m (10,000ft). Sanders (1955) took specimens throughout Astor district from 1,370–2,600m (4,500–8,500ft) in the second half of July.

Habits: The author only encountered this species on the hottest mountainsides on shale slopes. It is a strong flyer. The foodplants of the larvae are grasses, particularly *Bromus* spp., which thrive in arid regions.

155. The Chitrali Satyr

Karanasa digna
(Syn: *Hipparchia digna* and *Kanetisa digna*)
Wingspan 50–55mm

Description: This species combines some of the distinctive features of both *Eumenis mniszechii* and *Karanasa actaea*, having the same sharply-dentate sub-marginal line on the under hindwing as *acteae*, with a broad orange-yellow band across both wings as in *mniszechii*. There is only one ocellus in the forewing apical area, which is unpupilled, and the orange band is continuous, not interrupted as in *Eumenis shandura* or *persephone*. This band is even along its inner border, but somewhat crenulated along the outer border. The hindwing outer margin is scalloped, with the cilia brown along the veins and white in between. The underside forewing is orange-yellow, striated with dark brown across the costa and cell, and with a sharply dentate fine dark sub-marginal line. The under hindwing is whiter, thickly freckled with brown, and showing a denser brown band across the post-discal area, with a dark brown toothed line curving around the sub-marginal area. The under forewing ocellus is white-pupilled. The slightly larger and paler female is otherwise similar.

Status: Not common and erratically distributed from Chitral to Yasin in western Gilgit, flying from June to August between 2,100–3,350m (7,000–11,000ft). Collected in the Laspur valley in July.

156. The Tawny Satyr

Karanasa huebneri, var: *cadesia* and *pupillata*
Wingspan 42–50mm

Description: This species averages smaller than the Chitrali Satyr (*K. digna*), but has the same broad orange-yellow band across both wings, differing from *K. digna* in having two pupilled black ocelli, a slightly larger one in the apical region and the second between veins 2 and 3. The orange band is quite toothed along its outer edge and the band on the lower wing is lunar shaped, both bands tending to be outlined with blackish-brown borders along the outer edge. The underside forewing is largely orange-tawny with a narrow grey-brown marginal border and darker brown basally, the two ocelli not being any larger than on the upper side. The under hindwing is dark, being thickly freckled with dark brown, but showing a sharply-toothed post-discal dark border which is bordered on its outer edge with a paler, whiter area. The veins are outlined in white. The sexes are very similar in appearance.

Status: Not very common, but widely distributed from 2,700–3,660m (9,000–12,000ft) from the Safed Koh and upper Kurram valley through Chitral into Gilgit and Baltistan. Collected from Astor, and Ghizar valley in Gilgit, and the Deosai plateau in Baltistan, and from Utzen and Tarben nullahs, the Shandur plateau and the Laspur valley (Evans, 1932) and Baroghil valley around 3,660m (12,000ft) in Chitral. Flying in July and August.

157. The Turkestan Satyr or Shandur Satyr in Evans (1932)

Hipparchia moorei (syn: *Karanasa moorei*)
Wingspan 50–57mm

Description: Similar to *Karanasa huebneri* in appearance, though averaging slightly larger in size, and with the broad tawny bands on the upper wings a much paler yellow-tawny, rather than orange. There are two black ocelli on the forewing, the lower one being very small, and both without pupils. The yellow band is dentate along its outer border and on the inner border is sharply indented at vein 4, almost dividing the apical part of the border which contains the ocellus. The hindwing also has its yellow border toothed along the outer edge; the hindwing margin is not scalloped. The underside forewing is ochraceous, with the cell area more rufous orange and bearing dark brown cross lines. The apical ocellus is pupilled and the lower ocellus often absent. The under hindwing is thickly freckled with dark brown like that of *K. huebneri*, and also bears a toothed dark line encircling the post-discal area, with a whiter, less densely speckled area along its outer edge.

Status: Uncommon but occurring in Chitral and western Gilgit from 2,700–4,260m (9,000 to 14,000ft) in July and August. Collected Yasin and Ghizar valleys.

158. The Turkestan Satyr

Karanasa regeli or *Karanasa boloricus*

Description and Status: These two forms are part of a puzzling group of similar 'Satyrs', which Talbot in the *Fauna* series (Vol. 2, 1947), considered should probably be taken as one species. In appearance they are identical with *Karanasa moorei*, the Turkestan Satyr, but examination of the male butterfly's claspers shows that they are a separate species. Sanders (1955) collected *boloricus* above the Shandur Lake and near the summit of the Baroghil Pass, between 3,660–3,960m (12,000–13,000ft). He noted that the pale bands on the upper side were almost white in this form, whereas in *K. moorei* they were always pale ochraceous; however, there is some variation even in *boloricus* in the series of specimens in the Brit. Nat. Hist. Museum collection, some showing an ochraceous tinge. It is significant that Sanders (*op. cit.*) found *moorei* and *boloricus* flying sympatrically in the Baroghil valley and on the Shandur. Talbot gives the key for the genitalia of *K. moorei* as follows: 'Valve with dorsal edge angled about two-thirds from apex, being incurved above and below this point; side lobes obtuse'. Key for genitalia of *K. boloricus*: 'Valve with dorsal edge, about two-thirds from apex, with a small tubercle bearing a tuft of hair; edge of valve not incurved. Side lobes pointed'. *K. boloricus* has been collected also from Misgar in Hunza.

159. The Great Satyr

Aulocera padma
(Syn: *Satyrus padma*)
Wingspan 70–98mm

Description: As its name implies, this is a large butterfly, of a very dark brown or almost black colour on both wings, bearing a white diagonal band across both wings. In this species the band is fairly even in width throughout, but in the apical area of the forewing it is separated into irregularly spaced smaller spots between the veins. In *A. padma* the apical white spots do not surround the ocellus, there being only two on its outer edge. On the hindwing the white band is crossed by black veins, and extends down into vein 1b. The underside is a slightly paler brown with scattered fine white striations across the post-discal region. The discal white band is rather diffuse along its outer edge, but sharply defined with black along its inner edge. There is also an indistinct black post-discal band, lacking white striations.

Status: Very rare in Balochistan, occurring only on the Shingar in August and September. It is more common on the Safed Koh, and plentiful in the lower Kaghan valley and the Murree Hills. It has not been recorded from Chitral but occurs rather rarely in western Gilgit, in Yasin, and more plentifully around Astor. It is on the wing in the Himalayas from late May to the end of June and favours open rocky areas.

Habits: The larval foodplants are grasses (*Poa* and *Bromus* spp.) and, as far as is known, the larvae develop at different rates, with some hibernating and forming an early summer brood and others developing quickly, producing a second, late summer brood. The pupae are formed underground. The butterfly haunts warm rocky places in open areas, away from forest, and is very territorial, chasing away all passing butterflies, often returning to a favourite rock perch.

160. The Common Satyr

Aulocera swaha
Wingspan 60–70mm

Description: A slightly smaller version of *Aulocera padma*, being distinguished mainly by the diagonal band being more creamy white, almost yellow in some specimens, and this band narrows on the hindwing as it reaches the inner margin, where it does not quite extend into vein 1b. The creamy band in the apical region is split up into four separate white spots surrounding a barely-visible black ocellus. The hindwings are not quite such a dark brown as in *A. padma*, having an almost bronzy sheen. The underside has more fine white striations, especially in a wide band around the sub-marginal area, so it looks less dark underneath than *A. padma*.

Status: Found at slightly lower altitudes than the Great Satyr, and adapted to more forested areas. It is very common in Hazara district and the Murree Hills, and less commonly encountered in the Safed Koh, lower Chitral, and in Gilgit on the slopes of Nanga Parbat. It has also been collected from Campbellpur. It is on the wing from May to September and normally occurs from 1,800–3,000m (6,000 to 10,000ft). Collected TJR at Miandam in Swat on 22 July at 2,600m (8,500ft), and at Dunga Gali at 2,280m (7,500ft) on 15 May.

Habits: Bell describes the larvae as being brown with a rugose hairless skin. The larval foodplants are grasses including *Poa* spp. and *Bromus* spp. As stated above, this species is more closely associated with forested areas than *A. padma*, and occurs at lower altitudes.

161. The Striated Satyr

Aulocera saraswati
Wingspan 65–70mm

Description: Very like *Aulocera padma* in having a fairly straight, even, white diagonal bar across dark blackish-brown wings. It is smaller than *A. padma* and the white bar on the hindwing is quite broad and reaches the margin. The apical black spot on the forewing is surrounded by four white spots, the inner one in males being quite tiny. The underside is distinctive, as the hindwing bears a very broad white discal band, inwardly margined by a thin dark brown line and a less distinct narrow whitish band in the sub-marginal area, also margined on the inner side by dark brown. Overall there are more numerous fine white striations, and the underside fore and hindwings are more buffy yellow than dark brown as in the other two spp. found in Pakistan.

Status: This species is rare in Chitral and is the least common of the three species occurring in Pakistan. It is found at lower elevations than the Common Satyr (*A. swaha*), being found near Dewal at 1,800m (6,000ft) in the lower Murree Hills, and also from Campbellpur. It has also been collected as high as 3,000m (10,000ft) in Kumaon, India. It is only on the wing from late July to October.

162. The Common Argus

Erebia nirmala, var: *daksha*, and *materta*
(Syn: *Callerebia nirmala*)
Wingspan 45–50mm

Description: This is another genus of butterflies which are weak flyers with rather sedentary habits, giving rise in consequence to many slightly varying forms and some variation in taxonomic treatment. Like the entire genus, they have strongly-arched costa on the forewing, making it rather rounded in shape, with weak, floppy flight. The overall colour is dark brown above, with an apical ocellus which is double-pupilled and surrounded by a buff ring. The hindwing also bears a smaller, single-pupilled ocellus in between veins 1b and 2. The underside forewing is evenly coloured dark brown in the form *E. nirmala daksha* and dark maroon in the form *E. nirmala materta*. The underside hindwing is paler buffy brown with fine black speckling, and a very tiny pupilled ocellus in the tornal region.

Status: This is an extremely abundant butterfly at lower elevations from Chitral eastwards through Swat, Gilgit, and the Murree Hills. It has also been collected from Campbellpur. It is on the wing towards the end of the summer, between 1,800–2,700m (6,000–9,000ft). The form *daksha* is the most common in the Murree Hills, lower Chitral, and the main vale of Swat. The form *materta* is most common in Swat Kohistan and rare in the Murree Hills.

Habits: The females of the Arguses, genus *Erebia*, scatter their eggs on the ground near their larval foodplants, which are grasses. The larvae are spindle-shaped and green with darker longitudinal stripes. They have short processes or horns on their anal segment, and in some species, their skin is roughened with tiny tubercles. Most of them spend a long time in the larval stage, undergoing prolonged hibernation in the early stages of growth. They pupate on the ground at the base of grass, roughly drawing some blades together with silk and pupating inside this shelter. The pupae can be green or brown depending upon their surroundings. The butterflies, as stated above, have a weak, floppy sort of flight, never travelling far, and keeping close to the ground. They will visit flowers, and are particularly attracted to Compositae.

163. The Ringed Argus

Paralasa annada
(Syn: *Erebia annada* in Talbot, 1947)
Wingspan 60–70mm

Description: This species is typical of the genus in being a dark velvety brown on the upper surface, with the forewing costa strongly arched, and bearing on the forewing apex a large yellow-ringed black ocellus, usually with two small white pupils. It is distinguished from others of the genus by having the hindwing produced to a rounded lobe in the tornal area, and by having a broad discal band on the under hindwing extending into the tornal area, where there are two small ocelli. This discal band is frosted with fine white striations, and bordered by two indistinct sub-marginal and discal bands. The underside forewing is deep red-brown in the apex, turning to more ochraceous brown basally. Compared with the other *Erebia* Arguses, this species is rather larger and more handsomely marked.

Status: This species in included on the basis of Butler's (1886) identification of a series collected by Major Yerbury in the localities mentioned below. This is one of the lower elevation *Erebias*, favouring fairly dry, open slopes. Collected in Hazara district from Abbotabad, Kala Pani, and Mansehra, flying around open stony hillsides. There are two broods, the first emerging in mid-April and the second on the wing from late August through to October.

Habits: The eggs are scattered on grasses, and the larvae feed on grass. The larvae have two short projections on the last anal segment. Like all the Arguses, the adults have a typically weak, hopping sort of flight trajectory and they are attracted to damp patches on the ground.

(163b. The Pallid Argus

Paralasa scanda
[Syn: *Erebia scanda* in Talbot, 1947]
Wingspan 50–60mm

Description: This is very similar to *Paralasa annada* in all respects, except that the upper wings are a duller, slightly paler brown, and it is like the latter species in being rather variable, with some forms showing more white frosting or striations on the under hindwing, and generally the dry season brood is less well marked. Like *P. annada* there are small white dots on the under hindwing in the post-discal area, and the male possesses modified scent scales on the upper discal area.

Status: Both Talbot [1947] and Evans [1932] give the western distribution of this species as from Kashmir and Kumaon eastwards. There are no specimens in the British Natural History Museum from Pakistan, though there is a very good series from the western Himalayas within India, as far west as the Kulu valley, whilst *P. annada* seems to be more confined to the eastern Himalayas, from Darjeeling to Bhutan. Based only upon visual external examination of the many specimens in the British Museum, this author thinks it probable that the specimens collected by Major Yerbury from within present day-Pakistan were more probably *Paralasa scanda* [Butler, 1886] and therefore I include this species in the category of 'extralimital' until more information on the overlapping range of these two sibling spp. can be ascertained. Wynter-Blyth [1957] writes that *P. annada* frequents dry open slopes, whilst *P. scanda* prefers more shaded wooded hills, and the habitat where Major Yerbury collected *P. annada* is certainly dry and open.)

164. The Scarce Mountain Argus

Erebia kalinda
(Syn: *Paralasa shakti*)
Wingspan 45–50mm

Description: This species is distinguished by having rather paler ferruginous brown upper wings, with no ocellus on the hindwing, but on the forewing in the apical area a single pupilled ocellus ringed with buff, and an obscure discal band of deep ochraceous red, contrasting with the darker brown basal area. The underside of the forewing is ochraceous red with a large, single-pupilled and ringed ocellus, whilst the underside hindwing, with no ocellus, instead shows an arc of seven small white spots around the post-discal region.

Status: This Argus is rare and only found above 2,700m (9,000ft), up to 3,960m (13,000ft), from Chitral, through Gilgit including Yasin, Hunza, and Gurais. It is on the wing from June to August.

Habits: Like the previous species, the female drops her eggs on the ground near to grasses which form the larval foodplants. The larvae taper sharply at the head and tail, being thickest in the middle and they have two short projections on the anal segment. They pupate in mid summer and the butterfly hatches out within about four weeks, whereas the larvae overwinters as a very small caterpillar and then feeds rapidly in the spring and early summer.

165. The Mountain Argus

Erebia shallada
Wingspan 45–55mm

Description: Very similar to *Erebia kalinda*, with only a single pupilled ocellus in the apical region of forewing, no ocellus on the hindwing, but having an arc of small white dots on the under hindwing in the post-discal area, and usually more noticeable dusting of white scales in the discal area. The underside forewing is also a rather darker maroon than in *E. kalinda*.

Status: Locally common above 2,700m (9,000ft) in lower Chitral, Swat Kohistan, also in the upper reaches of the Neelum valley, Azad Kashmir, flying at lower altitudes than the Scarce Mountain Argus. It has been taken as low as 1,800m (6,000ft) in Himachal Pradesh, India. On the wing from June to August.

166. The Yellow Argus

Erebia mani (of Evans, 1932)
Wingspan 45–53mm

Description: Like *Erebia shallada*, with a single white-pupilled ocellus on the forewing and none on the hindwing, but differing in the presence of a broad, pale, yellowish-brown discal band on the forewing which contains the ocellus. The underside of the forewing is ochraceous red with the apical ocellus ringed buff, and the underside of the hindwing is greyish-brown dusted with a scattering of small white scales, and the same curved line of seven small white spots in the post-discal area.

Status: This is a high-elevation Argus found from northern Chitral through Gilgit and Hunza, flying from July to August. It is everywhere rare. It has been collected on the Shandur Pass.

The 'Rings' of *Ypthima* genus are mostly adapted to warmer latitudes and differ in their early life history from the Mountain Arguses, genus *Erebia*. They all have eggs, which are some shade of green, with their surface minutely patterned or pebbled. Their larvae are spindle-shaped and pubescent, that is covered with minute short hairs, and they have fairly long projections or horns on their anal segment, but no horns on their head as in the Wood Browns, genus *Lethe*. They pupate above ground though close to the soil, suspended hanging head downwards by the cremaster or tail claspers. Their larval foodplants are grasses, and the pupae hatch within about two weeks.

167. The Lesser Threering

Ypthima inica
Wingspan 30–34mm

Description: Like all members of the genus, this is a small brown butterfly with rather rounded apex to the forewing and a weak floppy flight. The Lesser Threering is uniformly dark brown on the upper side with a single prominent ocellus in the apical region, between veins 6 and 7, which is double-pupilled and ringed with yellow. The upper hindwing bears one small ocellus in the tornal area, lacking any white pupil. The underside hindwing shows three ocelli, one in the apical area and two tornal, the lower one always being minute and sometimes obsolete in dry-season forms. Both fore and hindwings on their underside are liberally striated with fine white and brown lines, especially denser white around the sub-marginal area. In the keys given by Wynter-Blyth (1957) and Talbot (1947), the apical and tornal ocelli, under hindwing, are described as not being in line, but in the author's opinion this feature is not easily discernible when compared to other *Ypthima* spp.

Status: Has been collected from Lahore (Rhe-Philipe, 1917) in June and November, where it must be considered very rare. A specimen was collected from Bannu early this century. It occurs more commonly farther east in Amritsar, and is widespread in the plains of India.

Habits: Like all the *Ypthima* species, this little butterfly lays its eggs on grasses, and the larvae pupate close to the ground surface among the roots of grasses, but hanging vertically, attached to a stem by the cremaster. The adults frequent open places where there is some grass growing, and they do not usually settle on flowers. They have a weak, floppy flight, settling often with wings closed, on some grass low down.

168. The Large Threering

Ypthima nareda
Wingspan 40–45mm

Description: There is no marked difference in this species between wet and dry season forms, and *Y. nareda* is chiefly distinguished by having prominent large bi-pupilled ocelli on both upper and hindwings, a single one in the forewing apical area and a single one in the hindwing tornal area. There is a scattering of white striations around the post-discal area of both wings, which highlights a sub-marginal dark brown band. The underside hindwing bears three ocelli, the apical one very large and ringed with yellow, and two smaller tornal ocelli each with a single pupil and ringed yellow. A darker brown sub-marginal band is prominent on both fore and hindwing undersides. The fine striations on the hindwing are evenly spread and rather buffy, less white than in *Y. inica*.

Status: It has been collected from the Ravi river vicinity near Lahore (Puri, 1931) and also from Campbellpur and Murree at lower elevations. The author found it fairly common in mid March in the Margalla Hills on the summit ridge around 600–900m (2,000–3,000ft). It no doubt occurs around Sialkot, but it is less common than *Y. asterope* in Pakistan, preferring Himalayan foothill country.

Habits: This is a bit stronger on the wing than the previous species, but has the same habit of settling very soon with wings folded and often concealed amongst grasses close to the ground.

169. The Common Threering

Ypthima asterope
Wingspan 30–37mm

Description: Smaller in size than *Y. nareda*, the upper forewing bears a large apical ocellus with two pupils ringed with yellow, and traces of a darker brown discal band in front of the ocellus and a narrow dark band along the sub-marginal area. The upper hindwing has a single small pupilled ocellus. The underside bears finer, less prominent, pale whitish striations very densely packed, and the apical ocellus on the forewing is surrounded by a paler ring of white scales, whilst the under hindwing bears two minute ocelli in the tornal region and a barely visible one in the apical region, which may often be entirely absent.

Status: Early records are of a specimen collected in Balochistan from near Loralai at 1,200m (4,000ft), and another from Karachi in 1886 (Swinhoe), but as it is widespread over the rest of Pakistan, it has probably been overlooked. Collected Rawalpindi, near Barakao, Nurpur Shahan, Abbotabad, Attock, Campbellpur, Khyber Pass, and Hasan Abdal. It favours dry rocky hills.

170. The Himalayan Fivering

Ypthima sakra
Wingspan 45–55mm

Description: This is the largest of the *Ypthima* spp. occurring in Pakistan, and is easily separated from the others by having on its underside two ocelli close together in the apical region of both wings, the lower one between veins 5 and 6 being always much larger. The upper forewing is dark brown without any trace of darker sub-marginal bands, and with a bi-pupilled apical ocellus, followed by a minute one close beneath. The hindwing upper side bears one small unpupilled apical ocellus, and three in the tornal area, the lowest one minute. The underside is evenly striated with fine buffy-white or creamish lines and all five ocelli pupilled, the apical and sub-apical ones conjoined by a yellow ring.

Status: Very common throughout the Murree Hills from May to October. Collected TJR on 15 September at 1,980m (6,500 ft) in Dunga Gali.

Habits: Sevastopulo (1946) records that the female lays her eggs singly on grass blades and that the eggs are greenish-white, barrel shaped with a flat top, and the surface minutely pebbled. The eggs collected by him hatched on the ninth day, and the larvae grew quickly. The full-grown larvae has a square-shaped head, buff in colour and divided into two short points. The body is pale buff with a brown dorsal stripe and sub-dorsal and lateral darker brown lines. The spiracles are black and the whole body is covered with a pubescence, and the last segment (anal) bears two longish processes. The pupa has its head produced into a triangular projection and is coloured buff, streaked and speckled with dark brown. It is suspended by the cremaster, hanging downwards. Sevastopulo reported that these larvae pupated on 13 June and the butterflies emerged on 26 June.

171. The Desert Fourring

Ypthima bolanica
Wingspan 35–40mm

Description: This is a medium-sized Ringlet, being a uniform dark umber brown on the upper side, with a prominent bi-pupilled ocellus in the apical area of the forewing surrounded by a yellow ring, and another, darker brown outer ring, with a frosting of white scales around the outer apical side. The hindwing bears a similar but smaller, single-pupilled ocellus in the tornal region. The underside is overall rather pale with much white and buff striation, except at the base of the forewing, which is uniformly hazel brown. The forewing apical ocellus is large and bi-pupilled, and the underside hindwing bears two small, pupilled ocelli in the tornal area, with three pupilled ocelli between veins 4, 5, 6, and 7, in a vertical line. The number of hindwing ocelli can vary in all *Ypthima* spp., and there are specimens in the Brit. Nat. Hist. Museum with only three ocelli on the under hindwing, but despite its common name, five ocelli on the under hindwing is usual.

Status: It occurs uncommonly throughout Balochistan and the NWFP between 900–1,800m (3,000 and 6,000ft), being on the wing from April to August. It has been collected from Quetta, the Bolan Pass, Bannu, and Peshawar in the NWFP, and from Campbellpur in Punjab.

Habits: Evans (1933) found that this butterfly, though very widely distributed throughout Balochistan and the NWFP, was always encountered singly, in contrast to others such as *Ypthima sakra*, which is often seen close to three or four others of the same species.

172. The Jewel Fourring

Ypthima avanta
(Syn: *Ypthima ordinata* of Marshall and de Niceville, 1886 and *Ypthima lisandra* in Talbot, 1947)
Wingspan 32–38mm

Description: There are wet and dry season forms of this species, which has led to their descriptions as separate species, *Y. ordinata* post monsoon and *Y. avanta* during the dry season. It is distinguished from other *Ypthima* spp. by having, on the underside, two sub-apical ocelli on the forewing and three tornal ocelli on the under hindwing. In the wet season form these ocelli are larger and more conspicuous, but very much smaller in the dry season form. Above, both sexes are dark umber brown with faint, darker, sub-marginal bands. In the male the single sub-apical ocellus is very faint and may be absent in the dry season form, but is quite conspicuous in the female. The underside of both sexes is pale whitish-brown, closely covered with dark brown striations, and both wings crossed by distinct discal and sub-marginal darker brown bands. There is a large double-pupilled and yellow-ringed ocellus in the forewing apex, and the under hindwing bears a row of three tornal ocelli, with two costal ocelli. Sometimes in the wet season form the under hindwing bears a row of six ocelli, there being four tornal ones in a row, the lower ones being very small.

Status: Occurs in the dryer outer Murree Hills from Abbotabad at 1,280m (4,200ft) to Tret at 900m (3,000ft), but has also been collected from Bhurban, Murree, and Nathia Gali, but usually not above 2,100m (7,000ft). It is on the wing in August and September, and much less commonly in February and March.

173. The Common Evening Brown

Melanitis leda
Wingspan 60–80mm

Description: As can be seen above, this is a large butterfly, with rather curved costa on the forewing and truncated angular wingtip, whilst the hindwing is slightly tailed, and incurved along its outer margin. On the upper surface it is dark brown, with two prominent black ocelli on the forewing tip, each with white centres and margined on the inner side by a pinkish ochraceous band. The hindwing upper surface bears a smaller black tornal ocellus ringed with fulvous, and often another smaller ocellus above. The underside can be rather variable, but generally it is a pinkish-buff or ochraceous, covered all over with fine dark striations, and imitating in colour the dead leaves amongst which it rests by day. Along the outer margin of both wings are a series of white-pupilled small ocelli, varying in size from minute in the apical region of the forewing to a larger one between veins 3 and 4. The under hindwing usually bears two larger ocelli between veins 2 and 3, and 3 and 4, with prominent white centres and ringed with ochraceous pink. Another large ocellus is prominent between veins 5 and 6. In Pakistan there is only one brood, emerging after the onset of the monsoon, so the above description corresponds to typical wet season forms in other parts of the subcontinent. Dry season forms often lack these ocelli.

Status: Strictly a plains species. On the wing only during the latter part of the monsoon and up to October, it is comparatively uncommon throughout north-eastern Punjab and again in lower Sindh in the better wooded tracts. Collected TJR on 12 July at Khanewal. Puri (1931) considered it common around Lahore.

Habits: Sevastopulo (1942) describes the eggs as pale silvery-green and spherical in shape. They are laid in twos and threes on the underside of a blade of grass. The larva has a square-shaped head, which is black and shiny and covered with hairs. Its head bears two dark purplish horns, which are also hairy, and there is an elongate white spot on the sides, above the jaws. Its body is bright grass green with a dark blue-green dorsal stripe, edged with a line of minute yellow dots and a double sub-dorsal yellow dotted line. The pupa is suspended by the cremaster from a pad of white silk, and is coloured jade green, being rather translucent. Its wing cases are a darker, diffused colour and there is a dark abdominal stripe. The pupa also bears two short anal processes. In their early growth stages the larvae are gregarious. The butterfly is unique amongst the Rhopalocera in its crepuscular habits in fact it can be seen actively flying after dusk has fallen. Only a few Hesperiidae have the same crepuscular habits. If they are flushed during the day they fly jerkily for a few seconds, soon alighting again on some dried leaf where their underside cryptic markings make them difficult to locate.

174. The Common Bush Brown

Mycalesis perseus
Wingspan 45–55mm

Description: This butterfly is the only representative in Pakistan of a very big genus of butterflies known as Bush Browns, and in Pakistan it only appears as a wet season form after the onset of the monsoon. On the upper side it is uniformly dark brown, with a small black ocellus in between veins 2 and 3. Males have a small black brand on the underside of the forewing midway along and on vein 1b. There is no ocellus on the hindwing upper surface, but both wings have two pale and narrow marginal borders. On the underside forewing there are small white-pupilled ocelli in the apical area and between veins 3 and 4, with five to seven small ocelli along the sub-marginal area of the hindwing, bordered outwardly with a paler band. The most conspicuous feature is a dark, fairly straight post-discal line across both wings, broadly bordered on its outer edge with a wider band of bluish-white, and another dark line in the basal area, similarly bordered with white, but along its inner, not outer edge.

Status: Collected by Puri (1931) from the Ravi river near Lahore in August and considered rare by him. There are no other records from either Lahore region or the Himalayan foothills, where it should occur. It has been collected by TJR in Khanewal on 17 October. Wynter-Blyth records it as one of the commonest butterflies in the wetter wooded parts of north-west India, and it is probably spreading into Pakistan with the development of irrigation and tree plantation.

Habits: According to Sevastopulo (1944), the female lays her eggs singly on grass blades. The eggs are yellow-green in colour and unsculptured (smooth on their surface). When full-grown the larvae have a dark brown head bearing two brown forward-pointing processes. The body is green, covered with short white bristles, and with a paler green lateral stripe. There are two processes on the anal segment coloured pink. The pupae are blue-green with a series of sub-dorsal white dots along the abdomen. They are suspended by the cremaster attached to a pad of white silk. Eggs collected by Sevastopulo laid on 28 March hatched on 31 March. Full-grown larvae pupated on 14 April, and the butterflies emerged on 20 April.

Chapter 10
Acraeinae

Diagnostically the Acraeinae sub-family is characterized by the adults having elongate wings which are sparsely scaled. The cell is fully closed in both fore and hindwings, and the males have their tarsal claws toothed or asymmetrical. The antennae are short and stout with a distinct club.

This sub-family is represented by only two species on the whole subcontinent, but they reach their biggest numbers in Africa, where over 200 spp. have been recorded. They are also represented in Neotropical countries. They are very similar to the Danaids in having their wings only thinly covered with scales, and in being protected from most reptile and bird predators by means of a noxious yellow oily secretion which is extruded from the leg joints whenever threatened. They also resemble the Danaids in possessing tough and leathery bodies, difficult to kill. Consequently the adults have a slow and weak flight and a fearless disposition; they can often be caught with the fingers. They are, however, small in size when compared with Danaids, and have slim abdomens with the hindwing rounded and lacking any anal fold. In both sexes the forelegs are degenerate and unfit for walking. After fertilization, females bear a horny pouch at the end of their abdomen like the Parnassinae.

The larvae are spiny, with six longitudinal rows of stiff branched bristles, and their heads are hairy, but lacking any horns or other projections. The larval foodplants are Cucurbitaceae, and Passifloraceae in the Tawny Coster, both families containing poisons.

175. The Tawny Coster

Telchinia violae
Wingspan 50–65mm

Description: The 'Costers' are quite small butterflies, unlike the Danaids. Both sexes are a dull brick red on the upper side, with a row of black spots almost conjoined in the post-discal region, and a broad black elongate spot end cell, with a smaller one mid cell on the forewing. The margin is narrowly bordered with black, the black extending slightly along the veins. The hindwing is broadly bordered with black, bearing tawny spots inside the margin, and there are smaller scattered black spots in the discal region and basally. On the underside the wings look more transparent and shiny, and the overall colour is a paler, yellower-tawny, with the hindwing black margin bearing larger spots which are whitish. The base of the hindwing is also white surrounded by three looped black borders. The thorax is black spotted with white and ochraceous.

Status: Though the Tawny Coster is very widespread throughout India, including the Himalayan foothills, it has so far only been recorded from Sindh, where Bell (1910b) considered it not uncommon in dry hilly areas of that Province. It should be looked out for in the Siwalik zone of the Punjab.

Habits: Bell (1916) describes the eggs as shiny yellow with seventeen longitudinal ridges. They are dome-shaped with flat bottoms. The female lays her eggs in batches of up to fifteen, on the growing shoot of the foodplant. In most of the subcontinent these belong to the family Passifloreae, especially *Passiflora foetida*, and *Modeca palmata*, but in Sindh, he found them on Cucurbitaceae. The larvae are cylindrical in shape, bearing two dorsal rows of branched spines with another shorter sub-spiracular row of branched spines. The head is small and covered with hairs, and the body is a greasy claret-brown colour. When full-grown the larva measures about 21mm in length. In their early stages they feed gregariously, but after the third instar feed separately. Haribal (1992) recorded an instance of cannibalism by a larva eating a pupa. The pupae are formed hanging freely by the tail claspers from the underside of a stem or leaf. They are pinkish at first, becoming whiter, and conspicuously spotted with black spiny spots having orange centres. Both the pupae and the larvae have a disagreeable odour. The butterfly frequently visits flowers, and has a slow fluttering flight, interrupted with gliding on wings outstretched. They usually keep close to the ground and settle often.

Daininae—Milkweed Butterflies

Classification

The Daininae possess many striking features such that they were formerly placed in their own separate family, but recent thinking has ranked them within the super-family Nymphalidae because of many shared characters. The greatest number of species of this sub-family occur in South-East Asia, though Pakistan has only four species, as most are lowland forest-loving insects, the climate here being too dry to favour natural forest. The males, like all the Nymphalidae, have forelegs (prothoracic) without claws at their tip, and in both sexes they are in fact degenerate and useless for walking. Their pupae are attached only by the tail, hanging vertically not upright, and held by a silken band as in the Pieridae and Papilionidae. Danaids are distinctive in all of them being large butterflies with very tough bodies, and protected by nauseous smells and taste. They are difficult to kill by the normal means used by human collectors with all other species, that is, by pinching the thorax, and because of the poisonous substances produced by their foodplants, they are able to absorb and sequester this both in the larval and adult stages, thus becoming highly distasteful to predators. They actually seek out plants from the families Asclepiadaceae, Apocynaceae, and Moraceae, all of which exclude a milky white sap when the leaf tissue is injured which contains the poisonous compounds pyrrolizidine alkaloids, which are cardiac glycosides (Brower and Brower, 1964). Consequently the butterflies tend to be slow in flight and not easily frightened, and the caterpillars are brightly marked with contrasting bands or rings of colour, easily recognizable to potential predators (aposematic colouring). The adult males all have special scent organs, in the form of twin extrusible hair-like brushes at the anal end, as well as small pockets on the under hindwing in some species (the Tigers, *Danais*, and *Tirumala* genera), which contain scent scales. Other species (the Crows, *Euploea* genus) possess scent brands on the forewing. The eggs of Danaids are white and higher than wide, barrel-shaped, with their exterior ribbed, and in some species further decorated with raised reticulations. The larvae when they reach full growth have bright contrasting bands of colour and are hairless, but decorated with long fleshy horns in pairs, usually two pairs thoracic and one anal, and they can wave these about when disturbed (*see*, Plate 4). They are fully developed in five instars and, depending on temperature and humidity, growth can be rapid, with newly-hatched larvae ready to pupate after ten days (Ackery and Vane-Wright, 1984). The pupae are thick and rather rounded at the abdominal end, with the head end square-cut. They are green or brown in colour, with striking decorations of gold or silver spots and streaks. Miriam Rothschild has shown that when their diet is carotene free this produces silver markings, and these metallic colours are produced by interference between cellular layers of varying density (Rothschild, Gardiner, and Mummery, 1978).

176. The Blue Tiger

Tirumala limniace
(Syn: *Danais limniacae*)
Wingspan 90–100mm

Description: Like all members of this family, which are highly distasteful to predators in the adult stage, these butterflies have a slow flight and confiding habits. This is a large butterfly, strikingly marked with pale greenish-blue streaks and spots against a black background. The thorax is black with conspicuous white spots, whilst the abdomen is olive brown. The underside hindwing bears a special rounded flap, coloured yellowish-brown, which protects the scent gland, which is also just visible on the upper side as a patch on vein 1c. The marginal areas of both wings have a series of small blue spots, with another irregular series in the post-marginal area, and in this row of spots, between veins 2 and 3, there are two oval blue spots one exactly above the other, and another spot between veins 3 and 4, all three in line, which serves to separate this species from other members of the Blue Tiger group. On the underside the forewing apical area is olive brown, as well as the post-discal area of the hindwing. The cell of under hindwing is very pale, almost white, with a thin black streak along its middle.

Status: Occurs widely over Sindh, Punjab, and the NWFP, but is comparatively rare in Sindh and only a straggler into Balochistan. It is most common in northern Punjab, particularly the Salt Range, and in all regions is only found on the wing during and after the monsoon season. Its preferred habitat is savanna or areas with scattered bushes, and it is less adapted to very dry open country than is *Danaus chrysipus*. In the Himalayan foothills it occurs up to 1000m (3,500ft).

Habits: The female lays her eggs singly, anywhere near the foodplant, but generally on young leaves. The eggs are pale yellowish-white with longitudinal ribs. The larvae are velvety smooth without hairs, but bearing long fleshy horns or tentacles, a longer pair on segment 3 which is movable, and a slightly shorter, rigid pair on segment 12. The head is pale green with two black rings, and the body is pale bluish-white with dark brown rings between each segment. The ventrum is described as whity-green, and the full-grown larva measures up to 37mm in length. The pupae are thick and blunt in shape, suspended from the tail claspers attached to a pad of silk, often on some twig or sheltered stem away from the foodplant. They are bright green, ornamented with metallic gold spots. The author has often found them from August up to as late as October in the northern Punjab, and their metallic markings sometimes make them easy to see. The larval foodplants are Asclepiadaceae, especially the vine *Dregea volubilis*, common in the Punjab Salt Range, and *Calotropis procera*. The butterfly (imago) is long-lived compared with most other families, with average life expectancy over sixty days (Haribal, 1994). This species also shows strong migratory tendencies. In flight they are, like all the Danaids, slow and fearless, but when alarmed or migrating have a strong swift flight.

177. The Common Tiger

Danaus genutia
(Syn: *Danais plexippus*)
Wingspan 75–95mm

Description: Though averaging smaller in size than *Danais limniace*, this is still a large butterfly, with the same rather slow, flapping flight and distasteful smell and taste to potential predators. This species is a rich orange-brown, with each vein and the wing margins prominently margined black and the apical area solidly black with a series of white spots and sub-apical elongated white spots. The hindwing upper surface bears a double row of small white spots around the margin, and on vein 1c there is a small scent gland visible as a swelling and partially covered by a small pouch, ringed black with chestnut centre. On the under hindwing this scent pouch has a small white centre, and the chestnut area between the black veins is paler in colour. The thorax is black with conspicuous white spots, and the abdomen yellowish-brown.

Status: This species is almost world-wide in distribution, and occurs in the Far East as well as Australia and North America, where it is known as the Monarch butterfly. In Pakistan it is widespread in Sindh, Punjab, and the NWFP, but in Sindh it only occurs in years when the monsoon rains are heavy. Odd specimens straggle into the main valleys of Balochistan and the northern foothills such as lower Chitral and Swat after the monsoon. In the plains it is only on the wing towards the end of the monsoon up to October. It is never as common as the Plain Tiger, *Danais chrysippus*.

Habits: The female lays her eggs singly on the underside of a leaf, and the larvae feed on the underside of leaves. The adult larvae are velvety black, marked with bluish-white and yellow spots, and along the sides a broad yellow band partly sub-spiracular and partly including the spiracles. The body bears three pairs of fleshy tentacles, the first, on segment 3, being the longest and movable. The second and third pairs are on segments 6 and 12 and are rigid. The tentacles are red basally and tipped black. The full-grown larva measures about 37mm in length. The pupa is formed under a leaf or twig, usually some distance from the foodplant. It is bright green, decorated with metallic golden spots and smaller black dots. It is stout and barrel-shaped. The larval foodplants are *Asclepias curasavica*, *Ceropegia lawii*, *Ceropegia bulbosa*, *Ceropegia macrantha*, *Cyanthum acutum*, *Marsdenia roylei*, and probably many other species of Asclepiadaceae. Compared with *Danais chrysippus* this species is more shade-loving and more likely to be encountered flying lazily close to bushes and undergrowth beneath trees, and it is less likely to visit flowers. Like its congener, the Common Tiger (*D. chrysippus*), the adult butterfly contains cardiac glycosides, and studies of the American Monarch show that it does so more consistently than *D. Chrysippus* (Rothschild *et al.* 1975).

178. The Plain Tiger

Danaus chrysippus
(Syn: *Danais chrysippus*)
Wingspan 70–80mm

Description: Smaller than the Common Tiger (*D. plexippus*), this familiar butterfly is distinguished from *D. plexippus* by the absence of thick black borders to the veins. Its upper surface is pale orange-brown, with black marginal borders and black forewing tips containing a sub-apical white band of elongated spots. There are three or four small black discal spots on the hindwing upper surface. The under wings are a paler ochraceous colour, with the forewing costal and cell areas darker red-brown. The under hindwing margin is black with a series of lunar-shaped white spots; the lowest of the discal black spots is white-centred, and in the male contains the not very conspicuous scent gland. The thorax is black with white spots. This widespread species exhibits several different colour morphs within the Indian subcontinent, and this polymorphism has extended to the population inhabiting east Africa (Smith *et al.*, 1997). One form collected from Campbellpur has a broader sub-apical band with more extended white spots in the forewing and an overall darker brown, less orange colouration. This has been named *D. chryssipus dorripus*.

Status: Without doubt this is the most widespread and commonest butterfly in Pakistan, found throughout the plains and lower foothill areas, being much better adapted to dry areas than any of the other Daininae species. Also, unlike the other Daininae species in Pakistan, it can be encountered in every month of the year. Stragglers turn up in lower Swat and Chitral also. This comparatively long-lived and confiding butterfly is constantly on the wing during daylight hours in search of nectar-bearing flowers.

Habits: The female lays her eggs singly on the underside of young leaves of the foodplant. The eggs are flask-shaped, cream-coloured and with longitudinal ribs. The foodplants are Asclepiadaceae of the Milkweed family, and the usual host plant is *Calotropis procera* in Pakistan, though elsewhere in India it has been recorded on *Asclepias curassavica*, and in the Punjab Salt Range, and foothills of Hazara and Punjab on *Cryptolepis buchananii*. Like the Common Tiger, the larvae, is velvety smooth, without hairs, and bears three pairs of black fleshy horns, two thoracic on segments 3 and 6, and one anal on segment 12. The first pair of horns are the longest, the other two much shorter. Its body is dark chocolate-brown or black, with narrow white bands between the segments and a sub-dorsal row of yellow spots, below which is another row through the spiracles of almost contiguous yellow spots, their upper edge being jagged. When fully grown the larvae measures about 38mm in length. It feeds on the edges of leaves as it gets larger. When ready to pupate it wanders some distance from the foodplant, and the pupa is coloured green or occasionally pale pinkish white, with a row of beads around segment 7, tipped with gold, edged black. The pupa, like *D. plexippus*, is rather squat and cylindrical in shape, its posterior end being hemispherical. It is suspended by a narrow neck from the tail, hanging head downwards. Courtship entails pursuit by the male everting his abdominal hair pencil, which secretes pheromones derived from alkaloids ingested by the larvae from their foodplant. If successful, the male will eventually force the female to alight, often by clasping her with his legs, and then continues to flutter over her, 'hair penciling' all the while. If receptive she will close her wings, enabling the male to alight laterally and attempt copulation.

(178b. The Plain Tiger

Danais dorripus, Klug, 1845
[Syn: *Limnas klugii*, Butler, 1886, p. 356]

Butler [1886] in describing a collection of Butterflies from Campbellpur and the Murree Hills, lists a single Dainaid from Campbellpur as *Limnas klugii*. This is not the Blue King Crow, *E. klugii*, but simply a variety of the Plain Tiger in which the forewing apex lacks the white-spotted black area.)

179. The Common Crow

Euploea core
Wingspan 85–95mm

Description: A large, rather long-winged butterfly, this species is a very dark velvety brown on its upper side, with a marginal row of small white spots curving round the apex, and a second sub-marginal band of slightly larger white spots on the forewing. The hindwing bears the same pattern, but these white spots are more elongated. The underside is a slightly paler, silky brown with traces of small white discal spots as well as the two marginal rows of spots. The male bears a small streak-shaped brand on the forewing between veins 1a and 1b. The thorax is black with conspicuous white spots.

Status. All the *Euploea* genus are forest-loving insects. Consequently this species has not been recorded from Sindh, but it is not uncommon in southern and eastern parts of Punjab, flying from about late July to early October, and only occurring in shady, tree-lined places. It has not been recorded from the Himalayan foothill areas. It has a seemingly slow and lazy flight, and shade-loving habits during midday hours.

Habits: The female lays her eggs singly upon the leaves of a variety of foodplants. These include many *Ficus* tree spp., such as *Ficus bengalensis, Ficus religiosa,* and *Ficus glomerata,* of the family Moraceae, as well as *Nerium odorum, Nerium oleander, Ichnocarpus frutescens, Holarrhena antidysenterica* of the family Apocynaceae, and *Hemidesmus indicus* of the family Asclepiadaceae. The eggs are yellow in colour, with longitudinal ribs, and flask-shaped with gently curved sides. The larvae are striped with black and red around each segment, with four pairs of fleshy horns or tentacles on segments 3, 4, 6, and 12, with the pair on 3 the longest. The pupae are barrel-shaped, like all the Danaines, with the rear half-hemispherical in shape. The pupa is shiny, smooth, and gold-coloured with some small black spots. Like all the Danaines, the Common Crow has a rather slow, lazy flight, and appears fearless. It generally keeps to shady places, flying close to the ground.

Libytheinae—The Beaks or Snout Butterflies

Classification

This is a very small family in terms of the number of species known which totals only ten in the whole world. However, they are widely distributed, with one species occurring over most of Europe, another covering most of North America, and the rest throughout the tropical regions of Africa and Asia. They are characterized by the greatly developed palpi which extend in a snout well beyond the eyes, hence their common name, 'beaks'. Males have their prothoracic legs atrophied, but these are only slightly under-developed in the females. All the adults are small with the forewing truncate and sharply angled below the apex. They are brown in colour above, with orange-tawny spots and bands. Underneath they are cryptically coloured. The larvae are green in colour, with paler longitudinal stripes, and cylindrical in shape, and smooth or with minute bristles (*see*, Plate 5). They do not have any projection on the head or anal end, though their heads are rather small, like those of Pieridae larvae. Their eggs are like Pierid ova also, being bottle-shaped and ribbed. Their pupae are like Nymphalids, being suspended, without any body band, parallel to the surface of the foodplant or the bark of a tree. The foodplants of the larvae are trees of the family Ulmaceae, in particular *Celtis* spp.

180. The Common Beak

Libythea lepita
Wingspan 45–50mm

Description: The Beaks are easily identified by their enormously pronounced palpi, sticking out like a beak from the head. They have the forewing apex square cut and the termen concave below, giving the wing tip an arcuate outline. The hindwing is toothed or slightly wavy. *Libythea lepita* is dark chocolate-brown above with an orange-red cell streak on the forewing, notched on the costal side, and with two whitish spots at the apex. On the upper hindwing there is a fairly narrow reddish-orange band post-discally. The underside hindwing is paler greyish-brown, unmarked, and the apical spots on the under forewing are white. The sexes are alike.

Status. Found mainly in the lower forested areas of the Himalayas, being rare in Chitral and Swat but common in the Murree Hills. It has also been collected from Campbellpur where it must be rare. It is on the wing from April up to September, flying between 900–2750m (3,000–9,000ft).

Habits: The eggs are bottle-shaped with longitudinal ribs, like Pieridae eggs. The larvae are smooth in outline and cylindrical, lacking any head or tail ornamentation or pronounced constrictions between the somites. They are pale glaucous green in colour, with narrow sub-dorsal white stripes and a broader sub-spiracular white band, as well as two rows of black spots along their sides. Their bodies are covered with short bristles. Their heads are comparatively small, shiny, and brown in colour. The larval foodplant is the tree *Celtis australis* (Family Ulmaceae), which grows between 1,500–3,000m (5,000–10,000ft) from Swat to the Murree Hills, but surprisingly Mohyuddin (1987) found larvae (wrongly identified as the Club Beak, *L. myrrha*) feeding on *Grewia* sp. and *Gossypium herbaceum* in Rawalpindi. The female lays her eggs at the tips of young leaves, and the larvae tend to eat only the soft tissues, feeding from the underside of the leaves, and leaving the midrib and leaf veins untouched. When disturbed they let themselves down by long silken threads. The larvae when not feeding rest in a characteristic posture with the front part of the body raised and the true legs bunched together off the leaf or stem surface, and with the head bent around to one side. The pupae are green in colour with yellow bumps along the dorsal surface, and rather short in shape, without the head case being produced into a 'beak'. They are held only by the tail cremasters, resting along the twig or tree surface. The butterflies are strong flyers but have a variable flight pattern, sometimes sailing, sometimes zigzagging rapidly, and are generally quite shy and wary. They do visit flowers and often settle on the tip of a leaf with wings closed, when they are difficult to see because of their cryptic colour and wing shape, resembling a dried leaf.

181. The European Beak or Nettle Tree Butterfly

Libythia celtis
Wingspan 35–50mm

Description: *Libythia celtis* is distinguished from *L. lepida* by having much broader orange streaks and spots on the upper surface and in particular by the presence of an orange-yellow spot between veins 1b and 2. The Common Beak lacks this spot. The orange post-discal band on the hindwing is also much broader. The underside is the same as in *L. lepita*, the Common Beak, but the lower edge of the hindwing cell vein is prominently white at its outer edge.

Status: So far only collected from Chitral, where uncommon and confined to the lower forested valleys. It is on the wing from April to August.

Habits: The larval foodplant is *Celtis australis*, known in Europe as the Nettle tree (Family Ulmaceae), and their life history is much the same as that of *Libythia lepita*, the Common Beak. The female lays her eggs singly on leaf tips and these are like Pieridae eggs, being rather tall and flask-shaped with prominent longitudinal ribs. The green larvae have yellowish-white longitudinal body stripes, and their bodies are covered with minute bristles. The adult butterflies frequent the vicinity of streams and, like the Common Beak, are swift on the wing, with irregular darting flight alternating with sailing. They keep to sunny glades in the vicinity of forested areas.

Chapter 11
Lycaenidae—Blues, Coppers, Hairstreaks, and Silverlines

Classification

This is a family comprising a huge number of species of worldwide distribution—an estimated 6,000 species worldwide, with greatest diversity in the tropics (Ackery, 1984). All of them are rather small in size, brilliantly coloured, and frequently showing marked sexual dimorphism. The adults are swift in flight but this is rarely sustained, as they settle after a short distance. They are in some respects rather homogeneous, in that the forewing has a reduced number of veins, usually eleven, in some genera only ten, with the forewing cell narrow and closed at the end, but the cell on the hindwing usually open-ended. The hindwing lacks a pre-costal vein (*see*, Fig. 1), and in both sexes the forelegs are functional, not degenerate, though they are always smaller in the male. Though the sexes differ in the upperwing colour pattern, the underwing pattern is consistent within genera and between the sexes and is the most important feature in aiding identification. The larvae are remarkably similar, being very small and woodlouse-shaped, that is, higher in the middle, curving downwards at the head and anal ends when viewed from the side, oval or elliptical in outline when viewed from above, and with the ventrum flat and the tiny legs more or less concealed under the body. The head is small and retractable under the first segment (prothorax), and the body is usually coloured green, but their colour does vary even within the same species according to the foodstuff, being pinkish and even purple-brown. Between segments 11 and 12 there is a mouth-like opening known as the honey gland, well developed in some species (but lacking in some groups), and this is associated with a fascinating symbiotic relationship between the larvae and various species of ants (Formicoidea). The ants tend the larvae, warding off potential predators, even in some cases carrying them down into their nest chambers when ready to pupate, whilst the larvae in return exude honey-dew secretions from this gland, which the ants greedily lap up. The genera, which are associated with ants in the larval stage, include *Azanus, Chilades, Catachrysops, Tarucus, Castalius, Lampides, Arhopola, Apharitis,* and *Spindasis*. The genera *Lycaenopsis, Celastrina, Everes, Argiades, Amblypodia,* and all the Hairstreaks (Theclidae) are never attended by ants. The eggs of the Lycaenidae are rather disc-shaped but with rounded edges and their surface beautifully sculpted with cellular indentations. The pupae are stout and rounded in outline; the thorax being humped (*see*, Fig. 2, Pl. 5). They are often dull-coloured, many are bone-coloured, and some are attached to the foodplant by a thin silken band and by their tail hooks. Others simply lie free on the ground amongst leaves, and some species pupate underground. The classification of this huge family has presented many difficulties amongst entomologists, and in this book I am following Cantlie (1962), but more recent taxonomic studies, particularly by the Japanese, have introduced many changes. Broadly speaking, the Lycaenidae can easily be separated by the non-specialist into two groups or sub-families: the Theclinae, with stout bodies and strong flying, including the Hairstreaks and Flashes, and the 'Weak Blues' or Lycaeninae, which includes such tribes as the Polyommatini (Blues) and Lycaenini (Coppers). There is so much variation in the early life history of the Lycaenidae that the reader is referred to the species accounts. Some feed entirely inside flowers, usually of the families Mimosaceae and Papilionaceae. The eggs are carefully laid inside the flower buds of the foodplant. Others, cared for by ants, are often carelessly laid anywhere, the newly-hatched larvae and even in some cases the eggs themselves, being taken by the ants to a variety of different foodplants, usually trees of the leguminous family. Others feed and pupate inside fruits and the eggs are laid on the newly-developing fruit buds or maturing flowers.

Pakistan has at least eighty-six Lycaenidae definitely recorded, and this book describes eighty-nine, which are most likely to occur. No doubt more species will be discovered when new studies can be carried out. The region, as can be expected, is particularly rich in the Palaearctic high-altitude adapted species, the early life histories of which have been little studied.

PIERROTS

182. Common Pierrot

Castalius rosimon
Wingspan 24–32mm

Description: Unlike the other Pierrots from our region, this species is mainly white on the upper wings, with broad black margins on both upper and hindwings, and only the base of both wings up to the cell an iridescent metallic blue. There are several square shaped black spots on both fore and hindwings, fewer on the hindwing. The female has the blue areas suffused with darker scales. The under forewing has a long black streak in the cell, with two widely-separated black spots, end-cell and between the base of vein 1a and 2. Both sexes are tailed.

Status: This is widespread in India but in our region only found in the foothill areas adjacent to the Himalayas from around Islamabad through the Margallas and Lehtrar valley, and most likely to be seen on the wing in December and January.

Habits: The egg is described (Bell, 1919) as hemispherical in shape, white in colour, with a reticulated surface, and is laid along young leaves or in the axis between a thorn and the stem. The woodlouse-shaped larvae are bright green with a roughish texture, and bear two yellow dorsal lines with scattered yellow spots. The larval foodplants are species of the family Rhamnaceae, most commonly the Ber, *Zizyphus jujuba*, and *Zizyphus mauritiana*, a widespread shrub in the foothills. They are regularly attended by ants. The pupa, according to Haribal (1992) is also green, with yellow bands on the abdomen, and a few black spots dorsally. It is formed along a leaf midrib, attached by the tail with a silken pad and also a loose silken band across the body.

183. Black Spotted Pierrot or Little Tiger Blue of Europe

Tarucus balkanica
(Syn: *Tarucus nigra*)
Wingspan 21–24mm

Description: This is a small Lycaenid, and is a rather dark, dull blue above, with two black discal spots on the upper forewing in addition to the end-cell black spot. The underside is dull white with a prominent black streak in the cell like *Castalius* spp., but with more small black spots on the forewing, some coalesced into continuous bands along the discal and sub-marginal areas. The hindwing underside has silver-pupilled spots in the tornal region. The female is dull brown above with two or three small dark tornal spots. Both sexes are tailed on the hindwing tornus.

Status: Collected from Sindh hilly tracts (Menesse, 1950), Balochistan in the Bolan Pass, Mach, and near Loralai, and in the NWFP from Peshawar. Also collected June 1993 by Rafiq Ahmad Rajput from Takht-i-Suleiman, and occurring widely across the dry foothills west of the Indus, always near *Zizyphus* bushes or trees.

Habits: No details seem to have been recorded, but the larvae are attended by tree ants and feed on the flowers of trees, especially the wild Jujube, *Zizyphus nummularia*.

184. Rounded Pierrot

Tarucus nara
(Syn: *Tarucus alteratus* or *Tarucus extricatus*)
Wingspan 23–24mm

Description: The male is dull violet-blue above with a black cell spot but no other discal spots on the forewing. Like the other *Tarucus* spp. it is tailed on the hindwing, and this species shows the shadows of the underside black spots on its upper surface. The underside is white with a prominent black streak in the forewing cell, and the black marks are elongated into streaks rather than rounded spots. The female is dull brown above.

Status: Occurs in much the same habitat as *T. balkanica*, in Sindh (collected Karachi) and the NWFP, but less common, and not found in Balochistan, and has been collected from Campbellpur (Attock district), where it is extremely common during May and June, but can be encountered on the wing most months of the year.

185. Pointed Pierrot or Common Tiger Blue of Europe

Tarucus theophrastus
(Syn: *Tarucus indica*)
Wingspan 22–30mm

Description: The male is a dark violet-blue with rather transparent wings showing the shadows of the underside spots. There is a rather inconspicuous black spot end cell. The female is quite pale whitish-brown on the upper side, the dark spots and streaks from the underside showing through prominently. The underside in both sexes is white with black streaks rather than spots.

Status: Collected from near Karachi, from Balochistan, and the NWFP in dry hilly areas, and Puri (1931) recorded it as very common around Lahore. Like its congeners, can be encountered on the wing most months of the year.

Habits: Like all the Pierrots, the larvae are attended by ants and are therefore not so host specific in their choice of foodplant. Bell (1919) reared them successfully from eggs on both Oleaceae (*Jasminum* spp.) and Rhamnaceae (*Zizyphus* spp.). Their eggs are turban-shaped and pale green in colour, covered with little tubercles in vertical rows and ridges arranged horizontally, both white in colour. The larva is pale green, covered all over with small white tubercles giving a shagreened appearance. There is a broad greenish-yellow dorsal stripe, with a red stripe down its middle. The tiny head is ochre coloured and normally hidden beneath the thoracic segment. The pupa is green, thickly speckled with black. Marshall and de Niceville (1890) quote an account by Mrs Wylly of the attendant black 'garden ants' driving the full-grown caterpillar down the stem of *Zizyphus* bushes and into a nest at the foot of the trunk, made by the ants themselves, wherein many larvae remain to pupate, covered lightly with earth, until they emerge, unmolested.

186. Balochistan or Mediterranean Pierrot

Tarucus rosacea
(Syn: *Tarucus mediterraneae*)
Wingspan 23–26mm

Description: The male is dull violet-blue above with rather transparent wings showing clearly the markings underneath. There is a black spot end-cell on the forewing, and a black tornal spot on the hindwing. The female shows more white on the upper side discal area of both fore and hindwings. The underside is white, with the black spots in the sub-marginal area of both wings forming a continuous line.

Status: Rather uncommon in western parts of Balochistan and not recorded outside of that Province. A huge series collected from Dalbandin in the Chagai. Collected on the slopes of Takht-i-Suleiman in July 1993 by Rafiq Ahmad Rajput.

Habits: The larvae are attended by ants, and their foodplants are members of the family Rhamnaceae (Buckthorn Family).

187. Spotted Pierrot

Tarucus callinara
Wingspan 24–26mm

Description: Rather a variable species according to geographic location. As its name implies, the underside bears black spots rather than streaks. The male above is dull violet-blue, often darker around the wing margins, and with a black spot end cell. The female is brown, with the base of the wings blue-scaled, and often-whitish areas discally.

Status: Has been collected from the NWFP, Balochistan but only from the Bolan Pass, and ascending into the Himalayan foothills up to 1,800m (6,000ft) in the Murree Hills. Not recorded from Sindh.

Habits: Much the same as for the other Pierrots. The larval foodplants, besides including *Zizyphus hysudrica* and *Z. mauritiana*, also include the Leguminous tree *Dalbergia sisso* ('Shisham'), the presence of ants on the host tree being a major factor determining the female's choice of location for egg-laying.

188. Himalayan Pierrot

Tarucus venosus
Wingspan 24–26mm

Description: The male is a rich violet-blue above

with a broad blackish border to both wings, and a prominent black spot end-cell on the forewing, as well as a black spot end-cell on the hindwing. The female is dull brown, not very transparent, but showing traces of black tornal spots on the hindwing. Underneath both sexes bear rather small black spots, and the sub-marginal spots are silver-scaled.

Status: Fairly common up to 1,800m (6,000ft), along the foothills around Abbotabad (Thandiani) and Murree, during July and August, and around Campbellpur (Attock) may be encountered in all months of the year.

189. Zebra Blue

Syntarucus plinius
Wingspan 22–30mm

Description: The male is slightly transparent above and rather a pale violet-blue, and the female is brown above with the base of the wings metallic blue and showing dark spots in the discal area. Underneath, both sexes are unlike any of the *Tarucus* spp., with paler brownish bands and streaks, broken into spots on the under hindwing. There are two metallic black-centred tornal spots on the under hindwing.

Status: In earlier records, only definitely recorded from Sindh, where it may be encountered during and immediately after the monsoon. Menesse (1950) considered it uncommon. Collected TJR from Rahimyar Khan district in southern Punjab.

BABUL BLUES

190. African Babul Blue

Syntarucus jesous
(Syn: *Azanus jesous*)
Wingspan 21–26mm

Description: The Babul Blues are not tailed, and *Syntarucus jesous* has the male violet-blue above with narrow black borders, and no black spot end-cell on the forewing, but two tornal black spots on the upper hindwing. The female is brown with blue scaling at the base of the wings. The underside of both sexes shows pale brown bands white-edged, with two black costal spots on the hindwing, a short black basal streak, and two black tornal spots.

Status: Fairly common and widespread in Punjab and northern Sindh. Not recorded from Lahore region. Collected TJR at Khanewal in November. The adults regularly visit garden flowers as well as roadside waste places, and, being well adapted to dry places, are very common in the plains.

Habits: The foodplants of the larvae are various species of tree Acacia, such as the 'Kikar', *Acacia nilotica*, and 'Jund', *Acacia senegal*. The adults generally lay their eggs high up near the flowering twigs, and the little larvae feed exclusively on the pollen anthers and stamens, hiding within the flowers. They are the normal 'woodlouse' shape, with the tiny head retractable under the first thoracic segment. Coloured green, they have a dorsal reddish band. The larvae are attended by ants, in Pakistan usually by *Camponotus* spp.

191. The Bright Babul Blue

Azanus ubaldus
Wingspan 20–25mm

Description: The male is bright violet-blue on the upper side with black borders, and no spot end cell. The female is a paler milky blue above with broad brown borders to both wings. Their underside lacks any black streak on the hindwing base as in *S. jesous*, but has the same pale brown bands and two black costal spots as well as two tornal spots.

Status: Occurs around Lahore, where uncommon, but ascends the outer Himalayas to 2,400m (8,000ft) and occurs commonly in the Murree Hills (collected Nathia Gali), and around Campbellpur where it is common. Has been collected from Dera Ismail Khan and Peshawar, as well as from Sindh, where it is much more common. As there are two broods, it is on the wing from April to December.

Habits: The life history of this species is similar to that of *Azanus jesous*, and in fact larvae of the two species can be found together. Like the former species, the foodplants are Acacias, mainly the 'Kikar', *Acacia nilotica*, and the larvae feed exclusively upon the yellow globose flowers. They are also attended by ants, which are rewarded by the secretion of droplets of a sweet liquid extruded from the pore on the top of segment 11.

192. Dull Babul Blue

Azanus uranus
Wingspan 20–25mm

Description: The male is a much duller mauve-blue on the upper side with traces of a tornal spot on the upper hindwing, and very narrow but clearly marked black borders. The female has broad black wing borders and quite extensive blue scaling on the upper side of both wings. The underside is similarly marked as in *S. ubaldus*, lacking the short black streak in the base of the hindwing.

Status: Much the same distribution as the Bright Babul Blue but far less common than the latter in Sindh and also in the NWFP. Has been collected near Attock Bridge, Hasan Abdal, Rawalpindi, and Peshawar. It also occurs in the Murree Hills but less commonly than *A. ubaldus*. It can be seen from July to about December.

Habits: Like the former species, the larval host plants are *Acacia* spp., especially *Acacia modesta* in the Murree foothills, *Acacia farnesiana*, and *A. nilotica* elsewhere, and the larvae are attended by ants.

TAILED CUPIDS

193. Shandur Cupid

Everes buddhista
(Syn: *Everes shandura* in Evans, 1932, also *Cupido buddhista* in D'Abrera, 1993)
Wingspan 21–25mm

Description: This is a difficult group taxonomically as they are rather variable. The Shandur Cupid lacks any tornal tail, and indeed the under hindwing shows no obvious orange tornal spots, both features otherwise typical of the Tailed Cupids. The male is a violet-blue above with wide black borders to the wings, and a small black end-cell bar on the forewing. The female is brown, without any tornal spots on the hindwing. The underside is pale greyish-buff with no black spot at the base of vein 1c, and all the spots are rather small and inconspicuous. The base of the hindwing is faintly striated with darker blue-green bars.

Status: Only recorded from the Shandur Plateau and Ispodagh in Chitral. Considered rare. On the wing June to August. Common in Western Tien Shan.

194. Chapman's Cupid

Everes argiades
(Syn: *Everes argiades diporides*, in Cantlie, 1962)
Wingspan 22–26mm

Description: The male is violet-blue above with more definite dark black borders to both upper and hindwings than *E. buddhista*, and with a tornal tail. The female is brown with two conspicuous orange-ringed black tornal spots. The underside is pale grey and bears small black spots more clearly marked than in *E. buddhista*, particularly the row of discal spots. The hindwing termen has several orange-bordered spots which are crowned with silver scales.

Status: Because of confusion within this genus in the past, the exact status of Chapman's Cupid in Pakistan is not clear, but it has been collected from Kashmir, and Mussoorie and Kumaon in India. Chapman's Cupid, *Everes diporides*, in Wynter-Blyth (1957) is really *E. hugelii*, and its distribution does not include Waziristan as listed by him.

Habits: Like the next species, the small green larvae are marked with a darker mid dorsal line, and their foodplants are a variety of creeping leguminous plants found in the hills, including *Lotus* spp. and *Trefoil* spp. According to Wynter-Blyth (1957) the adult butterflies are fond of flying round rocky hilltops and are not as closely associated with damp grassland as *E. argiades*. The larvae of *Everes* spp. are not attended by ants.

195. Tailed Cupid or Short-tailed Blue in Europe

Everes hugelii
(Syn: *E. argiades indica* in Evans, 1932), and *Catochrysops dipora* in Butler, 1886
Wingspan 20–28mm

Description: Male above shining violet-blue, with narrower black borders than in the two previous species, but usually showing two black tornal spots, more conspicuous than the rest of the border, on the hindwing. The female is dark brown with two conspicuous orange-bordered black tornal spots, and both sexes are tailed. The underside is pale grey-brown with relatively small, inconspicuous spots, and on the under hindwing the orange marginal spots are usually not dark-bordered above as in *E. argiades*.

Status: On the wing from June to October. Quite widespread in the Himalayas from Chitral eastwards through Gilgit, and common in the Murree Hills. Collected near Ziarat and Utzen nullahs in Chitral, common in Dunga Gali (where collected TJR), and Nathia Gali in the Murree Hills, and in Gilgit main valley, occurring from 1,500–2,400m (5,000–8,000ft). *Everes dipora*, listed by Wynter-Blyth (1957) as the Dusky Blue Cupid, is really *Everes hugelii*, as examination of the male genitalia have confirmed (Cantlie, 1962).

Habits: The larvae are of the usual woodlouse shape, green in colour with a darker mid dorsal line. They feed on small leguminous plants typical of grassy glades in the hills: Trefoil, *Lotus corniculatus*, Clovers, *Trifolium* spp. and Medicks, *Medicago lupulina*, and *Onobrychis* spp. The adult butterflies are fond of settling on damp earth, and are most likely to be encountered near streams.

HEDGE BLUES

196. Dusky Hedge Blue

Arletta vardhana
(Syn: *Lycaenopsis vardhana*)
Wingspan 38–44mm

Description: The Hedge Blues are larger insects than the Cupids, and none of them bear tornal tails. *A. vardhana* is particularly large, and the males are a distinctive pale silvery grey-blue with the forewing apex black, and a small black end-cell bar. The female is similar in colouration, but her discal area is white and the hindwing black border is broader. The under hindwing lacks any marginal spots, but the under forewing has conspicuous black spots in a sub-marginal band. Unlike the rest of this genus, there are no dark lunules on the under hindwing.

Status: Not uncommon in the Murree Hills and Neelum valley, Azad Kashmir. On the wing from May to early October at lower elevations (specimens collected October 11th from Kala Pani and from Murree through August to September), and favouring open grassy banks or forest glades. Collected TJR Mokhshpuri, Dunga Gali, at 2,830m (9,300ft).

Habits: The larval foodplants are species of Clover, *Trifolium repens*, and Trefoils, such as *Crotalaria medicaginea*. The larvae are only occasionally sought after by ants.

197. Common Hedge Blue

Celastrina puspa gisca
Wingspan 28–35mm

Description: This group of medium-sized blues are not tailed, and have no orange or metallic tornal spots on the hindwing, and usually the males are quite bright blue above with broad black borders, and in some species, white discal patches. *C. puspa* males are iridescent violet-blue with deep black margins and a white discal patch. The female is a paler, silvery-blue with broader black borders and a series of small white lunules on the hindwing sub margin. The underside is pale grey with quite coarse markings and marginal lunules on the under hindwing. The under forewing has the line of discal spots irregular, that between veins 5 and 6 being angled outwards from the others.

Status: Occurs rarely in the Murree Hills, flying between July and October, and found mainly in the outer foothill areas where there are open grassy slopes.

Habits: The larva has the typical 'woodlouse' humped shape of all the family, and is usually green but can be plum pink, with its body covered by rows of minute star-tipped white hairs. The larvae pupate on the ground or on a grass blade secured by a silken thread, and are pinkish in colour with darker blotching. They are only perfunctorily sought after by ants, according to Bell (1919). The foodplants of the larvae are various species of small Papilionaceae, such as Trefoils, Medicks, and Clovers.

198. Silvery Hedge Blue

Celastrina ladonides
Wingspan 38–45mm

Description: Larger than *C. puspa*, the Silvery Hedge Blue male has very narrow black borders and is iridescent dark blue with a silvery sheen, while the female has very broad black borders showing only restricted blue wing patches extending to the discal area. The underside is pale grey with very small black spots, and the under hindwing lacks marginal lunules, showing only very faint marginal spots.

Status: Occurs in the Murree Hills and, like the Common Hedge Blue, favours open grassy areas and bungalow gardens rather than true forest. It is on the wing from March to October and found up to the summit ridges of the Murree Hills.

199. Hill Hedge Blue or Holly Blue of Europe

Celastrina argiolus
Wingspan 25–32mm

Description: The male is a slightly darker silvery-blue than *C. ladonides*, with narrow black borders and no end-cell spot. The female is blue basally, with much broader black wing margins, widening at the apex, and with traces of marginal black spots on the hindwing tornal area, with paler blue lunules above. The underside is pale milky blue with rather obscure dark spots, and a sub-marginal ring of lunules on the hindwing.

Status: More widespread than *C. ladonides*, occurring from Chitral, Swat, and the Murree Hills. Collected by TJR in Dunga Gali, and particularly common around Khaira Gali, and preferring higher altitudes than the two previous species.

Habits: The western European population has been well studied, where there is only one hatching of adults each summer, the second brood spending the winter as pupae, and for this purpose choose as their foodplants non-deciduous trees or vines. In the Himalayas outside of Pakistan they have been recorded feeding on Cornaceae and Rosaceae, whilst in Europe the Ivy, *Hedera* spp., is favoured. Both *Cornus macrophylla* and *Hedera nepalensis* are common in mixed deciduous forest where this butterfly occurs in Pakistan. However, P. Mackinnon in Marshall and de Niceville (1890) collected larvae of this species at Mussoorie (India) on *Princepia utilis* (family Rosaceae), which is widespread in the Kaghan Valley and the Murree foothills. The eggs are bluish-white and button-shaped, with a slight depression in the top. They are laid singly, near flower buds inside which the newly-hatched larvae burrow to feed. The larvae are green in colour with a white lateral stripe extending from segments 2 to 10, each segment being deeply cut. The pupae are attached to the underside of leaves of the foodplant and are greenish-yellow in colour, varying to pale brown, and covered with short, colourless, bristly hairs.

200. Chequered Blue

Philotes vicrama
(Syn: *Polyommatus vicrama* & *Pseudophilotes* in Russia)
Wingspan 20–26mm

Description: In this species the white fringe of cilia is divided by black extending from the end of the veins, producing the chequered effect, and there are no tornal tails. The female is always brown, sometimes with some iridescent blue scaling basally, while the male is iridescent pale blue, with narrow, well-defined black borders and small black bars end cell. The underside is slightly darker grey than in *Everes* species, with numerous black spots, both discally, post-discally, and sub-marginally, on the hindwing. There are orange-crowned spots along the margin, and most of these underwing black spots are surrounded by white.

Status: Normally a very high-elevation species, and very widespread in the dryer northern Himalayan regions. Collected between 1,800 and 4,200m (6,000 to 14,000ft), from Chitral, through Gilgit around Punial and Ratu, extending into Baltistan at Dras. It also occurs rarely in the juniper forested zone around Ziarat and Uruk valley in Balochistan, and is on the wing between May and July.

201. Small Jewel Blue

Plebejus christophi
(Syn: *Lycaeides christophi*, or *Polyommatus christophi*)
Wingspan 28–35mm

Description: This species has given rise to some confusion amongst taxonomists in the past and has been placed amongst the *Lycaeides* and *Polyommatus* groups. *Plebejus christophi* is characterized by having very faint orange crowns to the marginal row of dark spots along both under fore and under hindwings, and with the hindwing tornal spots metallic-centred. The male varies from bright to dull violet-blue with a linear black border and no end-cell bar. The female is dark brown with wing bases metallic blue and up to three orange-crowned tornal black spots on the hindwing. The underside has quite large black spots in a sub-marginal band, but those on the under hindwing are small and faint, and the marginal spots and lunules, as well as their orange crowns, often rather obscure.

Status: Widespread in dry mountain tracts from Balochistan to the northern inner Himalayas. Collected from Quetta, Khojak Pass, and Saranan in Balochistan, and from Mastuj, Madaglash, and Harchin in Chitral. It has also been collected from Skardu, Baltistan, and from Hunza sub-sp. *samudra*. It flies from May to September and in Chitral at higher elevations from 2,400–3,000m (8,000–10,000ft).

Habits: The tiny eggs of this genus are white, flattened with rounded sides, and their surface covered with spiky reticulations. The foodplants are Papilionaceae, most probably *Astragalus* spp. in Balochistan. The larvae have the typical humped appearance, with the head hidden under the first thoracic segment. They are green with white sub-dorsal and sub-spiracular lines.

202. Balochi Jewel Blue or Zephyr Blue

Plebejus sephyrus
(Syn: *Polyommatus pylaon* or *Plebejus pylaon*)
Wingspan 28–34mm

Description: This small butterfly has the underside of both wings prominently marked with orange spots along the margin. The male is rather dark violet-blue above with black linear margins and usually traces of orange tornal spots on the forewing. The underside spots show through faintly, including a small end-cell bar. The female is dark brown with blue showing mainly on the hindwing base, and prominent orange-crowned tornal spots hindwing. The underside is grey, slightly browner in the female, and in both sexes the discal band of black spots is white-ringed, close to the margin, and even-sized and regular. There are usually no traces of metallic centres to the tornal spots, as in *P. christophi*.

Status: Occurs on the border regions of Balochistan around the Khojak Pass, Gwal forest, and Murgha-Mehtarzai Pass, mainly flying between 2,100 and 2,280m (7,000 and 7,500ft). It has also been collected from the Hanna valley, north-east of Quetta. Evans found it not uncommon from mid March to early May. It straggles to Sindh and was collected from Karachi on 1st December 1930.

ARICIA ARGUSES

203. Orange Bordered Argus or Brown Argus of Europe

Aricia agestis
(Syn: *Polyommatus astrarche* in Evans, 1932)
Wingspan 25–30mm

Description: In this species both sexes are brown on the upper surface with no traces of blue, but both wing margins on the upper side bear orange marginal spots or lunules. They could be confused with the females of some *Albulina* and *Polyommatus* spp., but under hindwing patterns are different from any of the latter females. *Aricia agestis* in both sexes has a small dark end-cell bar and no traces of metallic spots under hindwing, nor metallic green scales at the wing bases. The female is like the male but slightly larger, and both sexes on the underside are quite dark brown with black spots white-ringed, and a white flash on the under hindwing at the base of the discal region. There is an orange marginal band, usually more conspicuous in the female.

Status: Widespread from Chitral through Swat Kohistan and Hazara district flying from the foothills up to 2,700m (9,000ft) and on the wing from April to October. It also occurs in the Murree Hills where it is rather uncommon. Collected TJR Swat Kohistan, and from Murree in August. This species is also widespread in Europe, inhabiting heathland or moorland, in windswept hilly areas.

Habits: Well studied in the western European population, where the eggs are laid singly on the leaves of the foodplants, which in Europe are *Helianthemum* spp. or Rock roses, which do not occur in our area. Their eggs are white and button-shaped with a small depression in the top, and their surface intricately sculpted with a honeycomb-like pattern of indentations. In the Himalayas, the larval foodplant is *Erodium cicutarium* and probably also *Erodium bryoniifolium*, (family Geraneaceae), both of which are common in the foothills of northern Pakistan. The larvae feed, when small, between the upper and lower cuticles of the leaves, forming a sort of blister in the leaf which they are attacking. They are the usual woodlouse shape, pale green in colour with a brownish-purple dorsal stripe and faint pale lateral stripes. Each segment bears two small wart-like eminencies with projecting white bristles The pupae are usually rather pale creamy coloured with a black spot along the ventral side of each segment, and a reddish-purple dorsal stripe. Col. Lang in Marshall and de Niceville (1890), from whom the above larval description is taken, describes finding their pupae spun in a shelter between dried leaves of *Erodium* or *Artemisia*.

204. Streaked Argus

Aricia chiron (syn: *Aricia eumedon*, Esper)
Wingspan 28–32mm

Description: In this species both sexes are dark brown above with a small black bar end cell, but in the male no trace of orange lunules. The female bears four orange-crowned lunules on the upper hindwing. The underside is quite dark brown with the under hindwing showing metallic green scales basally, and with a prominent triangular-shaped black spot end cell, which is white-bordered and often with the white extending outwards to the termen. The orange spots along the hindwing margin are not very conspicuous as in *A. agestis*.

Status: Found in the inner dryer Himalayan ranges from Chitral through Gilgit, where it has been collected from Yasin and in Chilas district on the Babusar Pass. It also occurs in Baltistan and on the Deosai plateau, and is nowhere common.

Habits: Foodplants are probably species of *Erodium* (Geraneaceae).

205. Astor Argus

Aricia astorica jermyni
Wingspan 29–31mm

Description: In this species both sexes are alike, being dark brown above with a small black bar end-cell on the forewing, and traces of black spots around the termen in the female on the under hindwing. There are no orange spots or lunules on the under hindwing. The underside forewing is tinged coppery-yellow, and there is no trace of metallic green scales basally as in *A. chiron*. The under hindwing end-cell spot is quite thin and the outer half of this underwing is white.

Status: Widespread in Gilgit from the slopes of Nanga Parbat, Astor and Rupal, the Naltar valley, and also from Baltistan where it is quite common. It has not been recorded from Chitral. It flies from 2,400–3,000m (8,000–10,000ft).

206. Spotted Argus Blue

Turanana cytis
Wingspan 24–26mm

Description: In the genus *Turanana*, the underside forewing bears large white-ringed spots post-discally which are not in line but irregular in placement. The male *T. cytis* is pale blue with a diffused black border and, besides a black bar end cell, three more small black spots post-discally on the forewing. The female is brown with the same black spots on the forewing. The underside is leaden grey-brown with the black spots white-ringed and conspicuous on the forewing, that between veins 3 and 4 on the forewing being nearer the margin than the others.

Status: Only recorded so far from Chitral, where considered rare. It has been collected from as low as Drosh at 1,800m (6,000ft) and on the Shandur plateau, also from Mastuj at 2,400m (8,000ft).

207. Dark Blue Jewel

Vaccinia kwaja
(Syn: *Vaccinia hyrcana* in Evans, 1932)
Wingspan 28–30mm

Description: The *Vaccinia* 'blues', are mostly brown above, the males dusted with blue, and with very prominent black spots in the discal area of the underside forewing, which are almost touching. *Vaccinia kwaja* is a small butterfly, the male dull blue with a diffused black border and small tornal black spots visible on the upper hindwing. The female is dark brown with marginal dark spots visible on both fore and hindwings. The underside is grey-brown with a marginal band of orange lunules, rather dark and obscure, and the tornal spots metallic-centred. The under forewing has a prominent row of black spots post-discally, placed very close together, and the end-cell spot is a narrow bar, in contrast to the large round end-cell spot in *Vaccinia iris*.

Status: This species, fairly common in Iran, is rare in Pakistan and has only been collected from near Chaman on the Khojak and Bogra hills. It is on the wing from May to June and again in early September.

208. Jewel Argus

Vaccinia iris
(Syn: *Polyommatus iris*)
Wingspan 25–29mm

Description: This is another very small butterfly, with both sexes brown above, showing no blue. There are several black tornal spots on the hindwing, bordered above by obscure orange lunules, and a small black spot end-cell on the forewing. The underside is quite dark grey-brown with very large, white-ringed black spots on the forewing, four in the discal area, almost touching in a straight line, and a larger round spot end cell, in contrast to the thinner bar-like spot in *V. kwaja*.

Status: Occurs rather rarely in Balochistan on the Afghan frontier near Chaman and the Khojak and Murgha-Mehtarzai Passes, also in the upper Uruk valley and around Ziarat. There are two broods, one on the wing in April and May and the second in September.

GREEN UNDERWINGS

209. Greenish Mountain Blue

Agriades pheretiades
(Syn: *Polyommatus orbitulus walli*)
Wingspan 28–30mm

Description: Both sexes have black spots end-cell fore and hindwings. The male is rather a pale greenish-blue with dark borders diffused inwardly to grey-brown. The female is brown with no dark borders noticeable. The underside is pale grey on the forewing, darker brown on the hindwing, and with prominent black spots, white-ringed along the post-discal area of the forewing, and a prominent white spot end cell on the hindwing. Usually just the base of the hindwing is metallic green.

Status: An inhabitant of high alpine slopes and meadows. Widespread in the inner dryer mountain ranges, but rare in Chitral and more common in Swat Kohistan. Collected TJR on the slopes of Mankial (Swat), extending through Gilgit mountain ranges. Collected from Chilas and Astor.

Habits: Not much is known about the early life history of this species, but at the altitudes in which they live, like their congeners in the European Alps and Pyrenees, the females seek Primulaceae or *Androsace* spp. for laying their eggs, both of which families and genera are represented by many species in the northern mountain regions of Pakistan. The tiny larvae feed on the underside of the host plant leaves, hibernating as caterpillars in a rough cocoon of silk, or as pupae hidden in the base stems of the foodplant, not emerging till the following summer.

210. Alpine Argus

Agriades jaloka (Syn: *Lycaene orbitulus astorica* in Tytler, 1926, and *Polyommatus orbitulus jaloka* in Evans, 1932)
Wingspan 28–30mm

Description: This rather small Lycaenid is rather distinctive because of the presence of white spots on both upper fore and hindwings in the post-discal area. The male has the upper side rather dark brown, with the basal area of both wings thinly dusted with greenish-blue, whilst the female lacks this overlay of greenish-blue scales but has a white costal spot in the sub-apical area on the forewing. The post-discal white spots are conspicuous on the under forewing of females, but often less conspicuous in males. The most characteristic feature of both sexes, compared with *Agriades pheretiades*, is in the extent of white on the under hindwing in both sexes, generally resulting in complete merging of the post-discal white spots with the white border. There are prominent black centres to the post-discal white spots in the under forewing of both sexes, as well as end cell. These spots are white-rimmed. There is a large, conspicuous white spot end-cell in both sexes under hindwing, bordered by an irregular brown band.

Status: It is a high-elevation species, flying from 3,600 to nearly 5,100m (12,000–nearly 17,000ft). It has been collected from the Rupal Valley in Astor as well as in Chilas, southern Gilgit, and, just outside our limits, from Gurais (sub-sp. *A. jaloka marlene*). It is very common on the Deosai Plateau and has been collected from Chogo Lungma in Baltistan at 5,100m (16,800ft). Evans considered it very rare in Chitral. It is on the wing from mid July up to early September.

211. Mountain Blue

Albulina pheretes (sub-sp. *asiatica*, Elwes)
Wingspan 24–28mm

Description: This is a small Lycaenid, and the genus is characterized by having white spots on the under hindwing. The male of *Albulina pheretes* is iridescent dark violet-blue with a small spot end cell on the forewing, and black margins to both wings, that on the hindwing being broader. The female is dark brown without black borders. The underside is variable, but usually bears very small black spots post-discally on the forewing, and four white spots on the under hindwing in veins 6 and 7 and end cell. The underside of both wings is variably suffused with metallic green scales.

Status: Only recorded from northern Chitral at elevations over 3,600m (12,000ft), where not rare.

Habits: The next three species, characterized by a suffusion of metallic-green scales on the under hindwing, are all inhabitants of high alpine slopes, and though nothing has been recorded specifically, the family is known to feed in the larval stage on Papilionaceae, particularly *Astragalus* spp., and the butterflies regularly come to flowers to sip nectar.

212. Dusky Green Underwing

Albulina omphisa, Moore
(Syn: *Polyommatus omphisa*, Evans, 1932)
Wingspan 30–32mm

Description: The male is a pale shining violet-blue above with a fairly broad (up to 4mm) dark border on the forewing, and only the base of the hindwing dusted with blue. The female is plain brown above, and lacks any orange lunules as in *Albulina metallica*, *Albulina galathea*, and *Polyommatus florenciae*. Both sexes on the underside have the whole of the under hindwing an iridescent brassy or yellowish-green with white spots in the post-discal area. In the male the under forewing is brown, paler basally, and the dark spots on the underwing forewing are rather faint. Compared with *Albulina galathea*, *A. omphisa* is smaller, whilst the under hindwing is yellower, less blue-green than that of *A. galathea*. In both sexes the fringe of cilia on the wing margins is white and well developed.

Status: Cantlie (1962) gives the distribution of this species as Ladakh and Lahoul in India, but also in Kashmir, without locality, and describes it as not rare. In Pakistan it has been collected from the Rupal valley in Astor, from the Shyok valley in Baltistan, and, more recently from Hunza. It flies between 3,000–3,960m (10,000 and 13,000ft), and is on the wing from July through August.

213. Small Green Underwing

Albulina metallica de Niceville, (Syn: *Albulina Chitralensia*, Tytler)
Wingspan 26–30mm

Description: The male is a dark iridescent violet-blue above, with very broad black borders covering almost half the surface of both upper and hindwings. The female is dark brown, strikingly patterned with large, bright orange spots on both hind and forewings, excepting the apex. The underside is grey-brown on the forewing with a very small end-cell spot, but with a line of white-ringed black spots post-discally. The under hindwing is metallic green all over, sometimes of a bluish-green tinge, sometimes more silvery sage green, and with a row of white spots post-discally, and another white spot end cell.

Status: Widespread in the northern inner mountains from Chitral, where collected on the Baroghil Pass, Shandur plateau, and Madaglash valley, through Gilgit, where collected Astor, Chilas, and Baltistan, including the Deosai plateau. Not rare.

214. Large Green Underwing

Albulina galathea
(Syn: *Lycaena galathea*, sensu: de Niceville, Syn: *Polyommatus galathea*)
Wingspan 34–39mm

Description: As its name implies, this is a larger Lycaenid, with the male iridescent violet-blue above and with black borders about 2mm wide. The female is dark brown with smaller and fewer orange spots than *A. metallica*, and in some individuals these may be entirely lacking. The underside is grey-brown on the forewing and metallic green on the hindwing, with a row of white spots post-discally and another white spot end cell, patterned similarly to *A. metallica*, but the under forewing is a much paler grey and the iridescent green of the under hindwing is generally darker, less brassy than in *A. metallica*. In his key to this species, Cantlie (1962) writes that the under forewing is suffused with darker brown in the sub-tornal area, but this character is also evident in some specimens of *A. metallica*.

Status Not uncommon in Chitral, where collected from Utzen nullah and the hills above Shishi Koh valley. Also occurs in Swat Kohistan and Chilas district of Gilgit. Flying from 2,400–3,000m (8,000–10,000ft).

215. Western Green Underwing or Green-Underside Blue of Europe

Glaucopsyche alexis
(Syn: *Polyommatus cyllarus*)
Wingspan 28–30mm

Description: Though synonymized with *Glaucopsyche cyllarus/alexis* of Western Europe, the Pakistan population differs in being a smaller butterfly and the female iridescent blue and black bordered above, whereas in Europe females are largely brown above. *Glaucopsyche alexis aeruginosa* males are a powdery violet-blue above with a broad diffused black border, and females have the base of the forewing, as well as the base and tornal area of the hindwing the same iridescent blue, with the rest of both wings black. The underside is dark leaden grey-brown on the forewing, metallic green on the hindwing, with very large black spots post-discally on the forewing, but no markings at all on the under hindwing.

Status: An inhabitant of hotter, lower hills up to 1,800m (6,000ft) in Balochistan, where it is common for a brief spell in April around the Khojak Pass and the Gwal forest. Also occurs in North Waziristan. This species spreads into central Europe.

Habits: It is known that the larvae are not attended by ants, and that their foodplants are Papilionaceae, most probably *Asragalus* and *Sophora* spp. in Balochistan.

216. Gilgit Meadow Blue

Iolana gigantea
Wingspan 35–38mm

Description: This is a large Lycaenid, without any metallic green scaling on the underside. The male is bright iridescent blue, with no spot end-cell on the forewing and with thin line black borders to both wings. The female is the same rather dark iridescent blue basally on both upper wings with very broad dark borders, the borders on the hindwing showing darker black lunules tornally. The underside is pale greyish-buff, with relatively small spots post-discally on the forewing and no spot end-cell on the under forewing, the end-cell spot on the hindwing being quite small.

Status: Only collected from extreme northern Chitral and Gilgit (without location), where considered rare. More frequent further east in Ladakh and Tibet.

Habits: The foodplants are members of the Papilionaceae, especially *Colutea nepalensis* (Syn: *arborescens*), known as Bladder Senna in Europe. The eggs are laid singly and the larva lives and feeds entirely inside the seed pods.

MEADOW BLUES

217. Pale Jewel Blue

Polyommatus sieversi
Wingspan 30–32mm

Description: Males are rather dark violet-blue above with small spots end cell in both upper and hindwings and a fairly broad-diffused black border. The female has broad brown borders with iridescent blue bases on the upper surface, and the hindwing bears one conspicuous black tornal spot ringed pale blue with indistinct pale lunules around the termen. The underside is leaden grey, darker than in *P. loewii*, with no orange tornal spots, and the discal spot in between veins 2 and 3 is less prominent than that in *P. loewii*.

Status: Only recorded in extreme western Balochistan, from Khan Mehtarzai near the Afghan border. Very rare, but commoner in Afghanistan and Iran. Collected at 2,100m (7,000ft) on 24 May.

Habits: Most of the *Polyommatus* Lycaenids inhabit high alpine meadows, and nothing seems to have been recorded about their early life history. It is known only that the larval foodplants are Papilionaceae, most frequently Milk Vetch, *Astragalus* spp., and Clovers, *Trifolium* spp., and they are not attended by ants.

218. Large Jewel Blue

Polyommatus loewii
(Syn: *Polyommatus baroghila,* Tytler, 1926)
Wingspan 32–34mm

Description: Males are a darker lavender blue than *P. sieversi*, with the forewing end cell only showing the shadow of the spot underneath and narrower black borders than in *P. sieversi*. The female is brown with prominent orange spots on the hindwing tornal area, and with white to pale orange post-discal spots on both fore and hindwings. The underside is pale grey with elongated black spots in the discal area, and orange lunules not very distinct in the hindwing tornal area. Also, the lunules between the post-discal row of spots and the margin on the under forewing are fused together in a dark brown band. In *P. sieversii* this band consists of smaller, separate lunules.

Status: More widespread and common than the former species, occurring commonly in Chitral, where collected from the Utzen and Tarben nullahs, from Mastuj, and Chitral town. On the wing from May to August and found between 1,800–2,400m (6,000–8,000ft). Also widespread in Balochistan on *Artemisia* slopes, from Ziarat, Uruk valley, Ghundak valley, and the Khojak hills, and on the wing from April to June.

219. Brilliant Meadow Blue

Polyommatus sarta
Wingspan 35–38mm

Description: This is a very variable species. The male above is iridescent violet-blue, usually with narrow black borders, and always on the upper hindwing a conspicuous row of marginal black spots, and only a trace of a dark bar end-cell on the forewing. The female is dark brown, but beautifully marked with orange spots along both wing margins and often with a dusting of bright orange scales on the forewing discal area. The underside is quite dark grey-brown with orange marginal spots extending around the termen and usually a white triangular spot end-cell on the hindwing. The hindwing bears a marginal row of black spots below the orange.

Status: This very pretty species occurs all over Chitral, from Ziarat and the Lowari Pass, Drosh, and up to the Baroghil Pass, and on the Shandur plateau. It occurs in Yasin, and from Astor and Rupal in Gilgit. A high-elevation species, flying from 2,400–4,260m (8,000–14,000ft).

220. Dusky Meadow Blue

Polyommatus gracilis
(Syn: *Polyommatus devanic*a in Evans, 1932)
Wingspan 24–34mm

Description: The male is rather a dull blue above with black borders up to 2mm wide, the borders on the hindwing being more diffuse and bordered inwardly with separate marginal black spots. The forewing has quite a conspicuous black bar end cell. The female is dark brown with the wing bases dusted with blue and a faint row of paler spots on the hindwing margin. The underside is dark leaden grey with large white-ringed spots on the forewing, and orange spots on the hindwing more or less obscure.

Status: A high-elevation species, flying from 2,700–3,660m (9,000–12,000ft). Occurs widely in Chitral but rather rare and local, from southern side valleys up to the far north where collected from Baroghil Pass, Shandur Lake, and extra-limitally in Dras, Ladakh. Also from Yasin in Gilgit, and from the Deosai in Baltistan, and occurs in Afghanistan (Sakai, 1981). It is on the wing during July and August.

221. Balochi Meadow Blue

Polyommatus bogra
(Syn: *Polyommatus actis bogra* in Evans, 1932)
Wingspan 28–34mm

Description: The male of this species is a very bright, silvery-blue with narrow line black borders well defined, the veins outwardly black, and no spot end cell. The female is dark brown with traces of dark spots on the hindwing termen. The underside is dark leaden grey with white-ringed spots including a prominent end-cell bar and a small spot mid cell on the forewing. The under hindwing has well-defined marginal lunules and traces of orange spots tornally.

Status: Found only in Balochistan, where collected on the Shingarh Range, from Ziarat, and on the Khojak and Bogra hills near Chaman. On the wing from June to September. Uncommon.

222. Violet Meadow Blue or Common Blue in Europe

Polyommatus icarus
Wingspan 30–34mm

Description: The male is iridescent violet-blue without the veins black, and with a narrow line black border sharply defined. The female is variable, but mostly brown above with varying amounts of blue scaling at the wing bases, and a series of black marginal spots, crowned orange. These are less distinct on the forewing. Neither sex shows black spots end cell. The underside is silvery grey-brown, with all black spots rather small, and well-defined marginal orange spots along both wings. Four black spots on the under hindwing base form the corners of a square.

Status: A widespread and common species, occurring in gardens and cultivated areas in Quetta main valley, and on the wing from March to October. It occurs all over Chitral, locally in Gilgit, Hunza, and in the Murree Hills. This species extends to Europe and appears to have three broods a year.

Habits: This species has been well studied in the European population. The eggs are white, disc-shaped, with concentric rows of little bumps. The larval foodplants are mostly Trefoil, *Lotus corniculatus*, or species of Clover, *Trifolium pratense* in the Murree Hills, and *T. fragiferum* in Balochistan. The larvae when small feed on the underside of leaves and are inconspicuous, later they will eat the flowers and seedpods of their host plant. The pupae are green in colour like the larvae, but lack the dark mid dorsal stripe of the former. They are formed in a rough silken cocoon, low down amongst the stalks of the foodplant.

223. Common Meadow Blue or Eros Blue in Europe

Polyommatus eros
(including sub-sp. *P. eros stoliczkana*)
Wingspan 26–30mm

Description: This is another very variable species, but the male is usually silvery-blue without violet tinge and with a black border varying from narrow to medium width, and with no spot end cell. The female is brown above, usually lacking any trace of blue scaling basally, but showing marginal orange spots prominently. The underside is pale fawn and the marginal spots are very faint, without white patches surrounding. There are reddish-orange sub-marginal spots on the hindwing which vary in prominence, and usually some iridescent green scales around the hindwing base.

Status: This very widespread Blue occurs all over Balochistan, where it is not very common but has been collected from Hanna Lake, Bogra, Ziarat, and on the Shingar Range and Torghar Hills in Zhob. In Chitral it is commoner between 1,800–3,960m (6,000–13,000ft) from Drosh up to the Baroghil Pass, and in the NWFP from North Waziristan at Razmak, the Kurram and Khyber valleys, the Samana Hills, and into Gilgit and Hunza (sub-sp. *stoliczkana*), on the Batura Glacier, and Shimshal valley. The author collected it from Swat around Kalam and from the Murree Hills, where it is abundant, and often shares the same habitat as *Aricia agestis*.

Habits: The larval foodplants are alpine species of leguminosae, especially Milk Vetch, *Astragalus* spp. They are not attended by ants.

224. The Silvery Meadow Blue

Polyommatus florenciae (Tytler, 1926)
Wingspan 34–40mm

Description: The male is a clear, iridescent violet-blue without any grey wash, and having rather broad black wing borders. The underside is pale greyish-brown with conspicuous rows of orange lunules across the sub-marginal area of both fore and hindwings, and the under hindwing bearing a large white spot end cell and quite conspicuous black spots in a similar pattern to other *Polyommatus* spp. The female is dark brown, with both fore and hindwings frosted with blue scales basally, and bearing orange spots sub-marginally, three on the forewing and five on the hindwing.

Status: First described as a new species by Tytler (1926), based on two specimens collected from the Baroghil Pass in extreme northern Chitral. It has also been collected from Gazan, also in Chitral, at 6,000ft on 6 June. It was later collected from northern Hunza, and no doubt occurs in the Wakhan, Afghanistan, where Sakai (1981) did not do any collecting. Evans (1932) considered it very rare and nothing seems to have been recorded subsequently.

225. Lime Blue

Chilades laius
Wingspan 26–30mm

Description: This is a fairly small Lycaenid, and the male is dull violet-blue above with linear black borders. Neither sex shows any spot end cell on the upper wings. The female is blackish-brown above with the wing bases metallic blue and with the hindwing showing black spots around the termen which are framed above by white lunules, sometimes with a second, fainter, white row. The underside is grey-brown with the black spots white-ringed, and on the under hindwing those in the discal area not solidly black, but pale brown-centred.

Status: With the spread of Citrus orchards in Pakistan, this species, common in India, may be spreading, though limes (*Citrus aurantifolia*) are not grown much in Pakistan, except in lower Sindh. There are only two old definite records: one specimen collected in Quetta by Swinhoe and another by Rhe-Philipe (1917) from Lahore in October. Marshall and de Niceville (1890) also state that it had been collected from the 'North-West Provinces', without details, which at that date could include parts of Punjab.

Habits: The larva is normal in shape, being humped in the middle, with the small head shiny black and often tucked in under the first segment. Sevastopulo (1944) describes it as pale bright green in colour with a darker green dorsal stripe, and traces of a double dark green sub-dorsal line. It is clothed with a short-sparse pubescence. The larvae are quite cannibalistic both of pupating congeners and when other larvae are moulting their skins. The pupa is plain green with a dorsal and lateral conjoined line of brown spots, and is attached to the underside of a leaf of its foodplant (Marshall and de Niceville, 1890). The foodplants, as its name suggests, are Citrus trees, especially the Lime, *Citrus aurantifolia*, and Pomelo, *Citrus grandis*. The larvae are occasionally attended by ants, and Sevastopulo records *Compressus camponotus* which is common in Pakistan.

GRASS BLUES

226. Grass Jewel

Zizeeria trochilus
(Syn: *Chilades trochilus* in Puri, 1931 and *Freyeria trochilus* in Cantlie, 1962 and *Chilades trochylus* in Whalley, 1996)
Wingspan 15–20mm

Description: Like all the 'grass blues', this is a tiny butterfly, though the females are slightly larger. Neither sex is tailed. *Freyeria trochilus* is dark brown above in both sexes, with the hindwing termen bearing three or four black spots framed above by orange lunules. The underside is fawn-brown with two black spots at the base and mid costa on the hindwing, and all the other spots dark brown, not black. The hindwing termen has black spots framed with silvery-green and orange lunules above. There are no orange lunules on the under forewing.

Status: This species is widespread in Pakistan, being on the wing every month of the year in Sindh, though mainly near the sea coast and always in grassland, Found occasionally all over Balochistan as well as the NWFP, and also at any time of the year. It has also been collected from the Khyber Pass, and Idak and Miramshah in North Waziristan, and in July and September at Naghr in Chitral, where it was considered rare. In the Punjab it has also been collected from Multan, Lahore near the Ravi River, and from Attock Bridge and Campbellpur.

Habits: The larva is green in colour with a dark green dorsal line extending from segments 3 to 12. The tiny black head is usually buried beneath segment 2. It is covered with minute white star-shaped hairs, giving it a hoary appearance. The larvae are attended by ants, and the larval foodplants include *Heliotropium strigosum* and *Heliotropium bacciferum*, as well as *Lotus corniculatus* and other vetches. The pupa is pale green and densely covered everywhere except on the wing cases by fine white hairs.

227. Pale Grass Blue

Zizeeria maha
Wingspan 26–30mm

Description: This is a larger 'grass blue' than *F. trochilus*, and the sexes are different. The male is a bright silvery-blue with sky-blue tinges on the hindwing bases, and a clearly defined narrow black border. The female is dark blackish-brown with the base of the wings blue, usually a darker shade than the male. The underside is pale grey-brown with small black spots in a discal band on the forewing and an end-cell spot, as well as a tiny spot mid cell, and a tiny spot on the forewing costa in the discal region. The under hindwing usually has a spot at the base between veins 1a and 1b, and always a spot between 1b and 1c.

Status: Fairly common in the outer foothills including the Punjab Salt Range and Hazara district. Collected Campbellpur, Rawalpindi, Hasan Abdal, Peshawar, the Khyber Pass, and in the Murree Hills, including Thandiani. A straggler to Balochistan and collected from Duki, but fairly common in lower Chitral between 4,000–9,000ft, and on the wing from March to October. Adults haunt fodder fields and grassy patches.

Habits: The eggs are described by Sevastopulo (1947) as button-shaped, with rounded sides and flattened top, and are greenish-white. Their surface is intricately sculpted by white reticulations. They are laid singly on the underside of the leaves of the foodplant, which is the Indian Wood Sorrel, *Oxalis corniculata*. Other foodplants recorded include *Tephrosia* spp. and *Strobilanthes* spp. (Acanthaceae). The larvae are woodlouse-shaped, but rather thin. As they feed on the flowers of their host plant, they tend to be rather variable in colour. Sevastopulo describes them as having the head and first segment black and retractile, the rest of the body dark green with an indistinct darker dorsal line, and with each segment deeply cut, and clothed with long white pubescence. The pupae are yellow-green, formed on a carpet of silk on the under surface of a leaf. Those bred by Sevastopulo pupated on 17 March and emerged on 23 March.

228. Dark Grass Blue or African Grass Blue

Zizeeria knysa
(Syn: *Zizeeria lysimon* in Evans, 1932)
Wingspan 18–24mm

Description: A very tiny butterfly, in which the male is dark blue above with broad black borders, especially on the forewing. The female is wholly dark brown, occasionally showing traces of blue scaling basally. The underside is quite dark grey-brown with small but well-defined black spots. There is often a white streak on the hindwing between veins 4 and 5.

Status: This is a very common butterfly throughout Sindh and Punjab, and is on the wing throughout the year. Swinhoe (1887) records two broods in Karachi area, being plentiful during April and May and again in even greater numbers during November and December. It also occurs all over Balochistan, but rather rarely, and in Chitral it occurs in the lower valleys between September and October. In Punjab it has been collected from Campbellpur (now Attock) in May and June, and from Murree on 1st October.

Habits: The larvae are blue-green in colour, usually with two indistinct lateral white lines, and a shiny black head. They feed on the flowers of their host plant, which is usually a Vetch, *Zornia gibbosa*, but in the Rawalpindi area (Mohyuddin, 1987), they have been found feeding on flowers of the Puncture Vine, *Tribulus terrestris* (family Zygophyllaceae), and Polygonaceae, such as *Polygonum plebejum*, and Amaranthaceae, such as *Amaranthus spinosus*, and can vary in colour according to the foodplant. They are occasionally attended by ants. The pupa is pale green with a few short scattered hairs.

229. Persian Grass Blue

Zizeeria galba
(Syn: *Freyeria galba* in Cantlie, 1962)
Wingspan 20–24mm

Description: This is another tiny 'grass blue', the male being a rather dark violet-blue with narrow black borders. The female is brown. The underside is dark grey-brown with an almost continuous discal bar of spots on the forewing, and the under hindwing bears two costal black spots, a spot at the base of the cell, and two slightly larger tornal spots framed with silver.

Status: This butterfly, which is common further to the west in Iran and Afghanistan, has only been collected from Idak in Waziristan.

230. Lesser Grass Blue

Zizina otis
(Syn: *Zizera indica* Murray, 1874)
Wingspan 19–20mm

Description: This and the next species are the smallest butterflies found in Pakistan. The male is dark blue above with black borders, usually fairly prominent. The female is dark brown with iridescent blue scaling on the wing bases. The underside is a paler grey-brown, and the under forewing bears no costal spots nor any spot inside the cell as in *Z. galba* and *Z. knysa*. The under hindwing has a row of discal spots with the spot between vein 6 and 7 not in line with those below, being closer to the wing base.

Status: Probably overlooked in Pakistan, as this is a very widespread species in the whole of India, but has only been collected from southern Chitral at lower elevations.

Habits: The eggs are white and flattened, with rounded edges, and covered with reticulations. They are laid inside the flower buds of Vetches, *Lathyrus* and *Vicia* spp. The larvae are grass-green with a sub-spiracular yellow band and a darker mid-dorsal line.

231. Tiny Grass Blue

Zizula hylax
Wingspan 16–24mm

Description: With *Z. otis* these species have the distinction of being the smallest butterflies found in Pakistan. The male is dull blue above with a narrow black border on the hindwing and a broad, diffused black border covering most of the apex on the forewing. The female is plain brown without traces of blue scaling. The underside is a paler grey-brown with small but well-defined spots, white-ringed, and those on the margin quite sharp. The under forewing row of discal spots are all well separated and angled inwards. There are two small costal spots and an end-cell spot on the hindwing.

Status: So far this butterfly has only been recorded from southern Sindh, where it occurs only after years of good monsoon rains. It is more plentiful further east in the Rann of Kutch.

Habits: The larvae are green in colour and of the usual shape, with humped back. They feed on the flowers of Vetches. The tiny butterflies are weak flyers, keeping close to the ground herbage and grass and frequently settling, so that they are easy to overlook.

232. Gram Blue

Euchrysops cnejus
(Syn: *Catochrysops cnejus*)
Wingspan 25–33mm

Description: The *Euchrysops* genus are tailed, with quite prominent tornal spots often visible on the upper wing as well as the under hindwing, and these are orange-crowned and with metallic spots. *Euchrysops cnejus* is quite a small Lycaenid, and the male is pale violet-blue above with only thin line dark borders, except at the apex of the forewing, and on the upper hindwing two orange-ringed black tornal spots. The female is brown above with the base of the wings overlaid with iridescent blue scales. The underside is fawn with darker brown discal bars and no black spots on the under forewing, but two black costal spots on the under hindwing and two larger, black, tornal spots, orange-ringed and with metallic silver centres.

Status: Collected Campbellpur, and Lahore, and by TJR from Rawal Lake, Islamabad. Uncommon in the dry north-west, and more widespread in India. It is on the wing from May to October.

Habits: The eggs are turban-shaped, but depressed in the middle at the top. They are coloured light green with fine white raised reticulations. The larvae are either green or rose-pink coloured, with a darker mid-dorsal stripe and yellow lateral stripes running diagonally across each segment. The whole body is covered with minute bristly hairs, giving a shagreened or rugose effect. The larvae possess a small extendible organ on top of segment 12 from which they can emit a disagreeable odour. The pupa is green or rose-coloured with some black spots, and is usually formed on the underside of a leaf. The foodplants are leguminous trees such as *Butea frondosa* or *Acacia* spp., but they will also feed on the pods of leguminous crops such as *Cicer arietum*, *Lablab purpureus*, and *Phaseolus trilobus*. They are only occasionally attended by ants.

CUPIDS

233. Small Cupid

Euchrysops parrhasius
(Syn: *Euchrysops contracta*)
Wingspan 20–25mm

Description: As its name implies, this is a small butterfly. The male is a brighter blue than *E. cnejus*, with thin black wing borders and without the forewing apex black as in the latter species. There is only a single small black tornal spot on the upper hindwing. The female is dark brown with large areas of both wings overlaid with blue scales, and on the hindwing three or four black spots around the termen, of which only the largest is orange-crowned and the rest are white-ringed. The underside is pale greyish-fawn, with darker brown bars discally and two prominent black tornal spots crowned with metallic scales, only the largest orange-crowned.

Status: Common all the year round in Sindh, mainly in the southern regions, and it occurs all over Balochistan, but rather uncommonly. It has been collected from Peshawar but does not seem to have been recorded from the Punjab.

Habits: The larvae are attended by ants and the foodplants are generally leguminous trees. The tiny larvae feed upon the flowers.

234. Plains Cupid

Euchrysops pandava
(Syn: *Euchrysops parrhasius minuta*)
Wingspan 30–35mm

Description: Slightly larger than *E. parrhasius*, the male is bright blue above with clearly defined, 2mm-wide black margins and a single, rather obscure, black tornal spot. The female is dark brown above with an overlay of blue scales basally and like *E. parrhasius*, bearing three or four black tornal spots on the upper hindwing which are white-crowned. The underside is dark leaden-grey with two prominent black costal spots on the under hindwing and three small black spots basally, one in the cell and one in between veins 1b and 1c. There are two black tornal spots, the one between veins 1c and 2 being larger and orange-crowned. The form which flies in Pakistan has white discal bands prominent on the under hindwing.

Status: Occurs rather rarely in Sindh, mainly in the south-eastern region, and has been collected from Sibi, but does not extend into the high plateau regions of Balochistan. It also occurs around Lahore and in the plains regions of the NWFP.

Habits: The larvae are rather dull-coloured with a rough surface, and are either green or rose-coloured, with darker dorsal and sub-spiracular lines. The tiny head is hidden under the first thoracic segment. These caterpillars are always attended by arboreal ants. The foodplants are leguminous trees, and the larvae feed on or inside their flowers.

235. Forget-me-not

Catachrysops strabo
Wingspan 25–32mm

Description: The male is pale silvery-blue with black line borders, and on the hindwing one prominent black tornal spot between veins 1c and 2. The female is rather striking, with pale brown borders to both wings and most of the upper surface up to the discal region pale silvery-blue. The borders of both wings bear white marginal lunules and two orange-ringed black tornal spots on the hindwing. The underside is pale fawn with darker fawn discal bars, and on the under forewing there is a small pale costal spot just inside the discal bar. The under hindwing has two black costal spots and two black tornal spots, the one between veins 1c and 2 much larger and orange-crowned. This species, like the whole genus, is tailed.

Status: Locally common in Sindh and on the wing from November to March. It should occur in north-east Punjab as it has been collected from Jullundur and the Punjab Himalayas in India.

Habits: The larvae are green with a brown mid-dorsal line, which is wider at the forepart and anal end of the body. There is a short, white, diagonal line on the side of each segment. The adult lays her eggs on both leaf tips and flowers, and the larvae generally live and feed on the flowers and seedpods. The foodplants are variable, but always leguminous and usually climbing Vetches or vines, in the hills *Desmodium elegans*, and in the foothills *Ougeinia oojeinensis*. The pupa is light pink or creamy coloured with black specks, and covered with small stiff hairs. It is located on the ground, fixed to some object by the body girdle and tail pad. The larvae are generally attended by ants.

236. Pea Blue or Long-Tailed Blue of Europe

Lampides boeticus
Wingspan 24–36mm

Description: The male is very similar to *Euchrysops cnejus* in appearance but slightly darker violet-blue above. There are two black tornal spots on the upper hindwing. The female is brown, overlaid with paler, silvery-blue scales basally, and on the upper hindwing two white-ringed tornal spots, and two obscure white marginal lines bordering the upper wing margin. The underside is pale fawn with darker fawn bars rimmed with white and no spots on the forewing, but two orange-ringed black tornal spots, often with metallic crowns. Both sexes are tailed.

Status: This species extends in range from southern Europe across the Middle East and throughout the Indian subcontinent. It is particularly common in southern Sindh and widespread but less common in Balochistan. In both provinces it is found throughout the year, but in the Punjab it flies mainly during the winter months, and in the Murree foothills it flies mainly from March to July. It is common wherever it occurs.

Habits: The eggs are turban-shaped, with flattened top and bottom, and with a shining surface. They are light green in colour, covered with tiny white knobs or granulations. The larvae are the normal woodlouse shape, with the head hidden under the first thoracic segment. They are normally coloured light varying to dark green, sometimes with a purplish suffusion, with a double dorsal yellowish line and a sub-spiracular yellow line. However, Fraser (1911) describes larvae he collected and hatched from Hyderabad district in December, when there are few ants about due to the cold weather. These larvae were a dirty flesh colour with a dark crimson mid-dorsal line and short diagonal lines of the same colour on each segment, whilst their whole bodies were covered with short stiff black bristles. The spiracles were brown. On segment 11—where the honey gland is normally situated in the larvae which are attended by ants—there was only a bare patch in these Sindh specimens. They were not attended by ants because they sealed themselves inside a flower bud before it opened, by binding the calyx and leaf bud together with their silk. The young larvae normally feed within flowers, mainly inside the carpels when young, later feeding out in the open on the seed pods. The foodplants are either leguminous trees such as *Butea frondosa*, or shrubs such as *Crotolaria medicaginea* and *Crotolaria juncea* in the Murree Hills and Rawalpindi respectively. Also *Pisum sativum, Pisum arvense*, and *Trichodesma indicum* in Rawalpindi (Mohyuddin, 1987), and *Melilotus indicus* in Sindh. Before pupating there is much cannibalism between active larvae and those which are about to pupate. The pupa is formed on the ground, sometimes in a crack in the soil, and is secured by a silken band mid body with a silk pad at the tail. It is coloured dull or bright green with a smooth shining surface, and with a darker mid-dorsal line, plus a double sub-dorsal row of black specks. The larvae are usually attended by ants.

THE CERULEANS

237. Dark Cerulean

Jamides bochus
Wingspan 25–34mm

Description: The male is bright iridescent blue above with paler, milky-blue areas basally, and very broad black borders on the forewing reaching to the discal area. The female is a duller blue with the same broad black borders and, on the upper hindwing, dark marginal spots crowned by milky-white lunules. The female is tailed but not always in the male. The underside is brown with interrupted white bands, and one orange-circled black tornal spot on the hindwing. The basal area of the forewing is plain brown without bars.

Status: Recorded as occurring in Pakistan by Cantlie (1962), but there are no specimens from our region in the Brit. Nat. Hist. Museum collection. It is most likely to occur in the Murree foothills.

Habits: The turban-shaped eggs are pale greenish, their surface covered with raised white reticulations. They are laid on the flowers of leguminous trees, typically *Butea monosperma* and herbaceous *Crotalaria* spp. They spend their whole life until pupation feeding inside the flowers, upon the carpels and stamens. The larva is a dirty pinkish colour, varying to olive, with paler buff dorsal and lateral bands. Its body is covered with minute conical black hairs, with a few scattered longer white hairs. The pupa is generally formed on the ground attached to a dead leaf or earth clod in the usual way with a silken body girdle, and by the tail onto a silken pad. The pupa is a dirty yellow, varying to pinkish-brown, covered with minute black spots in rows along the dorsum and sides. The larvae are occasionally attended by ants.

238. Common Cerulean

Jamides celeno
Wingspan 27–40mm

Description: This is a tailed species and the male is a very pale bluish-white, with the upper hindwing showing a thin, thread-like black border, but that on the forewing widening at the apex. There are traces of small black spots on the hindwing termen, and the underwing pattern shows through on the upper surface as paler bars. The female is the same blue but rather duller and with broader black borders, and more pronounced black spots on the hindwing termen. The underside is pale brown with two narrow white marginal bands and on the under forewing four more vertical white lines discally, while the under hindwing bears six white lines.

Status: This species, like the Dark Cerulean, favours more humid scrub forest than is found in Pakistan but it has been collected by TJR from Islamabad gardens, and is also recorded from Pakistan (without locality) by Cantlie (1962). Like the Dark Cerulean, it must be considered rare in Pakistan.

Habits: The eggs are the normal turban shape with rounded sides, and are pale green in colour, covered all over with raised white reticulations. The female lays these singly on flowers or in the axils of very young leaves, and the foodplants are generally leguminous trees such as *Butea monosperma*, *Pongamia glabra*, *Abrus precatorius*, and *Derris indica*. The woodlouse-shaped larvae are green with a darker mid-dorsal line, and a row of lateral spots. Their bodies are described by Bell (1918) as shagreened by a covering of minute star-shaped tubercles, some milky-white, others green or brown. The larvae are dirty in yellow colour, sometimes with a pink flush, and covered with black spots. The larvae are usually attended by ants.

LINEBLUES

239. Common Lineblue

Nacaduba nora
(Syn: *Prosotas nora*)
Wingspan 18–25mm

Description: The Lineblues can be tailed or not, and are brown on the underside, crossed with white lines in pairs. The males are generally a rather dark shade of violet-blue and the females have much paler blue panels on the forewing. *Nacaduba nora* is a small butterfly, and the male is a dull leaden blue with narrow line borders of black. The female has wide brown wing borders, with the discal area of the upper forewing pale blue. Both sexes are tailed and have an inconspicuous small black tornal spot. The underside is pale brown with basal and discal bands consisting of double white outer lines and within two darker thin brown lines. The hindwing tornal spot is more conspicuous, often with metallic spots on its lower margin.

Status: Common in the Margallas and Murree foothills from May to October, and it has also been collected from Peshawar. It also occurs in Afghanistan.

Habits: The eggs are turban-shaped, bluish-green in colour, and covered with reticulations, forming cells, inside of which are rounded prominences. They are laid inside flower buds, and the larvae remain feeding inside the flowers. The larvae are the usual woodlouse shape, with the head hidden under the first thoracic segment. They are dark or light green, with a red dorsal stripe, bordered on each side by yellow, and also a sub-spiracular band of brownish-pink, bordered below by white or cream. Their bodies are also covered with minute shining bristles, both dark red and creamy. The larval foodplants are leguminous trees of the Acacia and Mimosa genera, in the Punjab Salt Range and Margallas, *Mimosa himalayana*. The pupa is rugose on its surface and pink in colour, with black blotched along the dorsum. The larvae are usually attended by ants.

240. Tailless Lineblue

Nacaduba dubiosa
(Syn: *Prosotas dubiosa*)
Wingspan 22–26mm

Description: Slightly larger than *N. nora*, this small butterfly, as its name indicates, has no tail. The male is a brighter violet-blue than *N. nora* but with similar thin black line borders and obscure small tornal black spot on the hindwing. The female is blackish-brown above with iridescent pale blue panels on the forewing, not extending into the cell as in females of *N. nora*. The underside is pale brown with similar pairs of white-margined darker bands as in *N. nora*.

Status: Rare in Pakistan and only found in the Murree foothills.

COPPERS

This group of distinctive butterflies are recognizable by the splendid coppery-red colours of the males, by having naked eyes, and by the legs in both sexes being very spiny.

241. White-Bordered Copper

Lycaena pavana
Wingspan 37–40mm

Description: All the Coppers are very bright coppery orange on the upper side with dark brown margins of varying width and the underside pale fawn with prominent black spots. *Lycaena pavana* is slightly larger than the Common Copper, *Lycaena phlaeas*, and the male is reddish-copper on the upper forewing with a brown border and two black spots mid and end cell, followed by an irregular discal band of spots. The upper hindwing is mostly brown with tinges of copper in the cell and basal area and, at certain angles, a purplish bloom along the upper margin of the hindwing. The female has the whole of her upper hindwing brown and on the forewing only the discal area and cell is orange-copper coloured. The spots follow the same pattern as the male. The underside is dull orange on the forewing, pale grey-fawn on the under hindwing, with a continuous sub-marginal row of black spots on both wings, those on the under hindwing bordered above by a band of orange and more black spots lining the inner side. The principle identifying feature, which gives rise to its common name, is a band of white on the under hindwing between the discal row of spots and the sub-marginal double row of spots. This feature alone distinguishes it from all other 'Coppers'.

Status: Rather rare in the Neelum valley, Azad Kashmir, but commoner further east in the Jhelum valley watershed. Rare in the Murree foothills, but this species may have been overlooked because of its similarity to the Common Copper. Flies mainly between 1,800–2,100m (6,000–7,000ft), but has been taken up to 3960m (13,000ft).

Habits: The larval foodplants are *Rumex* spp.

242. Common Copper, Small Copper in Europe

Lycaena phlaeas
Wingspan 26–30mm

Description: Smaller than *L. pavana*, the Small Copper male is similarly coloured above, being reddish-copper with brown borders, and black spots in the mid and end-cell as well as an irregular discal band of spots. The upper hindwing is brown with no suffusion of copper-orange in the cell as in *L. pavana*, but showing faint black spots end cell. The underside is brownish-fawn on the forewing apex and the whole of the hindwing, with very small black spots on the under hindwing and no white band or row of sub-marginal spots as in *L. pavana*. The female is similarly marked above and below, and both sexes have a marginal band of orange on the upper hindwing, as well as a few bluish-white spots around the sub-discal region.

Status: This species may almost be described as cosmopolitan in the northern hemisphere, as it occurs from Japan, China, across Asia, the whole of Europe, and into North America. In the Indian subcontinent it is easily the commonest of the 'Coppers'. Extremely common throughout the northern areas from Chitral, through Hazara district, Swat Kohistan, and the Murree Hills. On the wing from March to October, and found from 300–3,000m (1,000–10,000ft). It occurs occasionally in the foothill areas of the NWFP and has been collected from Peshawar and from Campbellpur (Attock district). In Balochistan it is uncommon but widespread, especially around Quetta gardens.

Habits: The female lays her eggs singly on the foodplants which, are Sorrels and Docks, *Rumex* spp. In the Murree Hills they have been found on *Rumex hastatus*. The eggs are green and flattened top and bottom, bearing raised white reticulations like a honeycomb. The larvae are green, covered with short brown bristly hairs and whitish warts. They usually have a pinkish sub-spiracular line. With humped back and legs hidden under the body, they are difficult to see, looking like some excrescence or gall on the surface of the leaf. The pupae are yellowish on the abdomen, with the wing cases darker olive, and speckled with black or dark brown. They pupate low down amongst the basal stems of the host plant, attached in the usual way by a body girdle onto a pad of silk. None of the Coppers are attended by ants.

243. Green Copper

Lycaena kasyapa
Wingspan 27–32mm

Description: This handsome Copper is dark reddish-copper in the male with a purple suffusion on both upper wings, and broad dark brown borders. The forewing has elongated black spots in the lower discal region. The female is dark brown basally on both upper wings with the same spots. The underside hindwing is iridescent verdigris green, unmarked, and the under forewing is green along the margin and base with the discal area orange, and with black spots discally.

Status: Common in Chitral from June to September, and very common Swat Kohistan, where collected TJR from Bahrein, Kalam, and the Jabba valley. Absent from the Murree Hills. It normally inhabits higher elevations than the Common Copper, flying from 1,800–3,000m (6,000–10,000ft). Marshall and de Niceville (1889) considered it very uncommon further east in Lahoul, Chamba, and Himachal Pradesh. Sakai (1981) did not find it in Afghanistan.

Habits: The foodplants are Docks, *Rumex* spp.

244. Golden Copper

Lycaena solskyi
(Syn: *Lycaena thetis* in Evans, 1932)
Wingspan 32–36mm

Description: This striking little butterfly is a brilliant yellow-orange copper on the upper surface in the male, without any spots, but having a narrow black border, which is indented into tornal spots on the hindwing termen. The female is a duller orange-copper with the hindwing overlaid with brown scales except for the margin, and with both wings spotted all over. The underside is orange on the forewing with a white border, and the under hindwing is dull white all over, and with both wings bearing spots all over.

Status: This beautiful Copper has been recorded from the far north of Chitral around the Shandur plateau, flying in July and August. It was collected in late July 1887 by Marshall near the Braldo Glacier in Baltistan at 3,600m (11,700ft), as well as near Satpura below the Deosai in late August. It also occurs very rarely in the more sheltered 'tangis' (ravines) around Ziarat, Balochistan. It is on the wing from June to the end of August.

245. Balochi Copper

Lycaena phoenicurus
Wingspan 28–35mm

Description: The male is purple on the upper surface with an overlay of brown scales and with darker brownish-orange borders, bearing a marginal row of black spots. The upper wings show shadows of the dark spots on the underside, and there is a tail on the hindwing tornus. The female is brown above with the hindwing margin orange and the costal area and margin of the forewing also orange. She is also tailed, and bears spots in two discal and sub-marginal bands. The underside is dull white on both wings, with a prominent dark orange area on the forewing discally. Both underwings are spotted all over.

Status: Rather uncommon in Balochistan, occurring in the upper Urak valley, near Malozai on the Speraragha Road, and near Wam on the Ziarat road. It is less rare further west in Afghanistan (Sakai, 1981) and in Iran. It is on the wing in May and June.

Habits: As far as is known, the larval foodplants are Polygonaceae, such as *Rumex hastatus* and *Rumex vesicarius*.

246. Purple Copper

Lycaena caspius
(Syn: *Lycaena hyrcanana* or *Lycaena athamanthis balucha*)
Wingspan 29–35mm

Description: The Purple Copper averages smaller in size than *Lycaena phoenicurus* though, as the size range given above indicates, there is seasonal variation. The male is purple-brown above with the basal areas more iridescent purple, and with broad orange-brown borders. Like *L. phoenicurus*, the upper surface shows shadows of the darker spots borne on the underside, and the end-cell spot is prominent. The female is a duller brown with a thinner scattering of purple scales basally, and with orange marginal borders more prominent, especially on the hindwing, and the end-cell spot also more prominent. Both sexes are tailed. The underside is dull white on both wings, but there is a marginal orange-red border, bordered outside and in with black spots, and also scattered discal spots across both wings.

Status: This rather diminutive Copper is locally common but erratic in occurrence. In Balochistan it has been collected from Gandak on the Ziarat road, and on the Murgha-Mehtarzai Pass, and is on the wing from May to June. In Chitral it is on the wing from July to August, occurring at higher altitudes from 2,700–3,000m (9,000–10,000ft), and has been collected in valleys such as the Tarben, Shishi, and Utzen. It is commoner further west in Afghanistan and Iran.

Habits: The larval foodplants are Polygonaceae, probably *Rumex hastatus*.

247. Red Copper

Thestor callimachus
(Syn: *Tomares callimachus* in Evans, 1932)
Wingspan 27–30mm

Description: Smaller than the previous species, *Thestor callimachus* males are bright coppery-orange above, with dark brown borders which are indented on the forewing but not around the hindwing. There are traces of dark spots on the forewing end cell and apex region, and neither sex is tailed. The female has broader brown borders to the wings, and most of the hindwing costal and basal area is also brown, as is the forewing apex. The underside hindwing is banded with grey and white, with an orange band discally margined both sides with dark dashes. The under forewing is white in the apex and costal areas, reddish-orange basally, and with bold black elongated spots costally.

Status: Only found in Balochistan around the Gwal forest area and on the Murgha-Mehtarzai Pass. On the wing from March to April. It is commoner further west in Iran and Iraq.

SAPPHIRES

248. Sorrel Sapphire

Heliophorus sena
Wingspan 28–33mm

Description: The Sapphires are among the most beautiful of the Lycaenids, and generally are ochre yellow on the underside with a broad red band on the hindwing. The Sorrel Sapphire is the least typical of this genus, being greenish-ochre on the underside and with a black-margined white band across both wings framing a broad marginal band of red lunules, outlined in black. The male is dark shining violet, with a black border on the upper hindwing comprising marginal black lunules, framed with red. The female is brown above and the red marginal band is continued on to the forewing. On the under hindwing there are two black spots, one on the costa and the other between veins 1b and 1c. Both sexes are tailed.

Status: Extremely common both in Chitral and Swat, flying from 1,200–2,100m (4,000–7,000ft), and on the wing from March to October. It is fairly common in late summer in the Murree Hills. Collected TJR from Miandam and Bahrein in Swat.

Habits: The larvae are the typical humpbacked woodlouse shape, green in colour with a darker green mid dorsal stripe, and darker, double lateral line. The larval foodplants are *Rumex* spp., typically *Rumex hastatus* in the Murree Hills. The pupa is clothed with a short colourless pubescence.

249. Western Blue Sapphire

Heliophorus bakeri
Wingspan 30–34mm

Description: This is a variable species with wet and dry season forms, in which the upper hindwing shows red lunules in the tornal area in the wet season form which are absent in the dry season. The male is a deep silky blue with clearly-defined broad black margins, rounded at the apex. Neither sex is tailed but there is a small tooth at the hindwing tornus, and, as indicated above, varying amounts of orange-red on the upper hindwing termen. The female is dark brown with an orange-red marginal band on the hindwing and a broad orange-red band across the forewing apex. The underside is ochre yellow with faint dark marginal border on the forewing and a prominent black tornal spot. The under hindwing has a broad red margin, not outlined in black as in *Heliophorus sena*.

Status: Common in Hazara district, especially in the lower parts of the Kaghan valley at Shogran and Shahran, where collected TJR. Flying between 1,800–2,700m (6,000–9,000ft) and on the wing from May to September. It is also fairly common in lower valleys of Chitral and Swat Kohistan. Collected by TJR at Kalam at 2,400m (8,000ft) in mid July.

Habits: The larval foodplants are *Rumex* spp., particularly the Common Sorrel, *Rumex nepalensis*.

250. Powdery Green Sapphire

Heliophorus tamu
(Syn: *Ilerda viridipunctata* in Marshall and de Niceville, 1890)
Wingspan 34–40mm

Description: The underside of this little butterfly at once identifies it as a Sapphire (*Heliophorus*), being ochre yellow on both under fore and hindwings, with only narrow faint black bars on the under forewing, and with black-rimmed red lunules along the under hindwing termen. The male is dark blackish-brown on both upper wing surfaces, with a powdering of metallic green scales at the base of both fore and hindwings. It appears very similar to the Green Sapphire, *Heliophorus androcles* (extra-limital), but the latter has more extensive iridescent green areas on both wings, and a less pronounced tornal tail than *H. tamu*. The female is brown above, less dark than the male, and with the hindwing termen toothed and bearing a wavy scarlet marginal line, larger in the tornal region, where the red lunules are margined black on their outer edge. The upper forewing also bears a broad scarlet bar across the discal area.

Status: Has been collected from Murree after the monsoon in August and September, when it was described by Butler (1886) as common. There are no subsequent records, and is most regularly encountered in the Eastern Himalayas, from Simla through to Bhutan.

LOBELESS HAIRSTREAKS

251. Balochi Hairstreak

Neolycaena connae
Wingspan 28–30mm

Description: This is a drab little butterfly; both sexes being dark brown, unmarked on the upper surface, and without any tail. The underside is dark grey-brown, but in the male the lower third of the under forewing up to the margin is black. There are fine white lines along the margin, and broken lines discally which are black edged. The female has the under forewing entirely dark grey.

Status: Rare in Balochistan but occurs locally above 2,100m (7,000ft) in the Zarghun mountain range, and on the Mehtarzai Pass as well as between Wam and Ziarat, again above 2,100m (7,000ft). It is on the wing in May and June.

252. Green Hairstreak

Callophrys rubi
Wingspan 20–32mm

Description: This little butterfly is a slightly warmer brown than the Balochi Hairstreak, with the hindwing margin slightly toothed. It has the entire underside bright grass green, sometimes showing faintly a row of white discal spots. There are no marks on the upper surface and the sexes are alike.

Status: Locally common and widely distributed at higher elevations in the juniper forest zone of Balochistan. Collected TJR Ziarat. Occurs upper Urak valley, and on the Shingar in Zhob district. It is on the wing in May and June. It has also been found locally in lower Chitral, with specimens collected in late April from the Kesu valley at 1,800m (6,000ft). This is the same species which occurs right across Europe, up to the boreal forest zone in Scandinavia.

Habits: Nothing seems to have been recorded about the Pakistan population of this Hairstreak, but based upon the European population, it is known that the eggs are turban-shaped with rounded sides and flattened top and bottom, and white in colour, covered with minute pits in concentric rows. The larvae are green, and the normal woodlouse shape, with yellow-green lateral and sub-spiracular lines. The foodplants are known to be quite variable, including Papilionaceae, Rosaceae, and Compositae. In Balochistan, where the author collected it, *Onobrychis* and *Astragalus* spp. (family Papilionaceae) were being visited by adults, but they did not appear to be interested in nectar. The tiny caterpillars start their lives feeding inside the flowers or flower buds of their host plant. The pupae are formed on the ground in a soil crevice or protected under a rock.

TAILED HAIRSTREAKS

253. White-Line Hairstreak

Strymon assamica
(Syn: *Strymon sassanides* in Wynter-Blyth, 1957 and *Strymonidia assamica* in D'Abrera, 1993)
Wingspan 28–35mm

Description: Larger than the Green Hairstreak, *Strymon assamica* is glossy brown above, with a prominent tail on the hindwing tornus, a narrow black border to both wings, and white cilia. The sexes are alike. The underside is a paler brown with a fairly straight discal white line across both wings, which is black-edged anteriorly. There are obscure black spots along the margin of fore and hindwings, the tornal spot being larger and orange-crowned, also a thin white marginal line outside of the spots, which on the forewing does not extend up to the apex.

Status: Quite common from Chitral, through Swat Kohistan and Hazara district, but has not been recorded from the Murree Hills. Collected TJR from Jabba valley, Swat. It also occurs quite commonly in central Balochistan at higher elevations in the juniper zone around Zarghun, Ziarat, and the upper Urak valley. Marshall and de Niceville (1890) recorded it from northern Kashmir and Baltistan. Flies between 1,200–3,000m (4,000–10,000ft) from May to August.

Habits: Nothing seems to have been recorded about the early life history of the Hairstreaks found in the hill zones of the subcontinent but, as far as is known, all the species found in Pakistan are associated with fairly mesic conditions, with the single exception of this species, *Strymon assamica*, which is more tolerant of dry areas. The larval foodplants of all the Hairstreaks are various tree or shrub species, especially *Prunus* and *Quercus*, and *Ulmus* genera. In Balochistan where this Hairstreak flies, the only widespread shrub which seems the likely foodplant is the wild Almond, *Prunus amygdalus*. The eggs of all the Hairstreaks are variable in shape, and can be hemispherical or flattened top and bottom with rounded sides, with their surface pitted or covered with minute rounded lumps or, in some species, small spines, and are generally white or green in colour. They are laid singly on the foodplant and the larvae do not attract ants. Typically the females lay their eggs on the leaf buds, where they remain throughout the following winter, and do not hatch until the next spring when the tree's young leaf buds are just bursting. The larvae are the normal humped shape, oval in outline when viewed from above, and can be variable in colour, often with each segment constricted before the next, and bearing concentric rings of contrasting colour. The adult butterflies tend to fly around treetops and to stay for long periods resting on tree leaves, and they do not often come to flowers.

254. Water Hairstreak

Euaspa milionia
Wingspan 30–34mm

Description: Male and female of this Hairstreak are dimorphic. The male is pale iridescent sky-blue above with broad black borders, that on the forewing covering the apex, and with a wide white band on the forewing upper discal area, and also on the upper hindwing. The female is a darker, duller blue, with the same broad black border covering the forewing apex, but no white panels on the hindwing and only restricted white patches in the forewing apical region. Both sexes are tailed. The underside is very dark chocolate-brown with a broad white band crossing both wings discally, and on the under forewing a white margin in which brown spots are placed. The under hindwing has a large black tornal spot, orange-ringed.

Status: Only found in Himalayan temperate forest near streams, and has been collected from Murree and Dunga Gali, where it is only locally common. It is less rare further east in Kumaon, India. Flies between 1,200–2,000m (4,000–6,500ft), on the wing from May to June, and appears to be single brooded.

Habits: Like all the Hairstreaks, the adults do not visit flowers, and this species seems to prefer shady ravines in the proximity of streams, and has a rather slow, weak flight. Nothing is known about its early life history. Mohyuddin (1987) found the larva of an unidentified tailed Hairstreak at Charra Pani in the Murree Hills at 1,200m (4,000ft), feeding on *Punica granatum,* which may have been this species.

255. White-Spotted Hairstreak

Chrysozephyrus ziha
(Syn: *Euaspa ziha,* **or** *Thecla ziha*)
Wingspan 35–38mm

Description: The male and female are alike in this species, above rather a grey-blue with the veins on the hindwing black, and on the forewing the apex and a broad margin brownish-black. There are two square white spots on the upper forewing, one just beyond the cell and the second between veins 3 and 4. The hindwing also has a thin, bluish-white marginal border, and both sexes are tailed. They can be distinguished from females of *Chrysozephyrus syla,* which occurs in the same areas, by the white spots on the forewing and the shiny blue base to the upper hindwing. The hindwing in *C. syla* is wholly greyish-black with only a light dusting of blue scales. The underside is dull white, with a brown discal bar on the forewing and paler bars along the margin, with two prominent black spots in the tornal region. The under hindwing has one to two tornal black spots, orange-crowned.

Status: Very rare in the Murree Hills, but has been collected from Dunga Gali and Murree. It is on the wing from May to July and flies between 1,500–2,000m (5,000–6,500ft).

256. Wonderful Hairstreak

Chrysozephyrus ataxus
(Syn: *Thecla ataxus* **in Wynter-Blyth, 1957)**
Wingspan 20–23mm

Description: The upper side of the male of this species is indistinguishable from males of *Chrysozephyrus syla,* but the underside is different, as is the female upper side. The male of *C. ataxus* is a shining metallic green with a black border widening at the forewing apex. The female is strikingly different, with the base of the forewings iridescent violet-blue, broad black borders covering the whole apex, and two prominent orange spots on the forewing, end cell and between veins 3 and 4. The underside is silvery-white in the male, with no marginal darker lines on the forewing, but brown bars end cell on both wings, that on the hindwing extending from the costa down to vein 4. The tornal black spot is orange-crowned. Both sexes are tailed. The female underside forewing is browner, with darker brown discal and sub-marginal bars separated by a broad silvery-white band.

Status: In most years it is rare in the Murree Hills, but has been collected from Dunga Gali and Nathia Gali, where it is on the wing from May to September. It flies from 1,680–2,100m (5,500 to 7,000ft).

257. Silver Hairstreak

Chrysozephyrus syla
(Syn: *Thecla syla* in Wynter-Blyth, 1957 & *Neozephyrus syla* in D'Abrera, 1993)
Wingspan 42–45mm

Description: Like *C. ataxus*, this is a sexually dimorphic Hairstreak, with the male a shining metallic green above and the female with iridescent blue on the forewing. The male shows bronze lights at certain angles, not so apparent in the very similar *C. ataxus*, also the black border is more even, not widening at the forewing apex as in *C. ataxus*. The female has broad black borders to the forewing and the upper hindwing is largely blackish-grey. The forewing end-cell and apex is pale silvery-blue, the rest of the forewing a darker blue. Both sexes are tailed. The underside is silvery-grey with darker brown sub-marginal and broader discal bars, outwardly bordered with white. There is also a brown bar end cell, and two tornal black spots, orange-ringed.

Status: Certainly the commonest of the tailed Hairstreaks, occurring from Chitral, where it has been collected from the Utzen and Ashreth valleys, through Swat Kohistan, the lower Kaghan valley, and the Murree Hills. Collected TJR Dunga Gali. It flies between 1,800–2,400m (6,000–8,000 ft), and is on the wing from May to September.

Habits: This species, like *Euaspa milonia*, may be found in the vicinity of small streams in thickly-wooded 'nullahs', and frequently comes down to settle on bits of rotting wood or moss-covered stones midstream. The butterflies avoid open areas, and typically are found near Oak trees. They never visit flowers.

258. Walnut Blue

Chaetoprocta odata
Wingspan 32–36mm

Description: The sexes are alike in this species, both being dark shining purple above, with broad black borders including the forewing costa, and with the hindwing tailed. The female has wider dark borders. The underside is greyish-white with a pale yellowish-brown discal band, and the sub-marginal band on the forewing ends in two black spots at the tornus. The hindwing has paler, white-edged bars, and sometimes one or two large tornal spots rimmed with dull ochre.

Status: Especially common in the lower Himalayan valleys where walnut trees (*Juglans regia*) grow, such as around Shahran and Shogran in the Kaghan valley, and in the Murree Hills. Also in the lower Chitral valleys such as Khilas and Shishi Koh, where it was reported (Leslie and Evans, 1903) to carpet the grass under walnut trees. Mainly on the wing from June to September, and found from 1,200–2,700m (4,000–9,000ft).

Habits: The female lays her eggs on the young leaves of the Walnut (*Juglans regia*), and, where it is plentiful, the adult butterflies can be seen at evening time, literally in thousands, fluttering around this tree. According to Wynter-Blyth (1957) the female lays her eggs in batches, and each egg, as it passes from the tip of her abdomen, becomes covered with a fan of brown scales which are naturally shed from a special brush of these scales at the tip. This conceals the eggs from potential predators such as minute Ichneumon wasps and mites. In a more detailed description by Marshall and de Niceville (1890), the female lays her eggs in five longitudinal rows, the whole batch laid along the stem of a walnut twig. There is only one brood a year, the eggs being laid generally from late May, up to July at higher elevations. The larvae are the typical woodlouse shape, of a pale rose pink colour, varying to yellowish-green, the surface pitted with minute tubercles, each bearing a short, bristly hair. The pupae are dark reddish-brown, covered with short hairs on their upper surface, and of typical Lycaenid shape, with rounded head, humped thorax, and abdomen slightly constricted posteriorly.

OAKBLUES

259. Pale Himalayan Oakblue

Narathura sanesa
(Syn: *Amblypodia dodonea* sensu: Evans, 1932 or *Arhopala dodonaea*)
Wingspan 38–44mm

Description: Resembling *Chaetoprocta odata* in appearance but averaging slightly larger, this species is a light purple-blue above with broad black borders. It differs from *C. odata* in the forewing termen having a wavy margin, and with a more wedge-shaped tail on the hindwing. The female has broader black borders and is slightly paler blue. The underside is ochraceous brown with a silky gloss and with darker brown mottling and spots, much darker than the underside of *C. odata*, and also darker than the underside of *Amblypodia rama*.

Status: Fairly common in the Murree Hills, and collected from Nathia Gali and, by TJR, from Dunga Gali. In Chitral uncommon but occurring in the Ashtreth and Utzen valleys. Two broods, and on the wing from May to June, and again from August to October. It is found in the same habitat as the Walnut Blue, *Chaetoprocta odata*, and also flies at the same elevations as the Dark Himalayan Oakblue, *Narathura rama*.

Habits: As far as is known, the larval foodplants are *Quercus* spp., in the Murree Hills, *Quercus incana* and *Quercus dilatata*. The caterpillars are like all the family, being oval in outline when viewed from above, but differ in being rather flatter, less humped than in most other genera. According to Bell, they are never attended by ants. The adult butterflies rarely come to flowers and spend much of the day flying around the crowns of oak trees, or resting on the underside of an oak leaf. They are attracted to patches of damp mud.

260. Dark Himalayan Oakblue

Narathura rama
(*Amblypodia rama* in Evans, 1932 and Wynter-Blyth, 1957) (*Arhopala rama* in Marshall and de Niceville, 1890)
Wingspan 38–40mm

Description: Very similar to the Pale Himalayan Oakblue, *A. dodonea*, but both sexes are a darker purple-blue on the upper surface, and with their black borders narrower than in *A. dodonea*. However, the female has broader black borders very similar in width to typical females of *A. dodonea*. There is a short, rather blunt tail on the hindwing, and the forewing bears a black notch or spot end cell. The underside is pale purplish or greyish-brown with a silky gloss and darker bronzy-brown bars and spots, which are generally less distinct than in *A. dodonea*.

Status: It is commoner than the Pale Himalayan Oakblue (*N. dodonea*) in Kashmir but apparently does not extend as far west as *Narathura sanesa*, and has been collected in the Murree Hills, from Thandiani, Bara Gali, and through to Murree. It is on the wing from May to September, often flying sympatrically with *Narathura dodonea*. Like the former, it prefers lower, warmer elevations and sticks to areas where its foodplant, the Silver Oak (*Quercus incana*), grows, but has been collected in India as high as 2,700m (9,000ft). It is on the wing from May to June, and again from late July to October.

Habits: Like the Pale Himalayan Oakblue, the larval foodplants are *Quercus* spp., and the larvae are less humped and more flattened in profile than typical Lycaenid larvae. The adult butterflies do not normally come to flowers, and spend a lot of the day resting amongst the oak leaves, being reluctant to fly, but in May and June they will swarm around moist places on the ground.

261. Tailless Bushblue or Tailess Oakblue

Arhopola ganesa
(Syn: *Panchala ganesa* and *Narathura ganesa*)
Wingspan 32–37mm

Description: Smaller than the previous species, and with both sexes similar in appearance. Male and female are iridescent prussian blue above with a very broad black border on the forewing extending down the costa, and with a black V-shaped indentation post cell, separating two small white areas between. The black border on the hindwing is much narrower and, as its name implies, it is tailless. The underside is pale grey on both wings except for the forewing base and discal area which is white, and a broad blackish-grey discal band and spots on the wing base and end cell.

Status: Locally common in the side valleys of lower Chitral, and collected Ziarat below the Lowari Pass. More frequently met with in the lower Kaghan valley, and common in the Murree Hills, and has been collected in much the same areas as the Pale Himalayan Oakblue. On the wing from July to September, and preferring lower elevations where Silver Oak (*Quercus incana*) grows.

Habits: As far as is known, the larvae are green with a reddish lateral stripe, and feed on young leaves of the tree host, but the preferred tree species is not known. The larvae of this species are attended by tree ants. The adult butterflies are not easily seen as they fly around treetops and do not visit low-growing flowers.

SILVERLINES

Unlike the Hairstreaks, these fast-flying butterflies frequently visit low-growing flowers, and in Pakistan are generally associated with fairly open areas, avoiding true forest. They all have their underwing surface yellow ochre with darker transverse bars frosted with silver, and on the hindwing two tornal tails. The larvae of all are rather unique amongst Lycaenid larvae in having two prominent blackish pillars on segment 12, with extrusible hairs or fine tentacles at their tip when the larva is frightened. Their bodies are less humped and more cylindrical in shape and their heads are comparatively large.

262. Yellow Silverline

Apharitis epargyros
Wingspan 30–34mm

Description: All the Silverlines have the hindwing with the tornus slightly lobed and with two tails, one short one at vein 2 and a longer one at vein 1b. Their underside is usually pale buff with broad, silver-centred and dark-bordered lines and spots. *Apharitis epargyros* is tawny orange above with black spots and bars in the same position as the silver bars underneath. The underside is yellow-buff, and on the under hindwing a basal bar of three conjoined spots, a discal silver bar, and a short post-discal bar which bends inwards at its lower end to meet the discal bar. Females are similar but with the markings less dark on the upper surface.

Status: In our area only found in Balochistan. It is fairly common above 1,800m (6,000ft) and has been collected from the Urak valley, the Khojak Pass, and on the Speraragha road.

Habits: The larvae of all Silverlines are assiduously attended by ants, and consequently their foodplants can be quite varied, depending upon the occurrence of ants.

263. Tawny Silverline

Apharitis acamas
(Syn: *Spindasis hypargyrus* sensu: Butler, 1886 and *Aphnaeus hypargyrus*, sensu: Marshall and de Niceville, 1890)
Wingspan 30–36mm

Description: Like the former species, the Tawny Silverline is orange-tawny above, with black bars and spots, but the forewing sub-marginal bar is more clearly separated from the marginal black band than in *A. epargyros*, and on the hindwing the basal dark bar is inclined to be obscure and not as sharp as in the latter species. The underside is pale creamy white, and the discal silver bar is turned outwards at its lower end to meet the sub-marginal bar at vein 4. As in *A. epargyros*, there are three basal silver spots, but the lowest one is elongated, not round. Like the former species, there are two tails, the lower one being much longer than the one above.

Status: Commoner than the Yellow Silverline in Balochistan, occurring everywhere and on the wing from May to August. It occurs rather rarely in Sindh in the hilly tracts, and has been collected from Campbellpur. It is common in Chitral at low elevations and on the wing from May to August.

Habits: The early life history was described in detail by Fraser (1910) from specimens collected near Hyderabad, Sindh. The egg is dome-shaped with a flat underside, white in colour, and with its surface finely pitted. The larval foodplants are species of *Cassia* (Papilionaceae), especially *Cassia fistula* and *Cassia auriculata*. The newly-hatched larva is mahogany red in colour. As it grows bigger, it becomes pale fawn, with fine double lines of darker mahogany red. On the first two thoracic segments there is a shiny black chitinous plate whose function is unknown, and on segment 12 two black pillar-like bodies, from the tip of which long extrusible tentacles emerge when the larva is frightened. The larvae are always attended by ants.

264. Common Silverline

Spindasis vulcanus
Wingspan 26–34mm

Description: Like the *Apharitis* genus but with more restricted orange-tawny areas on the forewing, and on the hindwing, orange scales confined to the marginal region. The underside is pale yellow-buff with dark-bordered silver lines, those on the base of the hindwing being separate spots, and the outer basal band of spots not extending downwards to vein 1c, as it does in *Apharitis* Silverlines. There is a black tornal spot on the hindwing underside, orange-crowned. As in all the Silverlines, the females are often larger, with slightly more rounded forewings.

Status: This is the most widespread and common of the genus in Pakistan. It occurs rather rarely around Lahore and Sialkot, but is common in the Himalayan regions from Chitral, through Swat, Hazara district, and the Murree Hills. It flies between 1,200– 2,700m (4,000–9,000ft) in the hills, and is on the wing from June to July. Collected from Kalam, Swat, at 2,400m (8,000ft) by TJR.

Habits: As in all this genus, the egg is dome-shaped with a flat bottom and pitted all over with deep cells. The egg is green with the raised cell walls white. The larvae are rather oblong in shape, having a parallel-sided body, being less elliptical than typical Lycaenid larvae. Their heads are rather large for a Lycaenid, and shiny black in colour. Their bodies are covered with minute tubercles, and are pale green in colour, with a darker green mid-dorsal line centred with a narrow white line. The full-grown larva is about 11mm-long. The larval foodplants are varied, according to Bell (1919), and include *Clerodendron* spp., Rhamnaceae, *Zizyphus spp.*, *Canthium passiflorum*, and Rutaceae. The larvae are always attended by ants and typically spend the day when not feeding sheltered inside a rolled-up leaf or leaves, held together by their own spun silk. Often several larvae of different ages shelter together inside the same leaf nest. When about to pupate, the larva spins an open-ended white silk cocoon between two held-together leaves.

265. Common Shot Silverline

Spindasis ictis
Wingspan 27–32mm

Description: The sexes are slightly different in this species, the male being dark brown above but overlaid with dull purplish-blue scales, whereas the female has less blue overlay, especially on the upper forewing. Both have an orange, well-defined patch on the forewing apex, that in the female usually being more extensive and triangular in outline. There are two tails on the hindwing, with a black tornal spot. The underside is reddish-buff with the usual silver bars well defined.

Status: Recorded as occurring in Pakistan at low elevations by Cantlie (1962), and in Leslie and Evans (1903) recorded as common in Chitral, near Naghr at 1,200m (4,000ft), and also up to 2,700m (9,000ft). Their specimens were subsequently assigned to *Spindasis elima uniformis*. Recorded as common in the Murree Hills (Hasan, 1994). There are no Brit. Nat. Hist. Museum specimens from Pakistan. This butterfly prefers low hill country or open plains and is rather erratic in occurrence, but its exact status is problematical because of confusion with *S. elima*.

Habits: Like the rest of the genus, the egg is dome-shaped with rounded sides, and is pitted. The larvae are covered with minute tubercles. Mohyuddin (1987) found the larva of this Silverline feeding on the leaves of *Loranthus longiflorus* at Kahuta in the Murree foothills. Nothing seems to have been recorded elsewhere about the larval foodplants. The adult butterflies seem to seek out dryer open areas and avoid jungle or forest.

266. Scarce Shot Silverline

Spindasis elima
Wingspan 28–35mm

Description: *S. elima* males are similar to males of *S. ictis*, with the forewing margin and costa blackish-brown but the rest of the upper surface on both wings shot with purple-blue. It differs from *S. ictis* in that the orange patch on the forewing is at most very small and obscure and generally entirely absent. The female is dark brown without any dusting of purple-blue, but with a more well-defined orange patch on the forewing apex, which can be distinguished easily from females of *S. ictis* in being roughly oval in shape, not triangular as in the latter. Both sexes usually have a prominent tornal lobe on the hindwing which has an orange spot. The underside is pale khaki-yellow with the usual silver stripes.

Status: This Silverline is fairly common in Chitral, on the wing in June and July, and found from 1,200–2,700m (4,000–9,000ft). A few were collected on the Balochistan/Sindh border in the Hab river valley in 1885 by Swinhoe, and it probably occurs erratically after years of good monsoon rains. It has not been recorded elsewhere from Balochistan.

THE ROYALS

The only difference between the genus *Tajuria*, here represented, and *Pratapa* is that the latter (Tufted Royals) have a brand in the males on the upper hindwing, with a corresponding tuft on the dorsum of the under forewing, which is lacking in *Tajuria*.

267. Peacock Royal

Tajuria cippus
Wingspan 31–37mm

Description: This is the only species of this family represented in Pakistan. All the Royals have the males brilliantly coloured with metallic, iridescent shades of blue, and they are easily recognized by the presence of two thread-like tails on the hindwing between a tornal lobe. *Tajuria cippus* males are a beautiful green-blue colour, with the apex and wing margins broadly bordered with black. The slightly larger female is much paler, not so iridescent, and with a milkier blue colour, bordered with brownish-black, wider margins and forewing apex. Her upper hindwing also bears a series of curved black lines around the post-discal region, with several black spots sub-marginally around the tornal area, which the male lacks. Both sexes have the underwings plain grey-brown, with an interrupted sub-marginal black line across both fore and hindwings, and on the under hindwing, two prominent tornal black spots, bordered above by orange.

Status: The Royals and Tufted Royals are predominantly inhabitants of moist forested areas, and they are a wholly Oriental genus, with only one species extending as far north-west as the Himalayan foothills. *Tajuria cippus* has been collected at Lehtrar, in the Murree foothills, where it is on the wing from June to August, flying between 450–1,200m (1,500–4,000ft).

Habits: The eggs are hemispherical in shape, their surface minutely rugose, and they are white in colour. They are laid singly on the foodplant, on either a leaf axil or stem, and the larva feeds exclusively on the flowers and young shoots of the host plant, which is *Loranthus longifolius* (Mistletoe family), a common parasitic plant found only in a narrow foothill zone of Pakistan. Mohyuddin (1987) records finding their larva in June 1961 at Kahuta. The larvae, like most Lycaenids, have a small head hidden under the first thoracic segment, and are woodlouse-shaped, but 'waisted', with segments 5 to 10 being narrower than the front and rear segments when viewed from above. Their colour is ashy grey and they are not hairy; rather, they have a shiny, oily-looking surface. Their body is patterned dark grey or maroon in patches, with short-curved black lines on the sides of segments 6 to 10. The pupa differs from other Lycaenid pupa hitherto described in that it is never attached by a mid-body silk strand, but hangs freely from the tail only, and for this purpose the last anal segments 12 to 14 are attenuated into a stalk. The pupa is brownish-grey in colour with light green markings. When it pupates, it does so on the upper surface of a leaf or stem, attached by the tail only, but the head is usually twisted downwards, touching the surface of the stem or leaf. The adult butterflies rarely come to flowers, preferring to fly around the crowns of trees, especially such as Siris, *Albizzia* spp., favoured by the parasitic *Loranthus*.

CORNELIANS

268. Cornelian

Deudoryx epijarbas
Wingspan 34–44mm

Description: This is a thick-bodied insect, males being tawny-red above with the forewing cell black and a broad, dark brown border on the forewing costa and margin. The hindwing is tailed, with an orange spot in the tornal lobe which is black-centred. Females are a bit duller and paler, the brown areas being more extensive. The underside is paler brown with darker brown discal bars which are white-edged,

PLATE 39A: MEADOW BLUES, LIME BLUE AND GRASS BLUES

A. *Polyommatus eros stoliczkana* - Common or Eros Meadow Blue, male B. *Polyommatus florenciae* - Silvery Meadow Blue B-1. Male B-2. Female C. *Chilades laius* - Lime Blue C-1. Male C-2. Female D. *Freyeria trochilus* - Grass Jewel D-1. Male D-2. Female E. *Zizeeria maha* - Pale Grass Blue E-1. Male E-2. Female

PLATE 39B: MEADOW BLUES, LIME BLUE AND GRASS BLUES

F. *Zizeeria knysa* - Dark Grass Blue **F-1.** Male **F-2.** Female **G.** *Freyeria galba* - Persian Grass Blue **G-1.** Male **G-2.** Female **H.** *Zizina otis* - Lesser Grass Blue **H-1.** Male **H-2.** Female **I.** *Zizula hylax* - Tiny Grass Blue **I-1.** Male **I-2.** Female **J.** *Euchrysops cnejus* - Gram Blue, male

PLATE 40A: EUCHRYSOPS CUPIDS, PEA BLUE, CERULEANS AND LINE-BLUES

A. *Euchrysops parrhasius* - Small Cupid **A-1.** Male **A-2.** Female **B.** *Euchrysops pandava* - Plains Cupid **B-1.** Male **B-2.** Female **C.** *Catochrysops strabo* - Forget-Me-Not **C-1.** Male **C-2.** Female **D.** *Lampides boeticus* - Peablue **D-1.** Male **D-2.** Female

PLATE 40B: EUCHRYSOPS CUPIDS, PEA BLUE, CERULEANS AND LINE-BLUES

E. *Jamides bochus* - Dark Cerulean **E-1.** Male **E-2.** Female **F.** *Jamides celeno* - Common Cerulean, male **G.** *Nacaduba nora* - Common Line-Blue **G-1.** Male **G-2.** Female **H.** *Nacaduba dubiosa* - Tailless Line-Blue **H-1.** Male **H-2.** Female

PLATE 41A: COPPERS AND SAPPHIRES

A. *Lycaena pavana* - White-Bordered Copper **A-1.** Male **A-2.** Female **B.** *Lycaena phlaeas* - Common Copper, sexes alike **C.** *Lycaena kasyapa* - Green Copper, male **D.** *Lycaena solskyi* - Golden Copper **D-1.** Male **D-2.** Female **E.** *Lycaena phoenicurus* - Balochi Copper **E-1.** Male **E-2.** Female

PLATE 41B: COPPERS AND SAPPHIRES

F. *Lycaena caspius* - Purple Copper **F-1.** Male **F-2.** Female **G.** *Lycaena callimachus* - Red Copper **G-1.** Male **G-2.** Female **H.** *Heliophorus bakeri* - Western Blue Sapphire **H-1.** Male **H-2.** Female **I.** *Heliophorus tamu* - Powdery Green Sapphire **I-1.** Male **I-2.** Female **J.** *Heliophorus sena* - Sorrel Sapphire, male

PLATE 42A: HAIRSTREAKS, OAK BLUES AND PEACOCK ROYAL

A. *Neolycaena connae* - Balochi Hairstreak, sexes alike **B.** *Callophrys rubi* - Green Hairstreak, sexes alike **C.** *Strymon assamica* - White-Line Hairstreak, sexes alike **D.** *Euaspa milionia* - Water Hairstreak **D-1.** Male **D-2.** Female
E. *Chrysozephyrus ziha* - White Spotted Hairstreak, sexes alike **F.** *Chrysozephyrus ataxus* - Wonderful Hairstreak
F-1. Male **F-2.** Female

PLATE 42B: HAIRSTREAKS, OAK BLUES AND PEACOCK ROYAL

G. *Chrysozephyrus syla* - Silver Hairstreak **G-1.** Male **G-2.** Female **H.** *Chaetoprocta odata* - Walnut Blue, sexes alike **I.** *Narathura sanesa* - Pale Himalayan Oakblue, male **J.** *Narathura rama* - Dark Himalayan Oakblue **J-1.** Male **J-2.** Female **K.** *Arhopola ganesa* - Tailless Bushblue or Tailless Oakblue, sexes alike **L.** *Tajuria cippus* - Peacock Royal **L-1.** Male **L-2.** Female

PLATE 43A: SILVERLINES, CORNELIAN, GUAVA BLUE AND FLASHES

A. *Apharitis epargyros* - Yellow Silverline, sexes alike **B.** *Apharitis acamas* - Tawny Silverline, sexes alike **C.** *Spindasis vulcanus* - Common Silverline, sexes alike **D.** *Spindasis elima* - Scarce Shot Silverline **D-1.** Male **D-2.** Female **E.** *Spindasis ictis* - Common Shot Silverline **E-1.** Male **E-2.** Female **F.** *Deudoryx epijarbus* - Cornelian, sexes alike

PLATE 43B: SILVERLINES, CORNELIAN, GUAVA BLUE AND FLASHES

G. *Varachola asocrates* - Common Guava Blue **G-1.** Male **G-2.** Female **H.** *Rapala manea* - Slate Flash **H-1.** Male **H-2.** Female **I.** *Rapala nissa* - Common Flash, *sexes alike* **J.** *Rapala iarbus* - Indian Red Flash **J-1.** Male **J-2.** Female **K.** *Rapala selira* - Red Himalayan Flash **K-1.** Male **K-2.** Female **L.** *Rapala extensa* - Chitral Flash, male

PLATE 44A: AWLS AND FLATS

A. *Bibasis sena* - Orange Tail Awl or Pale Green Awlet, sexes alike **B.** *Hasora alexis* - Common Banded Awl, sexes alike **C.** *Badamia exclamationis* - Brown Awl, sexes alike **D.** *Choaspes xanthopogon* - Indian Awlking **E-1.** *Celaenorrrhinus leucocera* - Common Spotted Flat, male.

PLATE 44B: AWLS AND FLATS

E-2. Female **F.** *Celaenorrhinus munda* - Himalayan Spotted Flat, sexes alike **G.** *Sarangesa purendra* - Spotted Small Flat, sexes alike **H.** *Seseria dohertyi* - Himalayan White Flat, sexes alike **I.** *Lobocla bifasciatus* - Marbled Flat, sexes alike

PLATE 45A: ANGLES, AND SKIPPERS

A. *Caprona agama* - Spotted Angle, sexes alike **B.** *Caprona ransonnettii* - Golden Angle, sexes alike **C.** *Gomalia elma* - African Marbled Skipper, sexes alike **D.** *Hesperia zebra* - Zebra Skipper, sexes alike **E.** *Hesperia galba* - Indian Skipper, sexes alike **F.** *Hesperia sertorius* - Brick Skipper or Red Underwing of Europe, sexes alike

PLATE 45B: ANGLES, AND SKIPPERS

G. *Spialia phlomidis* - Persian Skipper, sexes alike **H.** *Spialia geron evanidus* - Sindh Skipper, sexes alike **I.** *Muschampia staudingeri* (Syn: *Syrichtus poggei*) - Syrian Skipper, sexes alike **J.** *Hesperia alpina* - Mountain Skipper, sexes alike **K.** *Carcharodus altheae* - Tufted Marble Skipper, sexes alike **L.** *Carcharodus alceae* - Plain Marble Skipper, sexes alike

PLATE 46A: INKY SKIPPER, SCRUB HOPPER, PALM BOB, DEMON, ACE AND DARTS

A-1. *Nisoniades marloyi* - Inky Skipper, male from Murree **A-2.** Female from Swat **B.** *Aeromachus inachus* - Veined Scrub Hopper, sexes alike **C.** *Saustus gremius* - Indian Palm Bob, sexes alike **D.** *Notocrypta fiesthamelii* - Spotted Demon, sexes alike **E.** *Halpe homolea* - Indian Ace, sexes alike **F.** *Actinor suada* - Veined Dart, sexes alike

PLATE 46B: INKY SKIPPER, SCRUB HOPPER, PALM BOB, DEMON, ACE AND DARTS

G. *Taractrocera danna* - Himalayan Grass Dart, sexes alike **H.** *Taractrocera maevius* - Common Grass Dart, sexes alike **I.** *Padraona* (Syn: *Potanthus*) *dara* - Himalayan Dart, sexes alike **J.** *Astycus pythias* - Dark Palm Dart, male

PLATE 47A: PALM DART, CHEQUERED DARTER, SWIFTS AND HOPPER

A. *Astycus* (Syn: *Telicota*) *pythias* - Pale Palm Dart, male **B.** *Hesperia comma* - Chequered Darter or Silver Spotted Skipper of Europe **B-1.** Male **B-2.** Female **C.** *Baoris eltola* - Yellow Spot Swift, sexes alike **D.** *Baoris discreta* - Himalayan Swift, sexes alike **E.** *Pelopidas* (Syn: *baoris*) *mathias* - Small Branded Swift, sexes alike

PLATE 47B: PALM DART, CHEQUERED DARTER, SWIFTS AND HOPPER

F. *Baoris guttatus* - Straight Swift, sexes alike **G.** *Guttatus* (Syn: *Baoris*) *bevani*, Bevan's Swift, sexes alike **H.** *Gegenes nostrodamus* - Dingy Swift, sexes alike **I.** *Eogenes alcides* - Torpedo, sexes alike **J.** *Eognenes leslei* - Leslie's Hopper, sexes alike and two tornal spots, the one between veins 2 and 3, orange with a black centre.

Status: Widely distributed in the foothills along the Himalayas from lower Chitral, through Hazara district from Battagram to Mansehra, and common around Abbotabad and the Murree foothills. It has been collected from Campbellpur. On the wing mainly from August to October, and usually below 2,100m (7,000ft), but has been collected from Murree in October at that altitude (Sevastopulo, 1947).

Habits: This species has a rather untypical life history for a butterfly, of passing the larval and pupal stages entirely concealed within a fruit. In this respect it is more closely related to *Virachloa isocrates*, the Guava Blue, which also develops inside a fruit, than it is to the Flashes, *Rapala* spp. Bell (1919) describes the egg as hemispherical, but depressed at the top, coloured a shining white and covered with square-shaped cells. The female lays her eggs singly on the foodplant, on leaf stalks, flower buds, or the axil between leaves and twigs. The little larva, upon hatching, makes its way to the nearest fruit, into which it bores, and thereafter continues feeding inside a succession of fruits as it grows larger. The larva is a dirty oily green, with the front part of its body yellow, the middle part more brownish, and the anal segments 7 to 8 pale orange. The constrictions between the segments are well marked. The anal end is covered with a thickened shield, which acts as a shovel, enabling the larva to clean out refuse and frass from the tunnel as it eats its way into the fruit. This anal shield is coloured a dirty brown-green. When fully grown the larva is about 16mm in length. The favourite foodplant is the Pomegranate, *Punica granatum*, which grows wild in the Murree foothills. They have also been found inside fruits of the Indian Horse Chestnut, *Aesculus indica*, and *Sapindus marginatus,* and *Connarus wightii*. The pupa is often formed inside the last fruit it cleans out and is covered with minute erect hairs. It is a rosy brown colour, covered with black spots and smudges. The larva is occasionally attended by ants. Despite the adults being very rapid, strong flyers, they regularly visit flowers, when they can easily be caught.

GUAVA BLUES

269. Common Guava Blue

Virachola isocrates
Wingspan 40–44mm

Description: This is a large Lycaenid showing sexual dimorphism. The male above is dull brown, shot through with purple scales, the forewing costa and margin being black. The upper hindwing is wholly overlaid with shining purple, and the termen margin slightly toothed; there is a prominent tornal lobe and tail at vein 2. The female is fuscous brown above with no trace of purple, but showing dull orange patches end cell on the forewing and a prominent, black-centred, orange tornal spot on the hindwing. The underside is grey-brown with slightly darker discal band, white-bordered, and two black tornal spots. The male on the under hindwing has a tuft of dark brown scent scales mid dorsum. Females are noticeably larger.

Status: It is rather rare and local in the plains of Pakistan. The Zoological Survey Department of Pakistan has collected a good series from Guava orchards in Malir and Landhi, outside Karachi, where it is quite plentiful, but nowhere else in Sindh. In the Murree Hills it is more widespread—the author collected it at 1,830m (6,000ft) near Malach village, below Dunga Gali, and there are several specimens from Murree in the Brit. Nat. Hist. Museum. It ascends to 2,100m (7,000ft) in the Himalayas and is always associated with stone-fruit orchards, such as Pomegranate (*Punica granatum*) or Guava (*Psidium guajava*).

Habits: The eggs are laid singly on a variety of potential foodplants, preferably fruits bearing large stones or drupes, as the larva bores inside such fruit seeds and feeds exclusively on the fruit kernel, not the pulpy exterior. The eggs are laid near small flower buds, as well as on leaves and twigs, and the newly-hatched larva wanders until it comes across a developing fruit, whereupon it bores into the calyx. The larva is very ugly, being covered with stiff black bristles and having its body rather shiny and oily-looking. It has been described as coloured dark indigo blue, with yellow bands on the two anterior segments, and a dorsal square white patch on segments 7 and 8. Marshall and de Niceville (1890) describe it as blackish-brown in colour with flesh-coloured patches. The anal end is covered with a shovel-like plate which is an adaptation enabling it to clean out its tunnel from time to time. The larval foodplants include the Loquat, *Eriobtrya japonica*, family Rosaceae, the Guava, *Psidium Guajava*, family Myrtaceae, Leguminosae, *Tamarindus indica*, Loganiaceae, *Strychnos nux-vomica*, Rubiaceae, *Gardenia latifolia*, and *Punica granatum*. A fascinating adaptation of this species is the instinct of the larvae to fasten the fruit inside which they are feeding onto its stem by strong silken ropes, which they spin after emerging from the fruit. As the effect of their feeding commonly causes the fruit to wither or rot and fall, they are thus secured from falling to the ground, where they would be vulnerable to many more predators. E.H. Aitken (1886) writes of this habit. He took a pomegranate fruit, which bore evidence of several larvae having bored their way inside and placed it in an eggcup, and by the next morning the fruit was so strongly attached by silken thread to the rim of the cup that he could lift it with the eggcup still attached. The larvae are occasionally tended by tree ants of various species, which may also help to keep the inside of the host fruit clean. The pupa is formed inside the shell of the fruit, attached by a silken tail pad and body girdle in the typical Lycaenid fashion.

FLASHES

270. Slate Flash

Rapala manea
(Syn: *Rapala schistacea* in Evans, 1932)
Wingspan 30–33mm

Description: The flashes are variably coloured on the upper surface, from tawny-red to purple, with dark margins on the forewing, generally widening at the apex, and a tail at vein 2. *Rapala manea* is, as its common name suggests, a slaty blue above, with black costa and margin forewing. The female is paler dull purple and larger. Both sexes have a tornal lobe on the hindwing and are tailed. The underside is quite dark grey-brown with a thin darker discal line white-edged outwardly only. There is a faint bar end cell on the forewing and two prominent black tornal spots, the upper one orange-crowned.

Status: Listed by Cantlie (1962) as occurring in Pakistan, but without location. This Flash is widely distributed in the plains in more mesic areas of India and occurs in the Himalayan foothills where it has been recorded from Mussoorie. There are no Pakistan specimens in the Brit. Nat. Hist. Museum collection but it probably occurs occasionally in the Murree foothills.

Habits: The eggs are laid singly on flowers of the foodplants, which are variable and include *Spiraea sorbifolia* in the Himalayas, *Zizyphus* spp. in the foothills, Leguminaceae such as *Acacia caesia*, and *Quisqualis indica*, family Combretaceae, as well as *Antidesma acidum*, family Euphorbiaceae. The larvae which feed entirely upon the flowers, are attended by *Formica nigra* and *Crematogaster* spp. of ants, and probably the female's choice of foodplant is triggered by the presence of ants rather than its botanical affinities. The larvae are quite unique in appearance, being parallel-sided but almost trapeze-shaped in cross-section, because their flanks are covered by a sub-dorsal and sub-spiracular row of tubercles or tiny teeth, which are dull crimson bordered with white. The sub-dorsal humps are covered with bristly black hairs. The overall colour of the larvae is purple-brown (Haribal, 1992) varying to pale yellowish-green (Sevastopulo, 1940). The larvae usually fall to the ground before pupating on a pad of silk concealed under a dead leaf. They are normal in shape, their surface being covered with minute red hairs. They are pink in colour, speckled with black, the wing cases olive green.

271. Indian Red Flash

Rapala iarbus
(Syn: *Rapala melampus* in Evans, 1932)
Wingspan 33–38mm

Description: There has been confusion in the past both in the use of common names for the Flashes and in their taxonomy. *R. iarbus* is called the Common Red Flash in Wynter-Blyth (1957) and the Indian Red Flash in Cantlie (1962). *Rapala melampus*, formerly applied to the Common Red Flash, is preoccupied (Cantlie, *op. cit.*). *R. iarbus* males are red above and, in the Pakistan population, with no overlay of purple scales. The forewing costa is dark brown, as is the wing margin, but much less broadly than in the Cornelian (*Deudoryx epijarbas*). The hindwing is all red with a tornal lobe and tail at vein 2. The female is a duller, coppery red with broad, dark brown margins. The underside is pale grey with a very fine discal darker line and the two tornal black spots usually without orange crowns.

Status: This Flash occurs quite commonly in the Murree Hills, and has been collected from Dunga Gali (by TJR) and Murree. Also occurs Peshawar. It flies from 300–2,300m (1,000–7,500ft), and is on the wing from May to September. It has also been collected from Frere Gardens, Karachi, (Sindh) in the late 1960s by the Zoological Survey Department.

Habits: The eggs are button-shaped with a shining surface, light green in colour, and covered with shallow hexagonal cells. According to Bell, they are laid singly on *Rubus* spp., particularly wild Blackberries, and they are not much attended by ants, in contrast to the Slate Flash. However, Mohyuddin (1987) records finding the larvae on *Mangifera indica* in May near Rawalpindi, and on *Acacia nilotica* flowers at Lahore. The larvae are rather flattened when viewed from the side, and covered with a sub-dorsal and sub-spiracular row of tubercles or teeth, each tipped with a brush of bristly black or white hairs. The body colour of the larvae varies from ochraceous (Marshall and de Niceville, 1890) to transparent white with some green marbling (Bell, 1915). The pupa is brownish-pink covered all over with black specks, and there are minute hairs covering the first two segments. Like all the Flashes, the adults have a strong, swift flight and readily visit flowers.

272. Red Himalayan Flash

Rapala selira
(Syn: *Hysudra selira* or *Rapala micans* in Evans, 1932)
Wingspan 32–34mm

Description: In this species the male is shot with an overlay of purple scales, leaving a broad black border, and with a squarish-orange patch on the forewing. The upper hindwing has an orange-red band along the termen, both orange areas crossed by black veins. The female is dark brown above, with only a light dusting of purple scales and broader orange areas. Both sexes are lobed at vein 1a and tailed at vein 1c. The underside is ochraceous brown with a slightly broader dark discal band than in *R. iarbus*, and this is prominently white-edged outwardly.

Status: This is the commonest Flash occurring in Pakistan, from Chitral, Swat, and Hazara district, as well as the Murree Hills. It flies from 2,100–3,000m (7,000–10,000ft) in side valleys of lower Chitral, and from 2,400–2,700m (8,000–9,000ft) in the Murree Hills. Collected TJR Dunga Gali, and is on the wing during May and June.

Habits: Similar to the previous species, the foodplant being species of Blackberry, particularly *Rubus sanctus* in the Murree Hills.

273. Chitral Flash

Rapala extensa
Wingspan 23–33mm

Description: This is one of the smaller Flashes, with both male and female alike, being red above, without prominent black veins, and with a dark brown margin and costa on the forewing. There is no purple gloss in the male. The underside is pale grey, with a slightly darker, narrow discal band petering out on the hindwing below vein 2. There is a tornal lobe and the tail is not well developed. The bar end cell is very faint on the forewing.

Status: This Flash is common in Chitral at medium elevations. It has not been recorded elsewhere in Pakistan, but Sakai (1981) found it widespread in Afghanistan.

274. Common Flash

Rapala nissa
Wingspan 34–38mm

Description: This is a rather variable species. The male is steely blue above with broad dark borders on the forewing, and with an orange discal patch. The upper hindwing has a broad orange margin. The female is paler, and browner overall on the upper surface. The underside varies from dark ochraceous to reddish-brown, sometimes with a rosy overlay, and with the darker discal band quite prominent and white edged outwardly. The end cell bars are quite distinct.

Status: In Pakistan it has only been collected from Murree in August, where it was described as common (Butler, 1886). Further east, it has been collected in Kashmir, where it is uncommon, but it occurs commonly in the other north-western Himalayan regions of India.

Chapter 12
Riodininae (Erycnidae)—Judys and Punches

Classification

This is a family with comparatively poor representation in the Old World and the Indian subcontinent, but richly represented in South America with more than 900 species. They are similar to the Lycaenidae in many respects, particularly in the shape and size of the larvae. The butterflies are all small in size and stout bodied, usually with a tornal lobe and tail on the hindwing. They are distinguished from the Lycaenidae by the extraordinary sexual dimorphism in the development of the forelegs. In the male these are brush-like and tiny, but in the female they are jointed and with claws, fully developed for walking. The eggs are small, domed in shape and white, and the larvae are like the Lycaenidae in shape, but the head is not hidden under the prothorax, and they lack the honey gland. They are green in colour. Their pupae are also like those of the Lycaenidae, usually rather squat and broad, attached to the foodplant by a silken thread across the thorax and by the tail claspers. The foodplants are Asteraceae, Fabaceae (Leguminosae) Myrsinaceae, and Myrtaceae.

275. Common Punch

Dodona durga
Wingspan 30–40mm

Description: The Punches are mostly small brown insects with white or yellow spots and bands, and either a small tail or tornal lobe. *Dodona durga* resembles the other two species occurring in Pakistan, but can be distinguished by several characters. There is a small yellow spot at the base of the forewing inside the cell, absent in *D. dipoea* and *D. eugenes*. This species only has a tornal lobe on the hindwing, no tail, which the other two usually possess. On the upper forewing there are two ochraceous yellow bars, one mid cell and the second end cell. The other two species lack the yellow bar end cell. On the underside there are two black spots on the hindwing near the apex, a feature shared by the other species. The bars on the hindwing upper side are tawny-orange rather than ochraceous. The sexes are alike.

Status: Very common at lower altitudes in the Himalayan foothills from Chitral through lower Swat, the Murree foothills, collected TJR from Manga valley, and especially Lehtrar, where it swarms on damp patches. Found from 760–2,100m (2,500–7,000ft) and on the wing from April to September.

Habits: The eggs are shaped like a truncated cone, and the female lays them on the underside of leaves, usually in small batches, side by side. They are coloured white becoming mauve. The larvae are cylindrical in shape, and striped with alternate lines of yellow-green, blue, and olive-green, with the lower sides sparsely clothed with black bristles. The larval foodplants belong to the family Myrsinaceae, and in the Murree Hills the preferred host plants are *Maesa semiserrata* and *Embelia robusta*. The pupa is formed on a pad of silk, held by a body band of silk and its tail claspers. It is rounded in shape with the thorax slightly domed, and its head bears two rounded lobes. It is yellow-green in colour with blue dorsal and sub-dorsal lateral lines. Like all the Punches, the adult butterflies are fond of settling on damp patches on the ground, and visit flowers less frequently. They settle with their wings closed and often have a favourite leaf perch from where they aggressively chase conspecifics.

276. Lesser Punch

Dodona dipoea
Wingspan 35–45mm

Description: Despite its common name, this species is on average slightly larger than *D. durga*. Above it is slightly darker brown than the Common Punch, with narrower buff markings. It also lacks the yellow-buff bar end cell on the forewing, the tornal area of the hindwing lacks tawny-orange bands and the creamy bands in that area are rather obscure. There is a tornal lobe and often a slight tail. On the underside the pale yellow bars are markedly narrower than in the other two species of Punch, and on the under forewing the costal yellow spot alongside the end cell bar is separated and shifted in slightly, whereas in the other two the costal spot forms a continuous bar. The under hindwing has very obscure paler bands.

Status: Has only been recorded from our area from Murree, where it is rare.

Habits: There is little recorded about this species, but its habits and life history are believed to be similar to the Common Punch. The white, dome-shaped eggs are laid in small batches on the foodplants, which are members of the families Mysinaceae and Astercaeae. The caterpillars are broadest in the middle, tapering at both ends, and with a relatively small head like the Pieridae, and their bodies are covered with minute bristles. The pupae are similar in shape to those of the Lycaenidae, and are formed on a bed of silk, attached by a silk band around the thorax and by the tail claspers. The adult butterflies are rather weak flyers, never travelling far without settling, and regularly visiting damp patches or sucking moisture from leaves.

277. Tailed Punch

Dodona eugenes
Wingspan 35–45mm

Description: As its common name implies, this species has a prominent tail. The upper hindwing has tawny bars, not yellow, and the under forewing has the yellow costal spot forming a continuous bar with that end cell. The under hindwing is its most distinctive feature, there being a broad silvery-white bar from the costa across the mid cell, and the short bar before the apex is wider than the others are.

Status: Occurs rather uncommonly in the Murree Hills from Thandiani across to Murree, from medium elevations (Kala Pani at 1,500m [5,000ft]) up to 2,700m (9,000ft), and collected TJR Dunga Gali in late June. Flying from mid-May to September. It generally occurs at higher elevations than *Dodona durga*.

Habits: The larval foodplants of this species in India are bamboos (Arundinaria) and in the western part of their range, including Pakistan, their eggs are laid on grasses. The larvae are described as slug-shaped, of a pale emerald green colour, with two blue dorsal lines (Haribal, 1992). The pupae are green.

Chapter 13
Hesperiidae—Skippers

Classification

Skippers occur in all the major zoogeographic regions of the world, and, being rather drab and predominantly brown in colour, with swift, darting flight, they have always presented difficulties to the taxonomist, both in individual identification and in their phylogenetic relationships within the family Hesperiidae. The family Hesperiidae differs from all the other Rhopalocera (butterflies as distinct from moths) in several important respects, to the extent that most taxonomists believe they represent a more primitive group within the order, and that they exhibit some characters more moth-like than all the other butterfly families. Most authorities now consider that they should accordingly be placed in a separate super-family Hesperioidea (Ackery, 1984). Hesperiidae all have the wing venation arising directly from the cell and thereafter no veins are branched. All have the forelegs perfectly developed, and usually with the hind pair (metathoracic) bearing two pairs of spines on the tibia. In the adult stage their heads are large and wider than the thorax, with prominent eyes, and their antennae are often hooked at the thickened tip. Adults are usually small and brown in colouration, often with semi-transparent (hyaline) white spots, and they typically have a very fast, darting flight. The larvae are unique in that they are usually with a large head much wider than the first segment or 'neck', and they build a shelter for themselves from the time of emergence from the egg: if their foodplants are dycotyledonous, by sewing together two leaves or by folding a leaf segment around their bodies; if their foodplants are monocotyledonous, by making a tunnel of grass stems bound with their silk. Often the larvae tend to be shy and sluggish in habits. Usually they feed only at night, and their bodies are delicate, being semi-transparent when exposed, with the internal organs sometimes clearly visible through the skin. Some larvae are brightly coloured but usually they are green or whitish, and they never have fleshy spines as in other families such as some Lycaenidae and Nymphalidae. The pupae resemble those of moths in that they are smooth, without any protrusions, and are rather long in shape, and tapering. They also are developed inside a protective grass or leaf shelter, or in some species in a silken cocoon on the ground. In some species (the Awls), the pupa is covered with a white waxy secretion. Their eggs, however, can be very variable in shape and are difficult to characterize for the whole family, but generally they are dome-shaped and depressed in the middle at the top. They can be smooth or ribbed and often are white or red in colour. Compared to other families, the females lay rather large eggs for their size, but fewer in number. 'Skippers' are widely represented throughout the world, with an estimated 3,500 species worldwide (Ackery, 1984), and about 300 species (Haribal, 1992) recorded from the Indian subcontinent. Pakistan has about 40 species definitely recorded, with poor representation of the more tropical Oriental species, but with a great variety of the smaller Palearctic genera such as *Gomalia, Hesperia, Spialia,* and *Carcharodus*. As indicated above, identification within some genera is very difficult and there is still some disagreement among taxonomists. I follow Evans (1949), in his classical *Catalogue of the Hesperiidae* etc., in the British Museum collection. A more detailed diagnosis of the three main sub-families is given below.

Sub-Family COELIADINAE

These are rather larger than is typical for the whole family, with the hindwing lobed, and the antennae short and with the tip (apiculus) sharp-pointed and curved. The hindwing vein 5 is tubular, arising nearer to vein 6. The larvae are brightly coloured, and their foodplants are dicotyledonous, including Araliaceae, Asclepiadaceae, Combretaceae, Fabaceae, Malphigiaceae, Myristaceae, Myrsinaceae, Sabiaceae, and Zingiberaceae. In the Indian subcontinent, they include the Awls and Awlkings.

THE AWLS

278. The Orange Tail Awl or Pale Green Awlet in Haribal (1992)

Bibasis sena
Wingspan 45–50mm

Description: The Awls are the largest, and some of them the most colourful, amongst this rather drab family. They get their common name from the peculiar shape of the palpi which, comprising three segments, has the third or outer one thin, naked, and awl shaped (*see*, Fig. 8). *Bibasis sena* is plain olive brown above, unmarked, but with darker brown wing borders. The underside has prominent orange cilia on the hindwing termen, and a broad white band through the forewing in the mid-dorsal area and across the hindwing, which is diffused along its outer border. The forelegs are often orange and some hairs on the tip of the abdomen.

Status: Collected from Tret at 900m (3,000ft). Likely to occur Kahuta and the Margallas.

Habits: The eggs are dome-shaped with thirteen longitudinal ribs, and when first laid are honey coloured, later becoming paler and showing dark blotches. They are laid singly on the underside of the foodplants, which are Combretaceae. In the Murree Hills, most likely to be *Terminalia* spp. or *Combretum nanum*. The larva and pupa are illustrated in Wynter-Blyth (1957), and Bell (1925) describes the larva as a watery-looking light green, with a broad dorsal smoky black line, bordered by two light-blue dorso-lateral lines, and a bright yellow sub-dorsal line, and with velvet black patches covering the sides of each segment. They attain a length of 35mm when fully grown, by which time they are quite sluggish and feed only at night. They make a cell or shelter for themselves by folding over a leaf and binding this with silk. The adult butterflies appear never to visit flowers and are somewhat crepuscular in habits, active mainly in the early morning and keeping to shady forested areas.

279. The Common Banded Awl

Hasora chromus
(Syn: *Hasora alexis*)
Wingspan 45–50mm

Description: The male is dark brown above, unmarked, but the female has two small semi-transparent yellowish-white spots in the discal area between veins 2 and 3, and 3 and 4. The underside is brown with a purplish gloss, and a rather narrow, bluish-white discal band across the hindwing, which is diffused outwardly. There is also a blackish tornal patch on the hindwing.

Status: Collected TJR in Murree Hills, where not uncommon. Very rare around Karachi (Menesse, 1950). Also collected from Quetta, Balochistan (Evans 1933), where uncommon.

Habits: The egg is hemispherical in shape, with sixteen to eighteen longitudinal ribs, and is light green in colouration, becoming dirty white with age. The larval foodplants are *Ricinus communis*, *Pongamia pinnata* (Papilionoideae), and *Heymnia trijuga* (Meliaceae). The larvae are sub-cylindrical and hairy. Their colour is dark mauve, suffused with whitish-yellow dorsally, and with a dorsal mauve line and lateral yellowish-green line. They grow rapidly, becoming about 30mm in length before pupating—the adult butterfly in those specimens reared by Bell (1924) emerged in fourteen days from the time the eggs were laid. The adult butterflies have a skipping flight, and, unlike the previous species, will visit flowers and are not averse to flying in bright sunshine.

280. The Brown Awl

Badamia exclamationis
Wingspan 50–55mm

Description: The forewing of *Badamia exclamationis* is elongated and with the termen sloping sharply outwards. Both sexes are dark brown above, with the base of the hindwing ochraceous, and on the forewing yellowish-white semi-transparent spots, a smaller one in the cell, an elongated one between veins 2 and 3, and a third small triangular spot between veins 3 and 4. The female has these same spots larger. The underside is a lighter brown, with the tornus darker and an indistinct yellowish bar above.

Status: Very common in Murree Hills. Collected TJR in Dunga Gali at 2,400m (8,000ft). Collected from Lahore and occasionally straggles to the outer foothills. Very rare around Karachi (Menesse, 1950), collected Zoological Survey of Pakistan in late 1960s from Government Nursery, Karachi.

Habits: Detailed descriptions have been published of the early stages, both by Dudgeon (1895) and Bell (1925). The eggs are hemispherical, with a flat bottom and bearing thirteen longitudinal beaded ribs. They are light green in colour with the ribs white. The cylindrically-shaped larvae are yellow in colour with a yellow-black spiracular and dorsal band, sometimes with a black dorsal line. Their heads are heart-shaped when viewed from in front, yellow with a black band across the top, and their necks are banded black with a collar of two yellow bands. They attain a length of about 40mm. The larval foodplants are members of the Combretaceae, including *Combretum extensum* and *Terminalia belerica*. The larva binds the edges of a leaf together with its silk and shelters inside. The pupae have a square head, and vary from light yellowish-brown to dark yellow in colour, becoming chalky white, and they are wrapped up in a rolled leaf held together with thick white strands of silk. The adult butterflies are extremely fast, strong flyers, and when they settle, do so on the underside of a leaf, with wings folded tightly over their backs, resembling an exclamation mark with their large rounded head. They are mainly on the wing in early morning and rarely come to flowers.

THE AWLKINGS

281. The Indian Awlking

Choaspes xanthopogon
(Syn: *Choaspes benjaminii*)
Wingspan 50–60mm

Description: Both sexes dark brown above, unmarked, but with an overlay of bluish-purple hairs, and on the upper hindwing, a black border to the tornal area, with a bright orange outer border extending up the termen. The underside is shining grey-green with black veins, and the hindwing black tornal patch very prominent, with the same orange outer border.

Status: Presumed rare. Recorded from Kashmir (Evans, 1932), and flying between 1,000 and 2,400m (3,500 and 8,000ft).

Habits: The early life history of this Awlking was described by Mackinnon and de Niceville (1898) with coloured plates figuring two different types of larvae collected in the Mussoorie Hills, India. One form of caterpillar is black with a red head, bearing three black spots and the rest of the body with broad yellow inter-segmental bands and small blue spots on the side of each segment. The second form is also black with a red head, but the body is decorated with triangular yellow spots on the dorsum of each segment, and blue sub-dorsal spots on each segment, with a yellow spiracular band the whole length of its body. They attain a length of 45mm. While feeding they shelter inside a rolled-up leaf held together with their silk. The pupa is pale pink with black spots, and is covered with white waxy secretions. The larval foodplants in the Mussoorie region are *Sabia campanulata* and *Meliosma pungens*, family Sabiaceae. Both these plants occur in the Murree Hills. The adults are attracted to bird droppings as well as damp patches, and occasionally visit flowers. They are mainly active in the early morning and evening, and are on the wing from March to September.

Sub-Family PYRGINAE

THE FLATS

This sub-family includes the 'Flats' and the 'Angles' and, as their common name suggests, butterflies of this group generally settle with their wings fully open and laid flat. They are usually dark brown above with prominent white bands or spots. They are characterized by the forewing vein 5 not being curved at the base, not arising very near vein 4, but near vein 6 at its origin. The cell is either long or approximately two-thirds the length of the costa, and the antennae end in a point. The larvae are either brightly coloured, or uniformly green or white, slender in shape, and pilose (Bell, 1921). Larval host plants belong to the families Combretaceae, Fabacae (Leguminosae), Lauracae, Malvaceae, Myrtaceae, Rosaceae, Sterculaceae, and Verbenaceae. Caterpillars form shelters by folding and joining the leaves of their host plant.

282. The Common Spotted Flat

Celaenorrhinus leucocera
(Syn: *Leucocera leucocera*)
Wingspan 45–55mm

Description: The Flats of the genus *Celaenorrhinus* are velvety brown insects with white spots and bands prominent on the forewing, and usually smaller, dull yellow spots on the hindwing. They are chiefly distinguished from other Skippers by their habit of always resting with the wings open and flat, hence their common name. *C. leucocera* has a broad white semi-transparent discal band, with the spot in between veins 2 and 3 very large and almost square, and also a large spot end cell. There are three small apical spots, and on the upper hindwing a yellow spot end cell and fainter spots post-discally. The upper hindwing cilia (fringe) is yellowish. The antennae are white on the shaft. The underside has more prominent white spots on the under hindwing.

Status: Collected Murree Hills, only at Khaira Gali, where flying at 2,400m (8,000ft). Not as common as *C. munda*, but probably overlooked because of its similarity to the Himalayan Spotted Flat.

Habits: The female lays her eggs singly on the top or bottom of leaves of the family Acanthaceae. The eggs are dome-shaped with a shining surface, and greenish-white in colour. They are covered with very fine, close-together ribs, numbering fifty to sixty around the circumference. The larvae are dark olive green in colour, with two greenish-white lateral and spiracular bands. Their surface is covered with minute white star-topped hairs. They have translucent skin showing the internal organs beneath. When full-grown they reach a length of about 29mm. The pupae are a golden red-brown in colour and have their proboscis protruding from beyond the end of their wing cases. The larvae when small make a cell for themselves by cutting a tiny triangular flap at the edge of a leaf and binding this together with silk. As they grow larger they join whole leaves together and pupate inside a cell formed by two such leaves, which generally wither and fall to the ground with the pupa secure inside. The adult butterflies frequent shady places and settle on the underside of leaves, but will visit flowers. They are mainly active in the early morning and evening. The recorded foodplants are *Eranthemum pulchellum*.

283. The Himalayan Spotted Flat

Celaenorrhinus munda
(Syn: *Leucocera munda*)
Wingspan 45–50mm

Description: This species is very similar to *C. leucocera*, differing only in the absence of any yellowish spots on the upper hindwing, and with the cilia on the hindwing white, not yellow. The antennae shafts are chequered, with the clubs white. There is one small yellowish spot end cell on the under hindwing. The upper forewing white discal band extends into the costa, whereas in *C. leucocera* the white discal band does not usually extend beyond the cell, the costa being brown. Both this and the former species can be variable, and the colour of the fringe is not a reliable character; the presence of spots on the upper hindwing is the best distinguishing feature.

Status: Very common through the Galis and Murree Hills. Collected TJR Dunga Gali at 2,400m (8,000ft).

Habits: All the *Celaenorrhinus* 'Flats' in the larval stage make a shelter or cell for themselves, lined with their own silk, by cutting a corner of a leaf edge and binding this over. They also pupate in a similar leaf cell lined with silk. It is believed that the foodplants of the Himalayan Spotted Flat are *Impatiens* spp.

284. The Spotted Small Flat

Sarangesa purendra
Wingspan 30–35mm

Description: This species is dark brown, mottled on the upper hindwing with darker patches, and bearing small transparent yellowish spots on the forewing, a large square one between veins 1 and 2, a single spot in the cell, and a third above it in the costa, and three small apical spots. The fringe on the hindwing is obscurely chequered. The underside hindwing shows obscure yellowish spots, two on the costa, a row around the sub-margin, and a larger one end cell.

Status: Collected Sindh, both from Karachi district and Hyderabad (Bell, 1924). Commoner in Kutch, India.

Habits: The eggs are laid singly anywhere on a leaf stalk or leaf, and they are dome-shaped, with fifteen longitudinal ribs. They are coloured light green at first, becoming red before the larva emerges, and the ribs are white. The larvae are spindle-shaped, wider in the middle segments, tapering at both ends, and with their bodies covered with tiny feathered hairs. Their colour is dark olive green except for the first thoracic segment, which is white, and with the whole body showing a pulsating dark dorsal line. Their heads are large and slightly bi-lobed in front. The larval foodplants are members of the Acanthaceae, probably *Blepharis* spp. in Pakistan. The larvae make a shelter for themselves by turning over a triangular piece cut from a leaf edge, and, like all the 'Flats', rest inside this with their heads turned round on their side. When bigger they combine several leaves together with their silk threads, usually choosing one that is withered and two green leaves, and they pupate inside this leaf cell. The pupae are blunt and square in front, with a pointed tail, and fattest in the middle. In colour they are dark yellowish-brown, with the whole body covered by a bluish-white bloom. The butterflies usually fly low down and with a fast and jerky trajectory. They will settle on flowers, and bask on low leaves with the wings held out horizontally.

285. The Himalayan White Flat

Seseria dohertyi
Wingspan 45–50mm

Description: A striking species, with the upper side dark brown except for a broad white band on the forewing from vein 1 up to vein 3, and the hindwing brown at the base, bordered by a white area covering most of the wing except the margin, which is brown, and bearing a series of black spots between the white area and the margin. The upper forewing has small white spots apically and two larger spots between veins 3 and 4, and 4 and 5. The abdomen is white above except for the tip, which is brown. The underside is ochraceous on the forewing with a similar pattern of white spots, and the under hindwing shows two black spots in between vein 7 and 8, as well as the row of sub-marginal spots.

Status: Collected Murree, where uncommon, but less rare further east around Mussoorie.

286. The Marbled Flat

Lobocla liliana
Wingspan 45–55mm

Description: The Marbled Flat is dark brown above, with a broad white semi-transparent discal band or series of conjoined spots, and much smaller spots in the apical region of the forewing. The upper hindwing is unmarked, with a chequered border. The underside hindwing bears irregular bands dusted with greyish-white giving it a marbled effect. The under forewing is speckled with greyish-white in the apical area only. The termen of the forewing is slightly convex.

Status: Collected Murree, where rare. Commoner Mussoorie eastwards.

THE ANGLES

287. Spotted Angle

Caprona agama mettasuta
Wingspan 30–50mm

Description: The Angles have the termen of the hind and forewings crenulated or slightly toothed, and they are generally tawny brown above with obscure white patches. *Caprona agama* is larger than *Caprona ransonetti*, and a darker brown above. It has clear white spots on the forewing, one small spot base of cell and a crescent end cell, with three small elongated spots in the forewing apex. There are no white spots on the upper hindwing, but three rows of rather obscure darker spots. The underside is pale brown, frosted with white, and with fine black lines along the hindwing margin. Specimens from Pakistan are smaller and without the black spots on the under hindwing characteristic of this species in the Far Eastern parts of its range (Myanmar, Indonesia).

Status: Collected Nurpur Sharif, in the Margalla Hills and presumed more widespread in the Murree foothills further east around Tret and Kahuta.

288. Golden Angle

Caprona ransonetti yerburi
Wingspan 30–40mm

Description: As its common name implies, this species is a golden tawny-yellow above, with white semi-transparent spots on the forewing, usually two in between veins 1b and 2, a much larger one between veins 2 and 3, and a large one between veins 3 and 4. There are usually two small white spots end cell on the forewing also, and three conjoined apical spots. The upper forewing has an obscure row of white spots sub-marginally and both wings are irregularly streaked with darker brown. The underside is white with buffy-yellow borders and two or more rows of discal and post-discal, dark grey spots.

Status: Only recorded from Fateh Khan Rest House, Campbellpur, (now Attock) in the Punjab Salt Range. This forest-loving insect must be presumed rare and local in Pakistan.

Habits: The eggs are always laid singly on the top of leaves. They are dome-shaped with a flattened top, and pinkish in colour, with seventeen to eighteen longitudinal white ribs. The tops of the eggs are covered with curly white hairs shed from the abdomen of the butterfly. The larvae are yellowish-green in colour with large maroon-coloured heads and constricted necks. They are spindle-shaped, being fattest in the middle segments. Their bodies are also covered with minute white conical tubercles. The tiny larva makes a very characteristic shelter for itself. First it cuts a curved piece in the middle of a leaf, leaving one edge uncut to act as a hinge, then pulls it over to form a circular flap which is firmly held down all around the edge with silk, except for a small exit gap used when the larva emerges to feed. Inside the cell it rests on a bed of silk, and on the 'lid', not the leaf itself, so that it is upside down. As it grows bigger it makes fresh cells, turning over a triangular corner of a larger leaf. The foodplants belong to the Sterculaceae and, in Pakistan *Helicteres isora*, widespread in the foothills and the NWFP. The pupae are formed in the same leaf cell used by the larvae in their later stages of growth, and they are rather stout in shape with a square head, and coloured pale yellowish-green with a white bloom, speckled with black dots. The butterflies have a strong, fast flight, keeping close to the ground, and are quite tolerant of bright sunshine, visiting flowers and damp patches, and often basking with wings held flat low down on a bush leaf.

Sub-Family HESPERIINAE

THE SKIPPERS

These are typically rather small butterflies, and when at rest they hold their wings erect, not flat. They are characterized by the forewing vein 5 arising nearer vein 4 than vein 6, and their antennae are not pointed. The males have brands on their upper forewing, sometimes a conspicuous stigma (genera *Augias*, and *Pamphila*), but never with a costal fold (Ackery, 1984). The larvae are smooth or finely pubescent, with a large head and constricted neck (prothorax), and generally coloured greenish-yellow or white. Representatives are distributed worldwide. The larvae often construct tunnels of grass woven together for their shelter. Major larval foodplants belong to the families Arecaceae, Combretaceae, Cyperaceae, Liliaceae, Musaceae, Pandanaceae, Poaceae, Sapindaceae, and Zingiberaceae.

289. The African Marbled Skipper

Gomalia elma
Wingspan 25mm

Description: The Skippers are all rather small butterflies, dark brown above with chequered white spots and a paler, greener underside, and the pattern on the under hindwing is often crucial to identification. *Gomalia elma* is untypical of the rest of this group in that the upper forewing is more olive brown, without numerous white spots, there being only one bar of white spots across from veins 6 to 9. There is a thin black band across the middle of the forewing, margined with paler olive green on the outer edge. The upper hindwing bears a small white spot at the base of the cell, and a broad white band from veins 1 to 7. The underside is largely white with two straight edged, dull brown-greenish bands on the hindwing, and a third outer sub-marginal band.

Status: Collected Karachi, Sindh (sub-sp. *litoralis*), where it is comparatively uncommon and on the wing in July. Also collected from Quetta and Mach in Balochistan. Larvae found (Mohyuddin, 1987) on *Abutilon indicum,* Rawalpindi in August, and in September.

Habits: The eggs are dome-shaped but peculiar in structure, their surface being covered with irregular prominent rounded knobs, and the top bearing a circular lid which the tiny larva bites open when ready to emerge. The eggs are coloured white at first, darkening to reddish-orange before hatching. They are laid on both young and mature leaves, and the foodplants belong to the Malvaceae, particularly *Abutilon indicum*. The little larva makes itself a shelter by cutting a nearly circular piece from the edge of a leaf and turning this over to form a flap, the inside of which it lines with silk. As it gets larger it cuts larger sections of leaf, rolling these inwards, and it pupates inside a cell made from a folded-over whole leaf. The larvae have rounded bodies with flattened ventrum, tapering at the anal end, and with the head slightly bi-lobed and dark red-brown in colour. The rest of the body is pale greenish-white and covered with fine translucent hairs, some of them branched at their tips. They attain a length of about 20mm. The pupae are covered with a waxy white powder and are pale yellowish-brown in colour, with their surface also covered with fine white hairs. Bell (1924), in rearing this butterfly from eggs, reported a forty-day generation, from the time of egg-laying on 8 June, emergence of the larva on 10 June, pupation on 3 July, and emergence of the butterfly on 16 July. The butterflies have the typical swift jerky flight of Skippers, and visit flowers mainly in the early morning and afternoons, resting during the middle of the day with their wings half raised, the forewings at a higher angle than the hindwings. They frequent fairly open, dry, bush-studded areas.

290. The Zebra Skipper

Hesperia zebra
(Syn: *Syrichtus zebra* and *Spialia zebra*)
Wingspan 25mm

Description: Above dark blackish-brown with prominent white spots and chequered wing margins. The upper forewing has an elongated spot between veins 1 and 2 and a series of three more spots above, extending to end cell. There are small sub-marginal spots only in the upper part of the forewing and mid termen region of the hindwing, and a white spot end cell on both wings. The underside is distinctive, with three more or less straight-sided dark green bands across the hindwing, the rest being white. The forewing is also dark olive green with the upper surface pattern of white spots repeated.

Status: Collected north Punjab. The type specimen comes from Fateh Khan Rest house, in the Kalla Chitta Hills.

Habits: Nothing specifically recorded, but it is known that all the *Hesperia/Spialia* Skippers have similar spindle-shaped larvae which spend their time sheltering in a folded piece of leaf lined with their silk and securely held together with silk strands, and that they normally feed out of direct sunlight, sometimes only at night, and in the close vicinity of their leaf shelter. They pupate in a silk-lined cocoon on the ground or among the basal stems of a plant, secure in a wrapped-up leaf. The foodplants are probably members of the Rosaceae, in particular *Potentilla* spp.

291. The Indian Skipper

Spialia galba
(Syn: *Hesperia* or *Syrichtus galba*)
Wingspan 24–27mm

Description: This species resembles *H. zebra* above, being dark blackish-brown with numerous white spots. Both wings are prominently chequered along the margin, and on the upper forewing there is a crescentic white spot end cell and two more inside the cell, which distinguishes it from *H. zebra* with only one spot inside the cell. The upper forewing also has a prominent and continuous row of sub-marginal white spots. The upper hindwing has a white spot at the base of the cell, lacking in *H. zebra*. The underside shows clearly the three white spots along the cell and end cell, against a dark olive green background. The under hindwing is white with much broader and paler green bands than the Zebra Skipper

Status: Found locally throughout Sindh, including Karachi area, but not common, and rare in Punjab. Collected Lahore in October, where considered rare (Rhe-Philipe, 1917), and from Murree Hills. Recorded from Kashmir by Evans.

Habits: The egg is dome-shaped, covered with rounded knobs running in meridional lines. It is pale grass green in colour, with the tubercles white. The eggs are laid on leaves, stalks, or even on grass near at hand. The foodplants belong to the Sterculaceae, and in south India *Waltheria indica* has been recorded. The larvae are spindle-shaped, fattest in the middle, with large heads and rather narrow necks. Their body colour is glaucous green, with a darker, pulsating dorsal line and pale, almost white, spiracular line. Their heads are shining black, and the first thoracic segment, or neck, is orange. The whole of their bodies are covered with small white tubercles bearing white or brown hairs at their tips. Immediately upon hatching, the young larva makes a shelter for itself by cutting a small round portion from the edge of a leaf and turning this over upon itself, binding it down tightly with silk and lining the cell inside with more silk. It shuns bright light and feeds only in the vicinity of its cell, moving to bigger leaves as it grows. The pupae are formed in a similar leaf cell lined thickly with silk. The greenish-yellow pupa is covered with hairs except over the wing cases, and is also covered with a white waxy powder. The butterfly, like all the Skippers, is fast on the wing and often basks on the ground, but will regularly visit flowers, and, unlike the very similar *Spialia evanidus*, mainly flies during the spring and summer months.

292. The Brick Skipper or Red Underwing Skipper of Europe

Hesperia sertorius
(Syn: *Spialia sertorius carnea*)
Wingspan 30–32mm

Description: Above dark brown with a similar pattern of white spots, a chequered margin, and fairly prominent, small white marginal spots on both upper wings. There are two larger white spots between veins 1 and 2, which distinguishes this species from the Indian and Zebra Skippers, and no white spot at the base of the cell as in *H. galba*. The main distinguishing feature of this Skipper is the under hindwing, which has a brick red or tawny band on the outer half of the wing bordering the discal white spots.

Status: Widespread in Chitral, Shandur, Gilgit, Yasin, and Balochistan around Uruk valley and Ziarat. In the Himalayas, generally found above 3,000m (10,000ft) but at 2,700m (9,000ft) in Ziarat. Mohyuddin (1987) claimed larvae found on *Achyranthes aspera* (family Amaranthaceae) in July, at Rawalpindi, unusual for such a low altitude.

Habits: The larval foodplants are members of the family Rosaceae, including *Potentilla* spp., *Rubus* spp., and *Sorbaria* spp.

293. The Persian Skipper

Spialia phlomidis geron
(formerly called the Baluchi Skipper *Hesperia geron* [Syn: *Syrichtus geron*])
Wingspan 28–34mm

Description: The upper side is dark brown with the forewing white on the costa, and a crescentic white spot end cell, with a faint white spot base of cell and a larger white spot mid cell. Both wing margins are chequered and with prominent sub-marginal rows of small white spots. The underside varies from pale to dark greyish-green with irregular white patches on the hindwing and the upper forewing pattern of spots repeated. The under forewing is darker green around the cell and upper discal region.

Status: Collected Ziarat, Quetta, Kalat, Uruk, Wam, and near Loralai. Common May to September, flying between 1,500–2,100m (5,000–7,000ft).

Habits: The larval foodplants are members of the family Rosaceae, including *Potentilla* spp., *Prunus* spp., and *Spirea* spp.

294. The Sindh Skipper

Spialia geron evanidus
(Syn: *Spialia doris* in Evans (1949),
Syn: *Syrichtus evanidus*)
Wingspan 23–25mm

Description: Very similar in pattern to *Spialia phlomidis*, the Persian Skipper, differing only in the upper forewing costa being less prominently white and, on the upper hindwing, the spot base cell more prominent and the sub-marginal row of spots being more prominent than in *S. phlomidis*. The under hindwing between vein 7 and the costa is all white in the Persian Skipper, but shows separate white spots against a green background in *S. geron*.

Status: Collected from Campbellpur (Maj. Yerbury, 1886) flying from late May to June, and from Karachi and lower Sindh, spreading to Hab River, Las Bela, whence came the type specimen. Uncommon.

Habits: According to Bell (1927) in Sindh this butterfly is only on the wing during the winter months. It is probable that, like other *Spialia* species, its larval foodplants are members of the Rosaceae, particularly *Potentilla supina*, and possibly *Neurada procumbens*, both widespread in lower Sindh.

295. The Syrian Skipper

Muschampia staudingeri plurimacula
(previously called *Muschampia* [Syn: *Hesperia*] *poggei*, and the same species as *Syrichtus poggei* in Sakai and the Streaked Skipper [*Syrichtus plurimacula*] in W.H. Evans, 1949)
Wingspan 30–35mm

Description: There has been considerable confusion as to the taxonomy of this species, variously called the Streaked, Syrian, Turkestan, and Balochistan Skipper. I am following the British Nat. Hist. Museum nomenclature for this species. This Skipper is rather pale olive green on the underside hindwing, with obscure white streaks and patches, a pattern not repeated in the other spp. within *Hesperia* and *Spialia* genera. Typically, the white spot between veins 1 and 2 is extended as a streak to the termen, and the white spot between veins 3 and 5 is also an obscure, elongated white patch. The upper wings are dark brown and similarly patterned as the other Skippers, but the sub-marginal series of spots are rather faint and obscure on both wings, with the forewing costa white, and there is a faint spot on the upper hindwing base.

Status: Collected between 1,200–1,800m (4,000–6,000ft), Chautar (Chotair), Ragora, and Loralai, and the area between Dilkhuna and Ziarat, all in Balochistan. Two broods, on the wing April to August. Adults associated with the showy Labiate, *Hymenocrater sessilifolius*.

296. The Mountain Skipper

Hesperia alpina
(Syn: *Pyrgus alpina* in Evans, 1949)
Wingspan 25–30mm

Description: This Skipper is marked on the dark brown upper side with prominent white spots, like many of the other 'Chequered Skippers', and is chiefly distinguished by having a double white streak on the forewing costa, just above the end cell white spot. The spots below this are also quite large and prominent, and the spot in between veins 2 and 3 is just below the cell spot, whereas the spot between veins 3 and 4 above it is further out towards the margin. The underside is boldly marked with broad white patches and dark grey-green, varying to olive ochraceous, bands.

Status: Widespread and common in the far northern areas of Pakistan. Collected Chitral from Shandur plateau and Gharogar, and in Gilgit from Yasin, Rama lake above Astor, in Hunza from the Batura Glacier and Misgar, and in Baltistan from the Deosai plateau, and Chogo Lungma. Flying between 3,000–4,200m (10,000–14,000ft.), and on the wing from June to August.

Habits: It is probable that the larval foodplants are *Rubus* spp. and *Potentilla* spp., and that their pupae spend the whole winter as such, emerging as adults only in the following summer.

297. The Tufted Marbled Skipper

Carcharodus altheae
(Syn: *Carcharodus floccifera balucha* and sub-sp. *dravira*)
Wingspan 29–35mm

Description: The Marbled Skippers of the genus *Carcharodus* are olive-brown above with darker bands across both wings, giving a marbled effect, and white spots or streaks, some of which are translucent. *Carcharodus altheae* has comparatively large white spots on the upper hindwing in the discal area, rather obscure white spots on the upper forewing apex, and a white line across end cell. The underside is white with paler, greenish-grey bands. The under hindwing is crossed by two irregular dark bands, the discal band being slightly wider than the outer post-discal one.

Status: Sub-sp. *balucha* collected upper Uruk valley and Ziarat, Hannah Lake, Quetta, and Mastung in Balochistan. Sub-sp. *dravira* collected Chitral from Lowari Pass, Drosh, and the Shishi Koh valley. Evans (1910) mistook this sub-sp. for *Muschampia staudingeri*, and collected it from the Ashreth and Utzen nullahs in lower Chitral. He considered it rare in Chitral. It is on the wing from June to September.

Habits: All the *Carcharodus* species choose members of the Malvaceae or Labiatae families as their larval foodplants. *Carcharodus floccifera* caterpillars feed on *Marrubium vulgare* (Labiatae).

298. The Plain Marbled Skipper or Mallow Skipper of Europe

Carcharodus alceae alceae and sub-sp. *swinhoe*
Wingspan 26–34mm

Description: Similar in appearance to *C. altheae*, being greenish-brown above with darker brown marbling, and differing from the former species mainly in being browner above and having less distinct white spots, those on the upper hindwing discal region being smaller than in *C. altheae*. There is an obscure white spot between veins 2 and 3 on the upper forewing which is higher than it is wide, whereas in *C. altheae* this same spot is wider than it is high. The underside is paler greenish-brown with less white on the under hindwing costa than in *C. altheae*.

Status: Nominate sub-species widespread in Murree Hills, and Chitral around Drosh and Ziarat. Larvae found (Mohyuddin, 1987) in September on *Althaea rosea*, Peshawar and Marghazar, Swat. Sub-species *swinhoe* from Balochistan, not rare around Quetta gardens and cultivation in main valleys, with possible three broods. On the wing from March to October.

Habits: The larval foodplants are members of the Malvaceae, including the garden Hollyhock (*Althaea rosea*) and *Abutilon* spp.

299. The Inky Skipper

Nisoniades marloyi
(Syn: *Erynnis marloyi*)
Wingspan 28–30mm

Description: This widely distributed, mountain-dwelling Skipper has two distinct forms or sub-species in Pakistan, that occurring in Chitral being darker brown above without greyish frosting, and the Balochistan population being frosted on the upper forewing with light and dark bands. *Erynnis marloyi* is dark brown above, without any white spots except a band of small spots at the apex of the forewing. The Balochistan population (illustrated) has light and dark patches on the upper forewing caused by greyish-white scales, with two indistinct darker brown bands across the discal and post-discal areas. The upper hindwing is uniform brown, often of a more ochraceous brown, sometimes with small dark spots around the sub-marginal area. The underside forewing lacks the frosting of grey scales. As stated above, the Chitral population is plain dark brown on the upper forewing. In all populations the fringe is brownish. This is a variable species: specimens from Waziristan are darker, and I have collected specimens from Swat Kohistan with paler frosting.

Status: A larger and darker form is widespread in Chitral, and collected from Drosh and the Utzen nullah. A paler and patterned form occurs in central and northern Balochistan and has been collected from near Gandak on the Ziarat road, the Khojak and Mehtarzai Passes, Gwal Forest, the Shingai in Zhob, and from Razmak in Waziristan. They are fairly common in Balochistan, but more plentiful in Chitral. They are on the wing from March to August, flying between 1,500–2,400m (5,000–8,000ft) in Chitral and Swat. Collected TJR from Kalam, Swat. Found from March to May, flying between 2,000–2,400m (6,500 and 8,000ft), in Balochistan.

Habits: The author found this species readily settling on flowers and basking on the ground. The larval foodplants are believed to be spp. of creeping clovers and trefoils, such as *Lotus corniculatus*.

THE SCRUB HOPPERS

300. The Veined Scrub Hopper

Aeromachus stigmata
(Syn: *Aeromachus inachus*)
Wingspan 22–30mm

Description: As can be seen from above, this is quite a small Skipper, and all the Scrub Hoppers are small dark insects, distinguished by having on their undersides two rows of small pale spots, one marginal and one discal, and usually on the upper forewing only the discal row of spots showing. *Aeromachus inachus* is dark brown above, with a row of small creamy spots across the discal area of the upper forewing and a small spot end cell. The upper hindwing is unmarked. The underside is paler brown and the under hindwing has each vein outlined white from the outer half, with two rows of white spots marginally and discally, across both fore and hindwings. The male has a black brand on the upper forewing from mid vein 1b up to vein 3.

Status: Not very common in the Murree Hills, but locally common at Khaira Gali, around 1,500m (5,000ft).

Habits: The eggs are laid singly on grasses, generally on the underside of the leaf blade, and these form the larval foodplant. The eggs are hemispherical in shape, with nineteen tuberculate ribs, and the surface in between minutely pitted. They are dull white in colour, often developing a dark red ring around the middle (Bell, 1925). The larvae make a cell for themselves by loosely joining the edges of a grass blade together. They have a cylindrical body with a narrow neck, and the body colour is blue-green, with the head more yellow-green. There is a darker dorsal line and a white sub-dorsal stripe, and the spiracles are white (Sevastopulo, 1946). The pupa is formed upon a blade of grass, supported by a silk pad and attached by the tail and a silk girdle, the grass being slightly folded by cross-threads of silk. Its colour is grass green, and it is rather slender in shape, with the thorax keeled, and bearing white lateral lines. The butterflies are quite tiny and, with their swift jerky flight, difficult to see. They do settle on the flowers of leguminous creeping plants amongst the grass, keeping to grassy glades in the forest.

THE BOBS

301. Indian Palm Bob

Suastus gremius
Wingspan 32–42mm

Description: The Bobs are so named because of their dipping fast flight, with the greatest diversity of species found in the north-eastern parts of the Oriental region. The larvae feed on palms, which are not well represented in Pakistan. In all species the hindwings are unmarked, but the forewings usually bear, large semi-transparent white spots. *Suastus gremius* is quite large for a Skipper, being dark brown above, unmarked on the hindwing, and with one large white spot end cell and three in the lower discal area of the upper forewing. The underside is slightly paler brown, and this species is distinguished from all other Skippers by the under hindwing markings, which include four to five black spots sharply defined, the spot against the upper outer edge of the cell always being large. The under forewing has prominent white spots in the same pattern as the upper side.

Status: Lahore and Karachi, where occasionally occurs after good monsoon rains. Collected Zoological Survey of Pakistan in late 1960s from Karachi Sewage Farm. Also collected Murree, where rare. Larva recorded as feeding on leaves of Oriental Palm in Pusa, India.

Habits: The female lays her eggs singly on the tip or edge of palm fronds, the larval foodplants being species of palm, especially the wild Date, *Phoenix sylvestris*. However, ornamental palms are equally favoured, such as the Fish-tailed Palm, *Caryota urens*. Sometimes more than one egg is laid upon the same leaf, but on different blades. The eggs are hemispherical in shape, with thirteen broad meridional ribs. In colour the eggs are dark red-brown with the ribs white. The little larva is bright red in colour when it first emerges, but as it moults into its third stage (Bell, 1925) it become glaucous green with a darker mid-dorsal stripe. Its anal plate is large and broad and its body thickest in the middle, being flattened posteriorly. The head is brown with pale central and lateral stripes. The larva makes a cylindrical cell for itself by eating its way inwards from the edge of the palm leaf in two places and then rolling the freed section towards the leaf midrib. This is then lined thickly with criss-crossed strands of silk and the two ends are left open. The pupa is formed in the same way after cutting a whole leaf blade up to the midrib and binding it into a cylinder. The pupa is yellow-green in colour with the last segment reddish-brown, and the thorax a darker green with a reddish lateral spot. It is smooth and resembles the cocoon of a moth in shape, and is covered with a waxy powder. As the palm leaf withers and falls to the ground it often is swept away by rains, but the waxy secretion seems to repel water effectively and so the pupa survives flooding. As might be expected from its robust body, the adult butterfly is a very strong, swift flyer, but it does come to flowers and likes sunlight. It settles with the hindwings almost horizontal but the forewings raised.

THE DEMONS

302. The Spotted Demon

Notocrypta fiesthamelii
Wingspan 38–50mm

Description: The Demons are quite large Skippers, somewhat similar to the Flats in appearance. They are characterized by the presence of a broad, semi-transparent white discal band across the upper forewing, and some bands of white frosting across the under hindwing. *Notocrypta fiesthamelii* is dark brown with small, not very clear white spots across the forewing apex, and semi-transparent, large, conjoined white spots from end cell across the discal area of the forewing. The upper hindwing is unmarked and the fringe is brown. The under hindwing and the apical area of the under forewing have an overlay of white frosting, with a clear dark brown area across the discal region of the hindwing. The markings in this species are variable, and this description is from a Murree specimen.

Status: In Pakistan only recorded from Murree, where it is not uncommon in some years.

Habits: Bell (1927) describes the eggs as limpet-shaped and coloured a light brownish-orange, with barely any indication of meridional ridges. They are laid singly on the underside of the leaves of wild gingers (family Zingiberaceae), mainly *Zingiber officinale*, and also *Curcuma angustifolia*, which is wild in the Murree foothills, and possibly *Curcuma longa*, the cultivated Tumeric. The larvae are a light olive green colour, their skin being somewhat translucent, with their heads rather smaller than typical Skipper larvae, but still wider than their neck or first thoracic segment, which is whitish in colour, as are the spiracles. Sevastopulo (1947) describes the young larvae as having a black head, the rest of the body green without darker speckling. In habits they are sluggish, feeding mostly at night; when little, the larva makes an oblong tubular cell for itself from the edge of a leaf. Bell (1927) describes the pupae as being able to wriggle and jump around if molested and cut loose from their silken body moorings. They are formed in a channel or leaf cell made by drawing the sides of a leaf downwards, so that they lie on the underside. They are covered by a white waxy powder, and are grass green in colour, with a thin white tracheal line and black spots on the dorsal side of the abdominal segments. The adult butterflies are typically strong, swift flyers, and they do visit flowers, but mainly in the early morning and late afternoon.

(302b. The Indian Ace

Halpe homolea
[Syn: *Halpe egena*]
Wingspan 30–35mm

Description: The Aces are mostly forest-dwelling Skippers, confined to the areas of south and eastern India with high rainfall. The Indian Ace, *Halpe homolea*, is included as it is widespread from south to north-east India and may occasionally occur. *Halpe homolea* is a medium-sized Skipper, dark brown above with the central area paler on the upper hindwing, but small, semi-transparent spots on the upper forewing, one end cell, two to three small apical spots, and two elongated discal spots between veins 1 and 2, and 2 and 3. The under hindwing is dark brown with an obscure tawny-yellow row of sub-marginal spots and a wide tawny yellow discal band. The under forewing also has an obscure row of small sub-marginal tawny-yellow spots.

Status: May occur, but there are no specimens from Pakistan in Brit. Mus. collection.

Habits: The larval foodplants of the genus *Halpe* are all bamboos, and the female lays her eggs singly on the underside of bamboo leaves. The larvae are some shade of green, with their body the typical spindle-shape, with the head generally yellow in colour and higher than the thoracic segments, and the anal end being rounded and somewhat flattened. The larvae make a cell for themselves by joining the edges of a leaf near its tip. When they pupate they cut through the attachment of their leaf cell so that it falls to the ground, and their body, besides lying in a bed of silk within the shrivelled leaf, is also covered by a white waxy powder secreted by the larva just before pupating. The Aces are all strong flyers, and do come to flowers to feed. They rest with the wings folded over their back, the forewings drawn down between the hind.)

THE DARTS

303. Veined Dart

Actinor suada radians
Wingspan 34–42mm

Description: The Darts are medium-sized Skippers, without the prominent black-edged brand on the forewing of the male, so conspicuous in the Palm Darts. They are dark brown above, with a row of pale yellow or white discal spots, and the veins outlined yellow on the outer side of these bands. *Actinor radians* is dark brown above with semi-transparent pale yellow spots on both fore and hindwings forming an almost continuous post-discal band, and with large, pale yellow double spots end cell. The veins are outlined yellow from these spots outwards to the wing margin. There are ochraceous hairs on the base of the hindwing. The underside has the veins of the hindwing conspicuously outlined white and those of the under forewing outlined slightly yellower, with the pale yellow spot end cell prominent.

Status: Usually a low-elevation hill species found around 1,200m (4,000ft), but taken as high as 2,400–2,700m (8,000–9,000ft) in Chitral. Evans (1910) collected it in July from the Utzen valley.

304. Himalayan Grass Dart

Taractrocera danna
Wingspan 27mm

Description: The Himalayan Grass Dart is dark brown above with white, semi-transparent spots on the forewing in a discal band from vein 1 up to vein 8. Those between veins 3 and 4, and 4 and 5 are further out towards the margin, not in line with the others. There is an elongated white spot or streak in the forewing cell. The under hindwing has a large spot between veins 6 and 7, and a slightly smaller spot in the cell along its upper edge, and a post-discal row of spots between veins 2 to 5. The underside forewing apical region and the whole of the under hindwing are frosted with greenish-yellow scales.

Status: Collected from Kashmir to Kumaon (Evans, 1949), probably occurs Murree Hills.

Habits: The foodplants of the larvae are grasses and they will feed on almost any species. The larvae are green in colour and have the typical body shape, with narrow neck and tapering body at the rear. Their bodies are covered with minute hairs which have minute black tubercular bases, giving a speckled effect. They make a shelter from the grass blades, either by rolling the tip inwards or, cutting a section along the edge, and binding this with silk strands. The pupae are covered by a white powdery secretion, and are bone coloured.

305. The Common Grass Dart

Taractrocera maevius
Wingspan 24–26mm

Description: A smaller version of *Taractrocera danna*, the Himalayan Grass Dart, but differing in having fairly prominent white spots in the hindwing cell, and with a similar white streak along the upper forewing inside the cell. The under hindwing is different from *T. danna*, in that the veins are outlined white, and the white spot inside the cell extends right across it, while there is a smaller spot between veins 6 and 7, and a larger spot between veins 7 and 8, almost in line with the cell spot. When compared with *T. danna* this species also lacks the frosting of yellow scales prominent on the under hindwing of the latter.

Status: Collected Hasan Abdal and Peshawar, and not rare in the outer foothill region. Seen feeding on *Xanthium* flowers in August, Rawalpindi.

Habits: The eggs are laid singly on grass blades, and the little larvae upon hatching immediately make a cell for themselves by turning over the edge or the tip of a leaf and securing this down with silk threads. Their eggs are dome-shaped, with the surface covered by minute shallow cells, and are coloured dull yellowish-white. The larvae are slightly transparent and yellowish-green in colour, with a dark green pulsating dorsal line. Their bodies are covered with minute hairs, their bases darker, which gives a speckled effect. Their heads are slightly bilobed and, as in all Skippers, larger than the narrow neck. They attain a length of about 20mm. The pupae are formed after about one full month of feeding (Bell, 1925), and are enclosed in a tight cell made by rolling over a grass blade and binding the edges with silk. The pupae are pale yellow and covered with minute reddish hairs and are rather slim in shape. The butterflies frequently visit flowers or rest on the ground on a grass stem, and are rather weak flyers.

306. The Himalayan Dart

Padraona dara
Wingspan 29–32mm

Description: As can be seen above, this is a medium-sized Skipper, larger than the two previous Darts. On the upper side it is boldly marked with primrose yellow spots and streaks, there being a broad yellow streak covering most of the forewing cell and extending up into the costa. There are three spots discally from vein 1 upwards, and there are spots in between veins 4 and 5, and 5 and 6, which are further out towards the wing margin. On the upper hindwing there is a prominent spot between veins 6 and 7, but none between 7 and 8. The underside is largely yellow, with dark brown spotting on the under hindwing, and the under forewing pattern repeated as on the upper side, except for the apex, which is wholly yellow. The fringes are yellow.

Status: Collected Chitral and normally found above 1,800m (6,000ft), but also collected from Lahore and Rawalpindi, where not rare. Adults found on *Xanthium strumarium* in October (Mohyuddin, 1987) near Rawalpindi.

Habits: The larval foodplants are grasses and the eggs are laid singly, usually on the tip of grass blades. The newly-hatched larvae make a cell by joining the edges of the grass-blade tip together with silk strands. The larvae are covered all over with minute hairs and are light bluish-green, speckled with darker dots. There is also a broad whitish sub-dorsal band and a lateral white line along the whole length of their bodies. As the larvae grow bigger they live on the underside of grass blades, drawing the sides loosely together with silk, and the pupa is formed naked on the underside of a leaf, in a similar loose-folded, over leaf. It is grass green in colour with indistinct dorsal and darker lateral lines. The butterflies are strong flyers, but generally keep close to the ground, and they frequently bask with wings half-open, low down on leaves, as well as visiting flowers among low creeping herbs amongst the grass.

307. The Dark Palm Dart

Astycus pythias bambusae in Wynter-Blyth, 1957
(Syn: *Telicota ancila bambusae* in Evans, 1949)
Wingspan 33–35mm

Description: The Palm Darts are medium to small-sized Skippers, characterized by having tawny-yellow or orange discal bands, and in the males a prominent brand on the upper forewing, extending across the inside of the discal area as a grey streak outlined with black. *Astycus pythias*, as its name suggests, is a darker tawny-yellow above than the next species, from which its mainly differs by the upper forewing having the veins, which extend to the margin from the yellow discal area, only narrowly outlined tawny-yellow. In *Telicota augias* these veins are broadly edged yellow, so that by comparison the Dark Palm Dart has an uninterrupted brown border to the upper wing margin. The Dark Palm Dart varies from orange-tawny in the eastern part of its range to yellow-tawny in the west, with dark brown borders to both upper wings, and in the male the prominent grey brand outlined with black, already described above. The upper hindwing has the cell yellow and a broad tawny band across the discal region. The underside is paler yellow without such dark brown areas, and in the male the brand is invisible from underneath.

Status: Collected Lahore (Puri, 1931) and Murree, where rare due to paucity of bamboos, which form its larval foodplant.

Habits: The eggs are shaped like a high dome, unsculptured and coloured pinkish-white. The female lays these singly on the tip of bamboo leaves, the larval foodplant. The larva is the normal Hesperid shape, with large head, narrow first thoracic segment, and slightly humped body, the anal end being flattened and rounded. The head is rugose and pale fawn in colour with a dark brown central stripe. The rest of the body a translucent yellowish-green. Sevastopulo (1942) states that the trachea, silk glands, and gut are visible through its skin. The body is also covered with small fine hairs, longer over the anal flap. The spiracles show as yellow dots along the sides. Immediately after eating its way out of the egg, the young larva makes a cell for itself at the leaf tip by lying on the midrib and drawing the sides of the leaf blade together. As it grows bigger it makes exactly similar leaf cells using more of the leaf blade tip to fit its body, which finally attains a length of about 24mm. The pupa is cylindrical in shape, and according to Bell (1926) is dark reddish-brown in colour with ridged dorsum. Sevastopulo (1942) describes a pupa as being yellowish-chestnut in colour, with the head and wing cases tinged purplish. It is formed in a cell made by two to three bamboo leaves drawn together to make a cylindrical house, which is lined with white waxy powder secreted by the larvae before pupating. It is attached only by the tail inside this cell, the top or tail end of the cell being tightly sealed with silk. The stout-bodied butterflies are, not surprisingly, very swift, strong flyers, but have the habit of returning to the same resting spot on the top of a leaf.

308. The Pale Palm Dart

Astycus augias
(Syn: *Telicota augias*)
Wingspan 33–35mm

Description: Very similar to the previous species, *A. pythias*, and differing mainly in the upper side being a slightly paler yellow-tawny shade than in the former, except in the western population and on the upper forewing the dark border is interrupted by the intervening veins being broadly outlined yellow. In the male, the brand on the upper forewing has the central streak a slightly broader and darker grey than in *A. pythias*. The underside, as in the former species, is more uniformly tawny-yellow on the under hindwing, and with only the base and lower discal region of the under forewing showing dark brown.

Status: Collected Lahore by Puri in 1925, but this is surprising based upon its known distribution from Assam to Indonesia. However, with the great increase in sugarcane cultivation, its main larval foodplant, in the Punjab since Puri's time, it may spread into Pakistan.

Habits: The early life history of this Palm Dart is very similar to that of *Astycus pythias*, except that the normal foodplant of the larvae is sugarcane leaves (*Saccharum officinarum*), (Bainbridge Fletcher in Bell, 1926), and they are rarely also found on bamboos, keeping to more open country than *Telicota pythias*. Like the Dark Palm Dart, the adult butterflies are very swift on the wing, but they do settle frequently and readily come to flowers, and generally keep close to the ground when flying.

309. The Chequered Darter or Silver Spotted Skipper of Europe

Hesperia comma
(Syn: *Pamphila comma*)
Wingspan 29–35mm

Description: Above dark brown with orange-tawny areas on the upper forewing base and discal region, and with orange discal spots on the upper hindwing. The male has a black brand across the upper forewing up to the base of the cell, and this usually bears only a narrow grey inner streak. The female averages slightly larger, without the black brand and with more restricted orange-tawny areas on the upper forewing. The underside, from which it gets its common name, has prominent white spots, incurved on their outer edge, those on the under hindwing being conjoined between the base of the cell and between veins 4, 5, and 6. The under hindwing is overlaid with greenish-ochraceous scales, as also the forewing apex, making the white spots more conspicuous. The base of the under forewing is orange.

Status: Widespread in the Northern Areas from 2,700–4,200m (9,000–14,000ft). Collected Chitral on the Shandur plateau, Shishi Koh valley, and Laspur valley, also from Yasin and the Rupal valley in Astor district, both in Gilgit.

Habits: The early life history has been well studied in Europe, where it also occurs. The female lays her eggs singly, either on the pointed grass blades of Fesque, *Festuca* spp., or on the leguminous vetch, *Lotus corniculatus*. The larvae, however, feed only upon grasses. The eggs are pale creamy-orange in colour, dome-shaped with a flanged, broader, flat base, and micropyle hollow at the top. The larvae hatch the following spring, the eggs remaining undeveloped throughout the winter. They immediately make a cell for themselves by spinning together several blades of grass low down near the ground. The larvae have dark red-brown heads and olive green bodies with the spiracles black and traces of a yellowish supra-spiracular line. The pupa is formed in a similar leaf cylinder made from chewed and growing pieces of grass bound together with silk threads. It lies in a silken-lined bed inside this cell, attached both by tail hooks and by hooks on the top of its head. The butterflies have the typical skipping flight of the small Hesperiidae Skippers, close to the ground, and frequently settle on low-growing flowers.

THE SWIFTS

310. The Yellow Spot Swift

Baoris eltola
Wingspan 34–45mm

Description: The Swifts, as their name implies, are strong flyers; when settled they have their wings closed over their backs. It is a big genus of very similar appearing species, all being dark brown with pale, semi-transparent spots on the forewing. *Baoris eltola* is quite a large Swift, with olive brown upper parts, bearing oily yellow semi-transparent spots, that between veins 2 and 3 on the forewing being the largest and square in shape, and the upper hindwing having three spots, of which the top one crosses vein 5. The bases of both upper wings are clothed with ochre hairs. The under hindwing is more ochraceous in colour. Females have whiter, less yellow-appearing spots. The fringe is yellow.

Status: Collected by TJR Dunga Gali, and common in the Murree Hills up to 2,700m (9,000ft).

Habits: The eggs are similar to those of the *Hesperia* Skippers, being dome-shaped without prominent longitudinal ribs, and have a flanged, flattened base. They are pale green when first laid, becoming dull white. The larvae feed on *Saccharum* spp. and possibly other grass species, and are dull greenish-white in colour with rather translucent skin, showing the gut contents underneath and the tracheal tubes.

The pupa is described by Sevastopulo (1946) as being formed in a cell, made by folding over a leaf, thickly lined with white silk. It is supported by a silk girdle and tail pad. The pupa has a long shape and pointed head, and is a pale watery green in colour with a double white dorsal line and a thinner white spiracular line. The adult butterflies frequently return to the same leaf or twig to settle and have the typical fast flight, and seem to like the sunshine. They regularly visit flowers.

311. The Himalayan Swift

Baoris discreta himalaya
Wingspan 34–42mm

Description: Very similar to *Baoris eltola*, though averaging slightly smaller and with the semi-transparent spots white not yellow, and also the fringe being white. Like *B. eltola*, the semi-transparent spots are quite large, and there are three on the hindwing, and the spot between veins 2 and 3 on the forewing is large and square-shaped. Though the difference is small, the base of the wings are clothed with more greenish hairs than the ochraceous hairs on *B. eltola*. The difference in the fringe colour between these two similar species is, however, striking and a key character for separation.

Status: Collected in the Murree Hills, occupying much the same habitat as *B. eltola*, but uncommon.

Habits: The early life history of the Himalayan Swift is much the same as the other *Baoris* spp., and the female lays her eggs singly on grasses which are the foodplant. The eggs are dome-shaped with a flattened flange at the base, and the larvae hide in a cell made by folding over a leaf blade, emerging to feed mostly at night. The adult butterflies have been recorded on the wing as high as 3,000m (10,000ft) (Wynter-Blyth, 1957); they are fond of the sunshine and regularly visit garden flowers.

312. The Small Branded Swift

Pelopidas mathias
(Syn: *Baoris mathias*)
Wingspan 32–38mm

Description: This Swift averages slightly smaller than the two previous species and has smaller white spots on the forewing, usually with no spots on the upper hindwing, though in some individuals there are small spots between veins 3 and 4, and 4 and 5. The bases of the upper wings bear yellow hairs and the under hindwing is dusted with greenish-grey scales.

Status: Widespread Sindh, where common throughout the year. Also Punjab and NWFP. Collected Karachi and Hyderabad in Sindh, Multan and Lahore in Punjab, where fairly common from end of rains until beginning of cold weather (Rhe-Philipe, 1917). Also collected Dera Ismail Khan and Parachinar in NWFP, and Khuzdar in Balochistan (Swinhoe, 1887).

Habits: The eggs are high dome-shaped and nearly smooth on their surface, bearing a narrow shelving flat base, and are light green varying to pale yellow in colour when freshly laid. The larval foodplants are various species of coarse grasses, not bamboos. The larva is grass green in colour, with light yellow bands across its back and a whitish sub-lateral line. Its head is triangular in shape, with a narrow neck, and marked by a brown line between the head and neck. Bell quotes Forsayeth (1926), who bred these Skippers, as pupating on 3 September and the adult emerging on 13 September. The pupa is formed lying head upwards along a blade of grass, without any protective wrap-around cell. It is translucent green in colour and is attached to the grass by the tail claspers and a silk band around the thorax. The adult butterflies are very fast on the wing and frequently settle to bask on the ground as well as visiting flowers. They do not avoid bright sunshine, though they are most active early morning and late afternoon.

313. The Straight Swift

Parnara guttatus
(Syn: *Baoris guttata*)
Wingspan 32–38mm

Description: This is a relatively small Swift, with shorter antennae. The spots are white to creamy, and small. There is no spot inside the cell nor between veins 1b and 2. On the hindwing the series of small spots runs in a straight line from veins 2 to 5, decreasing in size as they ascend. This same pattern is repeated on the under hindwing, which is more ochraceous yellow.

Status: Widespread. Collected TJR Dunga Gali in August. Also occurs Thandiani, Khaira Gali, the Khyber Pass, Kohat district at Hangu, and Abbotabad and Hasan Abdal in Hazara district in October. In Punjab at Jhelum and Lahore, and in Chitral at Drosh. Larvae found (Mohyuddin, 1987) on sugarcane in Rawalpindi in September.

Habits: The eggs are smooth on their surface, dome-shaped with a narrow flanged base, and they are pearly white when freshly laid. The female lays them on the underside of grass blades, sometimes on sugarcane (*Saccharum officinarum*). The larvae are bluish-white in colour with their heads semi-elliptical in shape. The anal segment is broadly rounded and each segment is distinct, being marked by impressed lines along the rear margin. The larvae make for themselves a cell from blades of grass bound together very firmly, from which they emerge to feed in the evening. The pupa is covered with white waxy powder and is also formed in a tightly-bound leaf cell which is firmly closed at both ends, so that no moisture can get in. They are pale watery yellowish-green in colour, according to Bell (1926). The butterflies frequently settle to bask on the ground in the open, as well as visiting garden flowers even over 2,400m (8,000ft) in the Murree Hills.

314. Bevan's Swift

Guttatus bevani
(Syn: *Baoris bevani* and *Parnara bevani*)
Wingspan 32–36mm

Description: Again this Swift is confusingly similar to the other *Baoris* spp. It differs from *B. guttatus* in having a pale white spot between veins 1b and 2, though this may be absent in some male specimens. The spot between veins 3 and 4 is nearer to the one below it than to the one above. There is also a very small spot inside the upper part of the forewing cell but no cell spot on the hindwing, which is unmarked. The underside, as in the other *Baoris* spp., is more ochraceous, but the pattern of spots is repeated.

Status: Collected Murree Hills, more common in Kashmir. Very rare around Lahore in November (Rhe-Philipe, 1917). Also Karachi, where rare in most years. Collected by Zoological Survey of Pakistan in late 1960s from Karachi Sewage Farm.

315. The Dingy Swift

Gegenes nostrodamus
Wingspan 28–35mm

Description: Swifts of the *Gegenes* genus differ from *Baoris*, not only in their adaptation to a desert environment, but also in their generally paler colouration. In *Gegenes nostrodamus* both sexes are pale greenish to sandy brown above, without any pale spots, and on the underside they are even paler grey-green. The sexes differ, the female having a series of white spots on the under forewing, which are often visible also on the upper side. These spots are very small in the apical region and more conspicuous in between veins 1b and 5 in the discal region.

Status: Occurs Sindh where not rare (Menesse, 1950). Collected by Zoological Survey of Pakistan from Government Nursery, Karachi. Widespread but nowhere common, along Afghan frontier districts of Balochistan, and collected from Quetta and the Pishin valley. From the NWFP in Attock district, and Hazara district from Haripur, and Kala Pani. From the Punjab Salt Range around Campbellpur where there appear to be two broods, the first in June and July and then in October and November, when it is more common. Leslie and Evans (1903) found it fairly common at low elevations in the side valleys of lower Chitral. Fairly common around Lahore gardens October to November, (Rhe-Philipe, 1917).

Habits: The eggs are laid on desert-adapted grasses, generally the larger, clump-forming species, and the larvae when small shelter in a cell made by folding the tip of a grass blade backwards and binding this with silk, not by rolling the edge of the blade into a cylinder. When larger it often shelters on the underside of a grass blade without making any leaf cell. It has a triangular-shaped head, narrow neck, and humped body, with the anal segment flattened and rounded along its posterior margin, which is also fringed with visible hairs. The body colour is glaucous green with three thin, darker green lines along the sides of the body. The pupa is described as similar to *Baoris mathias*, being watery grass-green in colour and its body covered with a white waxy powder. It is formed on the underside of a grass blade in a loosely-made leaf cell, generally with its body fully exposed to view. The butterflies, like the entire genus, are very swift flying but frequently settle on the ground to bask, and will visit flowers, not shunning bright sunshine.

THE HOPPERS

316. The Torpedo

Eogenes alcides
Wingspan 38–44mm

Description: Above dark brown with semi-transparent white spots on the upper forewing, and the base of both upper wings dusted with sandy-grey scales. The spots in the apex are conjoined, and there is a double white spot inside the cell and, in the discal row, a double spot between veins 1b and 2, and two large spots between veins 2 and 4, and two much smaller spots near the margin between veins 4 and 6. The upper hindwing sometimes shows three faint spots between veins 2 to 5. The underside is paler-frosted with greyish-white scales.

Status: Not uncommon in Balochistan, around Ziarat, Uruk valley, and Kach from June to August. Also collected Chitral.

Habits: Nothing specifically seems to have been recorded. Larvae found on *Xanthium* sp. (Compositae), Rawalpindi, in June (Mohyuddin, 1987).

317. Leslie's Hopper

Eogenes leisliei
Wingspan 30–35mm

Description: Leslie's Hopper is slightly larger than the Veined Scrub Hopper, and does not have the two rows of pale spots on the underwings. It is dark brown above, with a white fringe, no spots on the upper hindwing, but three small, non-transparent, cream spots on the forewing apex and two larger creamy spots in the discal area, between veins 1 and 2, and 2 and 3. The underside is paler, being frosted with whiter scales, and on the under hindwing sometimes a spot between veins 4 and 5, more conspicuous in the female. The male also lacks the black brand on the forewing of the Veined Scrub Hopper.

Status: This species was named after Major Leslie (Leslie and Evans, 1903), who first discovered it in Chitral. Found usually at low elevations between 1,200–1,800m (4,000–6,000ft), but also taken as high as 2,700m (9,000ft) from the Utzen valley in June and July. It has also been collected from Malakand at 900m (3,000ft). On the wing from May to September.

Appendix 1
Bibliography

Ackery, P.R., (1975), 'A Guide to the Genera and Species of Parnassiinae (Lepidoptera: Papilionidae)', *Bulletin British Museum of National History*, Entomology, London, Vol. 31, No. 4., p. 105, and 15 plates.

Ackery, P.R., and R.I. Vane Wright, (1984), 'Milkweed Butterflies. Their Cladistics and Biology', Department of Entomology, British Museum (Nat. Hist.), 425pp.

Aitken, E.H., (1886), 'A note on the peculiar habits of the larva of *Virachola isocrates*', *Jour. Bombay Nat. Hist. Soc.*, Vol. 1, p. 216.

Antram, Charles B., (1924), *Butterflies of India*, Thacker Spink and Co., Calcutta, p. 226.

Baker, R.R., (1970), 'Bird predation as a selective pressure on the immature stages of the cabbage butterflies, *Pieris rapae* and *P. brassicae*', *Zool. Jour.*, London, Vol. 162, pp. 43–59.

———, (1984), 'The Dilemma: When and How to Go or Stay', in *Biology of Butterflies*, Vane-Wright and Ackery (Eds.), Royal Entomological Soc. of London, No. 11, Academic Press, pp. 279–95.

Barth, R., (1944), 'Die männlichen Duftorgane einiger *Argynnis*—Arten', Zool. Jarb. (Anat.), Vol. 68, pp. 331–62.

Bean, A.E., (1988), 'Observations on the occurrence and habits of the *Nacaduba* complex of the Lycaenidae (Lepidoptera), mainly from Pune district, Western Ghats', *Jour. Bombay Nat. Hist. Soc.*, Vol. 85, No. 2, pp. 332–63.

Bell, T.R., (1909), 'The Common Butterflies of the Plains of India', Part 4, *Jour. Bombay Nat. Hist. Soc.*, Vol. 19, No. 1, pp. 16–58.

———, (1909), Part 5, *JBNHS*, Vol. 19, No. 2, pp. 438–74.

———, (1909), Part 6, *JBNHS*, Vol. 19, No. 3, pp. 635–82.

———, (1910), Part 7, *JBNHS*, Vol. 19, No. 4, pp. 846–79.

———, (1910), Part 8, *JBNHS*, Vol. 20, No. 2, pp. 279–329.

———, (1911), Part 9, *JBNHS*, Vol. 20, No. 4, pp. 1115–36.

———, (1912), Part 10, *JBNHS*, Vol. 21, No. 2, pp. 517–44.

———, (1912), Part 11, *JBNHS*, Vol. 21, No. 3, pp. 740–66.

———, (1912), Part 12, *JBNHS*, Vol. 21, No. 4, pp. 1131–57.

———, (1913), Part 13, *JBNHS*, Vol. 22, No. 1, pp. 92–100.

———, (1913), Part 14, *JBNHS*, Vol. 22, No. 2, pp. 320–44.

———, (1914), Part 15, *JBNHS*, Vol. 22, No. 3, pp. 517–31.

———, (1914), Part 16, *JBNHS*, Vol. 23, No. 1, pp. 73–103.

———, (1915), Part 17, *JBNHS*, Vol. 23, No. 3, pp. 481–97.

———, (1916), Part 18, *JBNHS*, Vol. 24, No. 4, pp. 656–72.

———, (1918), Part 19, *JBNHS*, Vol. 25, No. 3, pp. 430–53.

———, (1918), Part 20, *JBNHS*, Vol. 25, No. 4, pp. 636–64.

———, (1918), Part 21, *JBNHS*, Vol. 26, No. 1, pp. 98–140.

———, (1919), Part 22, *JBNHS*, Vol. 26, No. 2, pp. 438–87.

———, (1919), Part 23, *JBNHS*, Vol. 26, No. 3, pp. 750–69.

———, (1920), Part 24, *JBNHS*, Vol. 26, No. 4, pp. 941–54.

———, (1920), Part 25, *JBNHS*, Vol. 27, No. 1, pp. 26–32.

———, (1920), Part 26, *JBNHS*, Vol. 27, No. 2, pp. 211–27.

———, (1921), Part 27, *JBNHS*, Vol. 27, No. 3, pp. 431–47.

———, (1921), Part 28, *JBNHS*, Vol. 27, No. 4, pp. 778–93.

———, (1923), Part 29, *JBNHS*, Vol. 29, No. 2, pp. 429–55.

———, (1923), Part 30, *JBNHS*, Vol. 29, No. 3, pp. 703–17.

———, (1923), Part 31, *JBNHS*, Vol. 29, No. 4, pp. 921–46.

———, (1924), Part 32, *JBNHS*, Vol. 30, No.1, pp. 132–50.

———, (1925), Part 33, *JBNHS*, Vol. 30, No. 2, pp. 285–305.

———, (1925), Part 34, *JBNHS*, Vol. 30, No. 3, pp. 561–86.

———, (1925), Part 35, *JBNHS*, Vol. 30, No. 4, pp. 822–37.

———, (1926), Part 36, *JBNHS*, Vol. 31, No. 2, pp. 323–51.

———, (1926), Part 37, *JBNHS*, Vol. 31, No. 2, pp. 655–86.

———, (1927), Part 38, *JBNHS*, Vol. 31, No. 3, pp. 951–74.

Boppré, Michael, (1984), 'Chemically Mediated Interactions Between Butterflies', in *Biology of Butterflies*, Vane-Wright and Ackery (Eds.), Symposium of Royal Entomological Soc. of London, No. 11, Academic Press, pp. 259–75.

Brower, L.P., and J.V.Z. Brower, (1964), 'Birds, butterflies and plant poisons: a study in ecological chemistry', *Zoologica* New York, Vol. 49, pp. 137–59.

Brower, Lincoln P., (1984), 'Chemical Defence in Butterflies', in *Biology of Butterflies*, Vane-Wright and Ackery (Eds.), Symposium of Royal Entomological Soc. London, No. 11, pp. 109–33.

Butler, A.G., (1886), 'On Lepidoptera collected by Maj. Yerbury in Western India', *Proc. Zool. Soc. of London*, pp. 355–91.

Cantlie, Sir Keith, (1962), 'The Lycaenidae Portion (Except the *Arhopala* Group), of Brigadier Evans', in *The Identification of Indian Butterflies 1932 (India, Pakistan, Ceylon, Burma)*, Reissued and Revised by the BNHS, Bombay.

Chai, Peng, (1996), 'Butterfly visual characteristics and otogeny of responses to butterflies by a specialized tropical bird', *Biol. Jour. of the Linnean Soc.*, Vol. 59, No. 1, pp. 37–67.

Chew, Frances S., and Roberts K. Robbins, (1984), 'Egg-laying in Butterflies', in *Biology of Butterflies*, Vane-Wright and Ackery (Eds.), Symposium of Royal Entomological Soc. London, No. 11, pp. 65–78.

Collins, N.M., and M.G. Morris, (1985), 'Threatened Swallowtail Butterflies of the World', International Union for the Conservation of Nature and Natural Resources, Gland viii, p. 402.

Courtney, Steven P., (1984), 'Habitat Versus Foodplant Selection' in *Biology of Butterflies*, Vane-Wright and Ackery (Eds.), Symposium of Royal Entomological Soc. London, No. 11, Academic Press, pp. 89–90.

Cumming, W.W., (1916), '*Vanessa Xanthonelaena* Found Occurring Below 2,000ft Altitude', *Jour. Bombay. Nat. Hist. Soc.*, Vol. 24, No. 4, p. 839.

D'Abrera, Bernard (1982), Butterflies of the World—Oriental Region – Part 1 Papilionidae, Pieridae, and Danaidae, Hill House, Victoria, Australia.

———, (1984), Butterflies of the World—Oriental Region – Part 2 Nymphalidae, Satyridae and Amathusidae, Hill House, Victoria, Australia.

———, (1985), Butterflies of the World—Oriental Region – Part 3 Lycaenidae and Riodinidae, Hill House, Victoria, Australia.

———, (1993), Butterflies of the World—Holarctic Region – Part 3 Nymphalidae (partim), Libytheidae, Riodinidae and Lycaenidae, Hill House, Victoria, Australia.

Davidson, J., and E.H. Aitken, (1890), 'Notes on the Larvae and Pupae of some of the Butterflies of the Bombay Presidency', *Jour. Bombay Nat. Hist. Soc.*, Vol. 5, No. 3, pp. 260–78.

Davidson, J., T.R. Bell, and E.H. Aitken, (1896), 'The Butterflies of the North Canara District of the Bombay Presidency', Part 1, *Jour. Bombay Nat. Hist. Soc.*, Vol. 10, No. 2, (3 plates, Danaiid and Nymphalid larvae), pp. 237–59.

———, (1897), Part 2, *Jour. Bombay Nat. Hist. Soc.*, Vol. 10, No. 3, (2 plates, Lycaenidae larvae), pp. 372–93.

———, (1897), Part 3, *Jour. Bombay Nat. Hist. Soc.*, Vol. 10, No. 4, (1 plate, Papilionid and Pierid larvae), pp. 568–84.

———, (1897), Part 4, *Jour. Bombay Nat. Hist. Soc.*, Vol. 11, No. 1, (2 plates, Hesperiidae larvae), pp. 22–63.

Dempster, Jack P., (1984), 'The Natural Enemies of Butterflies' in *Biology of Butterflies*, Vane-Wright and Ackery (Eds.), Symposium of the Royal Entomological Soc. of London, No. 11, Academic Press, pp. 97–104.

Dover, Cedric, (1922), 'Some Interesting Specimens of the Pierid Genus *Euchloe*', *Jour. Bombay Nat. Hist. Soc.*, Vol. 28. No. 4, pp. 1144–6.

Dudgeon, G.C., (1895), 'Description of the Transformations of *Badamia exclamationis*, Fabricius, a Hesperid Butterfly', *Jour. Bombay Nat. Hist. Soc.*, Vol. 10, No. 1, pp. 144–6.

Eliot, J.N., (1969), 'An Analysis of the Eurasian and Australian Neptini' (Lepidoptera: Nymphalidae)', *Bulletin Brit. Mus. Nat. Hist.*, Entomology, Supplement 15, p. 155.

Ellington, Charles P., C. van den Berg, A.P. Willmott, and A.L.R. Thomas, (1996), 'Leading-edge vortices in insect flight', *Nature*, Vol. 384. 19/26 Dec., pp. 626–30.

Eltringham, H., (1933), *The Senses of Insects*, Methuen and Co., London.

Ehrlich, Paul R., (1984), 'The Structure and Dynamics of Butterfly Populations', in *Biology of Butterflies*, Vane-Wright and Ackery (Eds.), Symposium of Royal Entomological Soc. of London, No. 11, pp. 25–40.

Evans, Brigadier W.H., (1910), 'Additions and Corrections to certain local Butterfly Lists, with Description of a New Species', *Jour. Bombay Nat. Hist. Soc.*, Vol. 20, No. 2, pp. 423–7.

———, (1924), 'The Identification of Indian Butterflies', Part 5 (Nymphalidae), *Jour. Bombay Nat. Hist. Soc.*, Vol. 30, No. 1, pp. 72–96.

———, (1932), *The Identification of Indian Butterflies*, 2nd revised edition, Bombay Natural History Soc., Diocesan Press, Madras, 454pp.

———, (1933), 'The Butterflies of Baluchistan', *Jour. Bombay Nat. Hist. Soc.*, Vol. 36, No. 1, pp. 196–209.

———, (1949), *A Catalogue of the Hesperiidae from Europe, Asia, and Australia in the British Museum (Natural History)*, British Museum (Nat. Hist.), B. Quaritch Ltd., London.

Feltwell, John, (1986), *The Natural History of Butterflies*, Croom Helm Ltd., Kent, 133pp.

Fink, Linda S., (1995), 'Foodplant effects on colour morphs of *Eumorpha fasciata* caterpillars (Lepidoptera: Sphingidae)', *Biological Jour. of the Linnean Soc.*, Vol. 56, No. 3, pp. 423–37.

Ford, E.B., (1945), *The New Naturalist—Butterflies*, Collins, London, 368pp.

Fraser, F.C., (1910), 'Biological Note on *Aphnaeus hypargyrus*', *Jour. Bombay Nat. Hist. Soc.*, Vol. 20, No. 2, pp. 528–30.

———, (1911), 'Notes on *Colotis* in Sind', *Jour. Bombay Nat. Hist. Soc.*, Vol. 20, No. 3, pp. 867–9.

———, (1911), 'A Note on *Polyommatus boeticus*', *Jour. Bombay Nat. Hist. Soc.*, Vol. 21, No. 1, pp. 287–9.

———, (1916), 'Biological Note on *Argynnis hyperbius*', *Jour. Bombay Nat. Hist. Soc.*, Misc. Notes 28, Vol. 24, No. 23, pp. 608–9.

Garcia-Barros, E., and J. Martin, (1995), 'The Eggs of European Satyrine Butterflies (Nymphalidae): External Morphology and its Use in Systematics', *Zool. Jour. Linnean Soc.*, Vol. 115, No. 1, pp. 73–115.

Gay, Thomas, Isaac David Kehimkar, and Jagdish Chandra Punetha, (1992), Nature Guides Series, 'Common Butterflies of India', World Wide Fund for Nature—India, Oxford University Press, Bombay, 67pp.

Hannyngton, F., (1911), 'Notes on the Life History of *Vanessa indica* and *kashmirensis*', *Jour. Bombay Nat. Hist. Soc.*, Vol. 20, No. 3, pp. 872–3.

Haribal, Meena, (1992), *The Butterflies of Sikkim Himalaya and their Natural History*, Sikkim Nature Conservation Foundation, Gangtok, Sikkim, 217pp.

Hasan, Syed Azhar, (1994), *Butterflies of Islamabad and the Murree Hills*, Asian Study Group, Islamabad, 68pp.

———, (1997) 'Biogeography and Diversity of Butterflies of Northwest Himalaya' in *Biodiversity of Pakistan,* Edits. Shazad Mufti, Charles Woods and Syed Azhar Hasan,

Pakistan Museum of Natural History, Islamabad and Florida Museum of Natural History, Gainsville.

Hashmi, Ali Asghar, and Afzala Tashfeen (1992), 'Lepidoptera in Pakistan', *Proc. Pakistan Congress of Zoology*. Vol. 12, pp. 171–206.

Häuser, Christoph L., C.M. Naumann, and W.G. Tremewan, (1985), 'On the Biology of *Parnassius charltonius* Gray, 1852 (Lepidoptera, Papilionidae)', in *Entomologists Gazette*, Vol. 36, No. 1, pp. 5–13.

Häuser, Christoph L., (1993), 'An Annotated Checklist of the Parnassinae (Lepidoptera: Papilionidae)', *Tidschrift voor Entomologie*, Vol. 136, pp. 137–46.

Ilse, D., (1956), 'Behaviour of butterflies before ovipositing', *Jour. Bombay Nat. Hist. Soc.*, Vol. 53, No. 2, pp. 486–8.

Johnson, S.D., (1994), 'Evidence for Batesian Mimicry in a butterfly-pollinated orchid', *Biol. Jour. of Linnean Soc.*, Vol. 53, No.1, pp. 91–104.

Kreuzberg, A.V.-A., (1985), 'Parusniki grupp *delphius, charltonius, simo* (Lepidoptera, Papilionidae), fauny SSR', in Ministerstvo kultury Uzbekskoj SSR (Ed.), *Issledovanija flory i fauny stednej Azii*, pp. 25–68, Tashkent.

———, (1987), 'Stenophagy in *Parnassius* (Lepidoptera: Papilionidae) of Central Asia and Altai', *Entomologists Gazette*, Vol. 38, pp. 95–102.

Kreuzberg, A.V.-A, and I.G. Pljushch, (1989), 'The Distribution, Ecology, and Biology of *Parnassius loxias* Phngler, 1901, (Lepidoptera: Papilionidae)', *Entomologists Gazette*, Vol. 40, No. 4, pp. 291–80.

Kunte, Krushnamegh, (1996), 'Strange Behaviour of Mottled Emigrant Males', *Jour. Bombay Nat. Hist. Soc.*, Misc. Notes no. 28, Vol. 93, No. 2, pp. 307–308.

Leslie, G.A., and W.H. Evans, (1903), 'The Butterflies of Chitral', *Jour. Bombay Nat. Hist. Soc.*, Vol. 14, No. 4, pp. 666–78.

Lewis, H.L., (1973), *Butterflies of the World*, Australasian Publishing Company, with George Harrap and Co., 312pp.

Mackinnon, P.W., and L. de Niceville, (1898), 'A List of the Butterflies of Mussoorie in the Western Himalayas and Neighbouring Regions', First part, *Jour. Bombay Nat. Hist. Soc.*, Vol. 11, No. 2, pp. 205–21, Second part, Vol. 11, No. 3, pp. 368–89 and Third part, Vol. 11, No. 4, pp. 585–602.

Magnus, D.B.E., (1950), 'Beobachtungen zur Balz und Eiablage des Kaisermantels *Argynnis paphia*, *Zeit. Tierpsychol*, Vol. 7, pp. 435–49.

Mani, M.S., (1986), *Butterflies of the Himalaya*, Dr W. Junk Publishers, Dordrecht, Series *Entomologica*, K.A. Spencer (Ed.), Vol. 36, 181pp.

Marsh, N., M. Rothschild, and F. Evans, (1984), 'A New Look at Lepidoptera Toxins', in *Biology of Butterflies*, Vane-Wright and Ackery (Eds.), Symposium of the Royal Entomological Soc. of London, No. 11, Academic Press, pp. 135–8.

Marshall, G.F.L., and L. de Niceville, *The Butterflies of India, Burmah, and Ceylon.*

———, (1883), Vol. 1, (Danainae, Satyrinae, Acraeinae).

———, (1886), Vol. 2, (Nymphalidae, Libytheinae, Riodinidae).

———, (1890), Vol. 3, (Lycaenidae).

———, Reprinted 1997, A.J. Reprints Agency, New Delhi.

Menesse, N.H., (1950), 'Butterflies of Sind', *Jour. Bombay Nat. Hist. Soc.*, Vol. 49, No. 1, pp. 20–24.

Mohyuddin, A.I., (1987), 'A Catalogue of Insects and Mites in the Reference Collection of PARC-CIBC Station up to 1986', Vol. 1, Lepidoptera, Pakistan Agric. Research Council and CAB International Institute of Biological Control.

Mosse, A.H.E., (1929), 'Note on the breeding of *Terias laeta* and *T. venata* and the probability of their being seasonal forms of a single species', *Jour. Bombay Nat. Hist. Soc.*, Vol. 33, No. 3, Misc. notes no. 40, pp. 727–9.

Moreshead, R.Y.A., (1925), 'Extraordinary display of "*Leucodice Soracte*" of the genus *Aporia*, *Jour. Bombay. Nat. Hist. Soc.*, Vol. 30, No. 4, p. 923, with plate.

Pinhiero, Carlos E.G., (1996), 'Palatability and escaping ability in Neotropical butterflies: tests with wild kingbirds (*Tyrannus melancholicus*, Tyrannidae)', *Biol. Journ. of Linnean Soc.*, Vol. 59, No. 4, pp. 351–65.

Reddy, T. Byragi, and C. Subra Reddi, (1995), 'Butterfly Pollination of *Clerodendrum infortunatum* (Verbeniaceae)', *Jour. Bombay Nat. Hist. Soc.*, Vol. 92, No. 2, pp. 166–73.

———, (1996), 'Pollination Ecology of *Duranta repens* (Verbeniaceae)', *Jour. Bombay Nat. Hist Soc.*, Vol. 93, No. 2, pp. 193–201.

Rhe-Philipe, G.W.V. de, (1917), 'The Butterflies of Lahore', *Jour. Bombay Nat. Hist. Soc.*, Vol. 25, No. 1, pp. 136–42.

Riley, N.D., (1929), 'Revisional Notes on the Genus *Heliophorus* (Lycaenidae), with descriptions of new forms', *Bombay Nat. Hist. Soc.*, Vol. 33, No. 2, pp. 384–402.

Roberts, T.J., (1991), *The Birds of Pakistan*, Oxford University Press, Karachi, 598pp.

Rothschild, M., J. von Euw, J. Reichstein, D.A.S. Smith, J. Pierre, (1975), 'Cardenolide storage in *Danaus chryssipus* with additional notes on *D. plexippus*', *Proceedings Royal Soc. London*, B 190, pp. 1–31.

Rothschild, M., B. Gardiner, and R. Mummery, (1978), 'The role of caretonoids in the "golden glance" of Danaid pupae (Insect—Lepidoptera)', *Jour. Zool. London*, Vol. 186, pp. 351–8.

Sakai, S., (1981), *The Butterflies of Afghanistan*, Tokyo, (in Japanese).

Sanders, D.F., (1955), 'Miscellaneous Notes on Indian Butterflies', *Jour. Bombay Nat. Hist. Soc.*, Vol. 52, No. 4, pp. 803–30.

Sevastopulo, D.G., (1938), 'The Early Stages of Indian Lepidoptera', Part 1, *Jour. Bombay. Nat. Hist. Soc.*, Vol. 40, No. 3, pp. 391–408.

———, (1939), Part 3, *JBNHS*, Vol. 41, No. 1, pp. 72–81.

———, (1939), Part 4, *JBNHS*, Vol. 41, No. 2, pp. 311–20.

———, (1940), Part 5, *JBNHS*, Vol. 42, No. 1, pp. 38–44.

———, (1941), Part 7, *JBNHS*, Vol. 42, No. 3, pp. 510–17.

———, (1942), Part 9, *JBNHS*, Vol. 43, No. 1, pp. 39–77.

———, (1943), Part 11, *JBNHS*, Vol. 44, No. 1, pp. 78–87.

———, (1944), Part 12, *JBNHS*, Vol. 44, No. 3, pp. 415–25.

———, (1945), Part 13, *JBNHS*, Vol. 45, No. 2, pp. 188–98.

———, (1946), Part 14, *JBNHS*, Vol. 46, No. 1, pp. 59–69.

———, (1946), Part 15, *JBNHS*, Vol. 46, No. 2, pp. 253–69.

———, (1946), Part 17, *JBNHS*, Vol. 46, No. 4, pp. 575–86.

———, (1947), Part 18, *JBNHS*, Vol. 47, No. 1, pp. 26–43.

———, (1947), Part 19, *JBNHS*, Vol. 47, No. 2, pp. 197–219.

———, (1948), Part 20, *JBNHS*, Vol. 47, No. 3, pp. 458–69.

———, (1944), 'Note on the courtship of *Euploea core core*', *Proc. Royal Entom. Soc. London*, No. 19, pp. 138–9.

———, (1948), 'Local Lists of Lepidoptera from the Punjab and U.P.', *Jour. Bombay Nat. Hist. Soc.*, Vol. 47, No. 4, pp. 586–93.

———, (1973), 'The Food-Plants of Indian Rhopalocera', *Jour. Bombay Nat. Hist. Soc.*, Vol. 70, No. 1, pp. 156–83.

Silberglied, Robert E., (1984), 'Visual Communiction and Sexual Selection among Butterflies', in *Biology of Butterflies*, Vane-Wright and Ackery (Eds.), Symposium of Royal Entomological Soc., London, No. 11, pp. 207–23.

Silberglied, R.E., and O.R. Taylor, (1978), 'Ultraviolet reflection and its behavioural role in the courtship of sulfur butterflies, *Colias eurytheme* and *C. philodice*', *Behav. Ecol. Sociobiol.*, Vol. 3, pp. 203–43.

Singer, Michael C., (1984), 'Butterfly-Hostplant Relationships: Host Quality, Adult Choice and Larval Success,' in *Biology of Butterflies*, Vane-Wright and Ackery (Eds.), Symposium of Royal Entomolgical Soc., London, No. 11, pp. 81–7.

Smith, Colin, (1989), *Butterflies of Nepal (Central Himalaya)*, Prof. C. Majupuria (Ed.), Tecpress Service L.P., Bangkok, Thailand, 352pp.

———, (1993), *Illustrated Checklist of Nepal's Butterflies*, T.C. Majupuria (Ed.), Rohit Kumar, Gwalior, India, 127pp.

Smith, D.A.S., D.F. Owen, I.J. Gordon, and K L. Ninian, (1997), 'The butterfly *Danaus chrysippus* (L.), in East Africa: polymorphism and morph-ratio clines within a complex, extensive and dynamic hybrid zone', *Zoological Jour. of Linnean Soc.*, Vol. 120, pp. 51–78.

Spencer Smith, David, and S. Azhar Hasan, (1997), 'A Preliminary Survey of Diversity and Distribution of Butterflies of Northern Pakistan: Gilgit to Khunjerab', in *Biodiversity of Pakistan*, Edits. Shazad Mufti, Charles Woods and Syed Azhar Hasan, Pakistan Museum of Natural History, Islamabad and Florida Museum of Natural History, Gainsville.

Stewart, R.R., (1972), Flora of West Pakistan, an Annotated Catalogue of the Vascular Plants of West Pakistan and Kashmir, E. Nasir and S.I. Ali (Eds.), Fakhri Printing Press, Karachi.

Sustare, Dennis, (1979), 'Physical Principles of Light', in *The Behavioral Significance of Color*, E.H. Burtt (Ed.), Garland STPM Press, New York, pp. 456.

Swinhoe, C., (1887), 'On the Lepidoptera of Karachi and its Neighbourhood', *Jour. Bombay Nat. Hist. Soc.*, Vol. 2, No. 1, pp. 269–80.

Talbot, G., (1939), *The Fauna of British India, Butterflies*, Vol. 1, (Covers Papilionidae and Pieridae), Taylor and Francis Ltd., London, p. 600.

———, (1947), *The Fauna of British India, Butterflies*, Vol. 2, (Covers Danaidae, Satyridae, Amathusiidae, and Acraeidae), Taylor and Francis Ltd., London, p. 506.

Tinbergen, Nikolaas, (1953), *Social Behaviour in Animals*, J. Wiley, London.

Tytler, H.C., (1926a), 'Notes on Some New and Interesting Butterflies from India and Burma', Part 1, *Jour. Bombay Nat. Hist. Soc.*, Vol. 31, No. 2, pp. 248–60.

———, (1926b), 'Notes on Some New and Interesting Butterflies from India and Burma', Part 2, *Jour. Bombay Nat. Hist. Soc.*, Vol. 31, No. 3, pp. 579–90.

Vane-Wright, R.I., and P.R. Ackery (Eds.), (1984), 'The Biology of Butterflies', Symposium of the Royal Entomological Society of London, No. 11, Academic Press, p. 429.

Varshney, R. K., (1979), 'Revised Nomenclature for Taxa in Wynter-Blyth's Book on the Butterflies of Indian Region', *Jour. Bombay Nat. Hist. Soc.*, Vol. 76, No. 1, pp. 33–40.

———, (1985), 'Revised Nomenclature for Taxa in Wynter-Blyth's Book on the Butterflies of the Indian Region—Part 2', *Jour. Bombay Nat. Hist. Soc.*, Vol. 82, No. 2, pp. 309–21.

Whalley, Paul, and Richard Lewington, (1996), *The Mitchell Beazley Pocket Guide to Butterflies*, Revised Edition, Mitchell Beazley Publishers, Reed International Books, London, p. 168.

Williams, C.B., (1938), 'The Migration of Butterflies in India', *Jour. Bombay Nat. Hist. Soc.*, Vol. 40, No. 3, pp. 439–5.

Wynter-Blyth, M.A., (1957), *Butterflies of the Indian Region*, Bombay Nat. Hist. Soc., p. 523.

Yerbury, Major J.W., (1886), 'List of butterflies received from Major Yerbury', Miscellaneous Notes, *Jour. Bombay Nat. Hist. Soc.*, Vol. 1, No. 4, p. 219.

Young, L.C.H., (1905), 'The Common Butterflies of the Plains of India', Part 1, *Jour. Bombay Nat. Hist. Soc.*, Vol. 16, No.4, pp. 570–79.

———, (1906), 'The Common Butterflies of the Plains of India', Part 2, *Jour. Bombay Nat. Hist. Soc.*, Vol. 17, No. 2, pp. 418–23.

———, (1907), Part 4, *Jour. Bombay Nat. Hist. Soc.*, Vol. 17, No. 4, pp. 921–7.

Appendix 2
Glossary of Technical Terms

Allopatric Taxa that do not overlap geographically

Androconia Specialized scent scales on the wings of certain male butterflies

Annular In the form of a ring

Antennae The long 'feelers' arising from between the butterfly's eyes

Apex Referring to that area of a butterfly's wings near the tip

Apiculus Applied to the hook at the tip of the antennae of Hesperiidae

Aposematic Colouring which has a protective role

Base Referring to that part of a butterfly's wings near the junction with the body

Bi-fid Cleaved in two, usually referring to the claws or spurs on a butterfly's leg

Bi-pupilled Having two paler-coloured centres within the ocellus or ring mark on a butterfly's wing

Brand A raised patch of specialized scent scales on the wings of certain male butterflies

Carinate Forming a ridge

Cell An elongate area extending from the base of both fore and hindwings, enclosed by veins, from which radiate other veins up to the wing costa and termen

Chrysalis The pupa of a butterfly

Cilia Hair-like fringe along the wing margin

Claspers In larvae, applied to the anal prolegs; in adults, applied to the organs of the males' genitalia

Contiguous Touching or joining, often referring to the arrangement of spots on the underwing

Costa The upper margin of both fore and hindwings

Cremaster An anal process of the pupa by which it is attached to its substrate

Cryptic (adj.), Crypsis (noun) Colouration which helps to conceal from enemies

Cryptic colours Those which conceal by making the butterfly resemble its surroundings

Dentate Appearing toothed

Diapause Equivalent to a state of physiological suspension akin to hibernation

Dimorphic Having two different forms or colour markings between individuals of the same species

Discal Pertaining to the middle area of the wing adjacent to and including the end of the cell and before the margin

Ecdysis The act of moulting or shedding of the outer skin by the larva

Emarginate Notched; usually referring to an obtuse or rounded section cut from the wing margin

Fringe The hair-like fringe on the outer margin of the wings

Genitalia The external organs of reproduction

Haemolymph The equivalent of blood in vertebrates, which in insects flows throughout the body cavity, not confined within veins and arteries

Hastate Shaped like a spearhead, narrowly diverging at the base

Haustellum The sucking mouthpart of the adult butterfly, also called the proboscis

Hermaphrodite Any individual in which the two sexes are combined

Hyaline Glassy, clear like glass

Imago The emerged perfect adult butterfly

Instar The larval stage between each casting of skin

Integument The natural covering of a body. That which encloses

Irrorated Sprinkled with minute dots

Larva The grub or 'maggot' stage; in butterflies, the caterpillar

Lobe The rounded extension in the tornal area of the hindwing in some butterfly species

Lunule A crescent-shaped spot

Marginal When applied to markings on the outer edge of the wings, the area between the the termen and discal region

Mesic Quality of an organism which requires constant moisture or of a site which provides it

Mesothorax The middle of the three segments comprising the insect's 'chest', and which in butterflies bears the forewings and one pair of legs

Micropyle The orifice at the top of a butterfly's egg for exchange of gasses. Literally 'a tiny opening'

Myrmecophilous Liking or semi-dependant upon ants

Nerve or Nervule   A rib or veinlet of the framework of the wings

Nullah   Vernacular name for valley

Obsolete   Almost absent, or at least very indistinct

Ocellus   An eye-like marking or spot on the wing

Ochraceous   A dull pale yellow or buffy-yellow colour

Osmeterium   Protrusible fleshy forked process, capable of being projected from a slit in the pro-thoracic segment of Swallowtail Butterfly larvae, which produces a strong aromatic smell

Palpi   Cushion-like process in front of the butterfly's head, curving up between the eyes

Phragis   Horny pouch formed at tip of abdomen of female Parnassians after fertilization

Pilose   Covered with soft hairs

Polymorphic   Occurring in several forms

Porrect   Extending forward horizontally

Pre-costal vein   The short vein emerging from vein 8 near the base of the hindwing. Present in all butterfly families except the Lycaenidae

Pro-thoracic legs   Those belonging to the first segment of the 'chest', and important in classification of butterfly families

Pubescent   Surface hairs which are short, soft and straight

Pupa   The resting stage of an insect. In butterflies also called the chrysalis (q.v.)

Quadrate   Shaped like a square

Rhopalocera   The family of Butterflies, literally 'club horns'

Sagittate   Arrowhead shaped

Scalloped   When the wing margin is indented with rounded tooths

Segment   In insects, a division of the body

Senesce   To grow old and wither, particularly applied to plants, and important for Neptini butterflies

Seta   Insect's bristle or hair

Setose   Bristly

Somite   Applied to one or more of the segments of a caterpillar's body

Sphragis   A hardened 'plug' at the tip of a female Apollo butterfly's abdomen deposited by the male after fertilization

Spinneret   A small perforated organ on the caterpillar's head through which it extrudes its silk

Spiracle   Paired respiratory openings along the side of the caterpillar which connect with the tracheal tubes. Sometimes conspicuously coloured

Stigmata   Literally, a mark made on the skin by branding; applied to the distinctive patches of special scent scales on the wings of certain male butterflies

Striated   Closely marked with short fine lines

Stridulating   Mechanically produced noise in insects made by rubbing specially adapted body parts together

Sub-apical   Near the wing but just inside the apex

Sub-marginal   Just inside the outer margin

Sympatric   Taxa that overlap geographically at least in some part of their range, especially applied to different species

Tangi   vernacular name in Balochistan for canyon or narrow ravine

Tarsus   Last four segments in the adult butterfly's legs forming the last apparent joint

Termen   The outer edge of a wing

Tornal   Particularly applied to the hind wing posterior distal extremity

Tornus   The lower corner of the wing at the junction of the termen and dorsum

Truncate   Cut off, especially at the wing tip

Tubercle   A small warty projection or knob, particularly applied to the skin of larvae

Xerophytic   A plant or one of its parts which shows a capacity to withstand drought

Appendix 3
Gazetteer

This Gazetteer differs from the two previous Gazetteers, which the author compiled for books on Pakistan's Mammalian Fauna and Avian Fauna. Many of the less well-known butterfly species are confined to Central Asian regions, often in high elevation, semi-arid habitats. This Gazetteer concentrates on giving co-ordinates for as many as possible of these far northern localities mentioned in the text.

Detailed maps of regions in Pakistan are difficult to obtain, and those available generally do not show the names of minor valleys and other physical features, tending only to show towns and villages by name. In preparing this Gazetteer I have been greatly helped by persons with local knowledge who are acknowledged in the beginning of this book. I have tried to give co-ordinates of all the places mentioned in this account of the various species status, excepting places outside of Pakistan. The reader should appreciate that many places recorded on the specimen labels of collectors have been spelled differently from modern spelling, and where possible I have shown alternative spellings commonly used in earlier publications. It should be realized that some of the more remote northern locations would only have been known to early butterfly collectors by the spoken name, which in the process of their transliteration, often by British army officers, became Anglicized in spelling. For example, at the turn of the twentieth century, when much collecting was carried out by army personnel stationed in Chitral, detailed maps of that region were not available. Usually the local languages from which place names originated are difficult to reproduce in English with respect to exact vowel sounds and 'a', 'e', and 'u' are sometimes substituted in the spelling of the same place. Users of this Gazetteer should also note that on maps of the region, valleys are often named *Gol* in the northern areas, or *Nullah*. Another problem in preparing a gazetteer of Pakistan is the duplication of place names. Ziarat, which means a shrine, is used for two well-known places in Chitral and Balochistan. Kala Pani, which means literally 'Black Water' and implies a place of desolation, occurs both on the slopes of Thandiani in Hazara district and in Astor district in Gilgit.

		Co-ordinates	
Name	**Approximate Location**	**Latitude (N)**	**Longitude (E)**
Abbottabad	Hazara District	34° 08'	73° 12'
Askole	Village on Braldu River, northern Baltistan	35° 40'	75° 51'
Astor	District in south-east Gilgit	35° 22'	74° 52'
Attock Bridge	North-west Punjab boundary	33° 54'	72° 14'
Ayun	Between Drosh and Chitral towns	35° 43'	71° 46'
Babusar Pass	Between Kaghan Valley and Chilas	35° 09'	74° 06'
Bahrain or Bahrein	Northern Swat	35° 14'	72° 59'
Baltoro Glacier	South of K2, Baltistan	35° 44'	From 76° 10' 76° 40'
Bangol Valley	North-eastern Chitral, valley on west bank of Yarkhun River	36° 31'-39'	From 72° 40' 72° 47'
Bannu	Town in Bannu District, NWFP	33° 00'	67° 02'
Barakao or Bharakao (old spelling)	Murree foothills	33° 44'	73° 09'
Baroghil Pass and Valley, also called Brughal	Northern border between Chitral and Wakhan	36° 31'	73° 23'
Battagram	Village in Hazara Tribal Area	34° 40'	72° 59'
Battakundi	Village in northern Kaghan Valley	34° 56'	73° 46'

184 THE BUTTERFLIES OF PAKISTAN

Name	Approximate location	Co-ordinates Latitude (N)	Longitude (E)
Batura or Batora Glacier	North-western Hunza	36° 32'	74° 20'
Beori Nullah	South-east Chitral	35° 29'	71° 48'
Bogra Pass	12 miles east of Chaman	30° 50'	66° 40'
Bolan Pass	Extreme north of Sibi District	29° 53'	67° 14'
Braldo Glacier	North of Skardu on border of Baltistan	36° 04'	75° 55'
Bunyan	Teep Nullah, Swat Kohistan	35° 21'	72° 41'
Burawai	Kaghan Valley	34° 57'	73° 52'
Burzil Pass	Eastern boundary of Deosai	34° 51'	75° 07'
Campbellpur (renamed Attock)	Attock District	33° 47'	72° 22'
Chagai	Town in western part of Chagai Desert	29° 18'	64° 44'
Chaman	Town on Afghanistan border	30° 55'	66° 25'
Chauter or Chotair	Village on road between Ziarat and Loralai, Balochistan	30° 52'	67° 55'
Chichele Pass or Ochhili Pass	Chitral, Wakhan eastern border	36° 07'	72° 28'
Chilas	Town in south-western district of Gilgit	35° 25'	74° 06'
Chiltan Hills	South-west of Quetta	30° 02'	66° 52'
Chitral Town	South-central part of State	35° 50'	71° 47'
Chogo Longma or Lungma	Glacier descending from Arandu, Baltistan	36° 44'	75° 10'
Dalbandin	Central Chaghai District	28° 53'	64° 26'
Darkot Pass	Between north-west Yasin and Chitral	36° 38-44'	73° 17'
Dashkin Forest	Northern Astor District	35° 28'	74° 48'
Dera Ismail Khan	South-eastern NWFP	31° 48'	70° 54'
Doiyan or Doyan	South of junction between Indus and Astor Rivers	35° 32'	74° 42'
Doubanni Mountains	East of Gilgit town	35° 10'-57'	74° 39'-47'
Dras	Extralimital in Eastern Ladakh		
Drosh	Southern Chitral	35° 33'	71° 47'
Duki	South of Loralai, Balochistan	30° 06'	68° 34'
Dunga Gali	Murree Hills	34° 03'	73° 22'
Faisalabad	Formerly Lyallpur	31° 25'	73° 07'
Farsat Pass	Between Rupal and Lilam valleys, south of Nanga Parbat	34° 59'	74° 33'
Fateh Jang	South of Kalla Chitta Range	33° 34'	72° 39'
Fort Monro	Dera Ghazi Khan District	29° 54'	69° 59'
Fort Sandeman, see, Zhob	Northern Zhob District	31° 21'	69° 28'
Garogarh or Gharogar	Village in Bangol Valley, tributary of Yarkhun River	36° 34'	72° 43'
Gandak or Ghundak Valley	North-east of Quetta	30° 21'	67° 08'
Ghizar Valley	Running westwards through lower Yasin, Gilgit	36° 10'	From 72° 45' 73° 44'
Gurais	Neelum Valley near Kashmir Cease-fire Line	34° 37'	74° 51'
Gwal Forest	North-east of Quetta	30° 30'	67° 10'
Hab Valley	Sindh, Lasbela border	From 24° 52' 25° 55'	From 66° 42' 67° 10'
Haldi	East bank of Hushe River	35° 66'	76° 37'
Hangu	Kohat District	33° 32'	71° 04'
Hannah Lake	East of Quetta	30° 05'	67° 57'
Harchin	On Laspur river, south of Mastuj	36° 04'	72° 31'
Haripur	Southern Hazara District	33° 59'	72° 59'

Name	Approximate location	Co-ordinates Latitude (N)	Longitude (E)
Hasan Abdal	Attock District	33° 48'	72° 42'
Hyderabad	City in southern Sindh	25° 24'	68° 22'
Idak	North Waziristan	32° 58'	70° 12'
Islamabad city	North of Rawalpindi	33° 43'	73° 05'
Ispodagh, or Ishpadrogh, or Ishperu Dok	Extreme north-east Chitral	36° 46'	72° 53'
Jabba Valley	East bank tributary of Swat River, near Bahrain	35° 20'	72° 40'
Jamrud	North-west of Peshawar	34° 00'	71° 19'
Jhelum city	On Jhelum River at foot of Salt Range	32° 57'	73° 44'
Kach	East of Takhatu Range	30° 26'	67° 21'
Kadir Gali	East Bank, Kaghan Valley	34° 41'	73° 42'
Kalam	Swat Kohistan	35° 31'	72° 35'
Kala Pani	Hazara Dist. on way to Thandiani	34° 11'	73° 21'
Kala Pani	Southern Astor District	34° 53'	74° 55'
Kamri Pass	Mountains south-west of Astor river	34° 44'	74° 57'
Karachi city	Arabian sea coast	24° 00'	67° 00'
Karumbar Pass	Gilgit-Chitral border in north	36° 51'	73° 43'
Katichula Pass	Eastern side of Deosai Plateau	35° 40'	75° 66'
Kawas	Village on road to Ziarat	30° 29'	67° 33'
Keru Lungma or Kero Lungma	Glacier Valley joining Chogo Lungma, Baltistan	36° 48'	75° 20'
Khaira Gali	West of Murree town	33° 59'	73° 23'
Khanewal	Southern Punjab	30° 18'	71° 56'
Khan Mehtarzai Pass	On railway line to Muslim Bagh	30° 42'	67° 28'
Khanozai	On railway line to Muslim Bagh	30° 37'	67° 18'
Khapalu	Village on Shyok river, Baltistan	35° 11'	76° 22'
Khilas Valley	Chitral	not located	
Khojak Pass	From Chaman into Kandahar	30° 52'	66° 28'
Khuzdar	South-east Kalat Division	27° 53'	66° 36'
Kine Chish Pass	Between Swat Kohistan and Chilas	not located	
Koch Boozla or Koch Borla	Baltistan above south bank of Indus	35° 55'	75° 25'
Lahore	Punjab capital on north-east border	31° 33'	74° 19'
Landhi	East of Karachi	24° 52'	67° 13'
Larkana	District Headquarters, north-west Sindh	27° 32'	68° 12'
Laspur Valley	Running south from Mastuj, Chitral	36° 12'	72° 28'
Lehtrar Valley	Eastern side of Murree foothills	33° 42'	73° 26'
Loralai town	North-eastern Balochistan	30° 23'	68° 38'
Lowari Pass	Between Dir and Chitral	35° 22'	71° 47'
Lower Topa	North-east of Murree town	33° 55'	73° 23'
Lutko Valley	North-east Chitral	36° 05'	71°20'-45'
Lyallpur, now Faisalabad	Central Punjab	32° 25'	73° 07'
Mach	Western Sibi District	29° 52'	67° 21'
Machiara Valley	Tributary of Neelum River on west bank	34° 31'	73° 37'
Madaglash Valley	North of Drosh on east bank of Kunar River	35° 47'	72° 02'
Makhrah peak or Makrah	Above Shogran on watershed of Neelum Valley	34° 33'	73° 29'
Malach	Village below Dunga Gali	34° 03'	73° 21'
Malakand Pass	Between Malakand Agency and Swat	34° 34'	71° 57'
Malezai or Malozai	Village on Zhob-Mekhtar Road	30° 41'	69° 23'
Malir	Eastern suburb of Karachi	24° 59'	67° 13'
Manga Valley	Murree foothills	33° 47'	73° 22'
Mankial Peak	Swat Kohistan, north of Jabba Valley	35° 19'	72° 44'

THE BUTTERFLIES OF PAKISTAN

Name	Approximate location	Co-ordinates Latitude (N)	Longitude (E)
Mansehra	Hazara District	34° 19'	73° 12'
Margalla Hills	West of Islamabad	33° 48'	73° 10'
Marghazar or Murghazar	Lower Swat	34° 39'	72° 19'
Mastuj	Northern Chitral	36° 17'	72° 31'
Mastung	Northern Kalat State	29° 48'	66° 52'
Mehtarzai Pass see, Khan Mehtarzai			
Miandam	Swat Valley	35° 04'	72° 34'
Mingora or Mangora	Main town in Swat	34° 47'	72° 22'
Mintaka Pass	Northern boundary of Hunza	36° 59'	74° 52'
Miranjani Mountain	North of Nathia Gali	34° 06'	73° 25'
Miramshah or Miranshah	North Waziristan	33° 01'	70° 04'
Mirogram or Miragram	Yarkhun Valley, northern Chitral	36° 32'	72° 47'
Misgar	Northern Hunza	36° 46'	74° 46'
Mokhshpuri or Mukhshpuri Mountain	North of Dunga Gali	34° 04'	73° 26'
Murgha Mehtarzai	14 miles east of Khanozai, Balochistan	30° 43'	67° 34'
Mushkin	Gilgit in Astor Valley	35° 30'	74° 44'
Naghr	Southern Chitral on Kunar River	35° 29'	71° 44'
Naltar Valley	North-east of Gilgit town	36° 07'	74° 14'
Nathia Gali	Murree Hills	34° 04'	73° 24'
Neelum Valley and River	Formerly Kishenganga River	34° 20'-43'	From 73° 28' 74° 51'
Nurpur Shahan or Nurpur Sharif	Margalla Hills	33° 44'	73° 07'
Nushki	Eastern end of Chagai	29° 33'	66° 02'
Orghutz or Uguthi	North-west border of Chitral with Afghanistan	36° 01'	71° 23'
Ormara	Mekran Coast	25° 13'	64° 40'
Parachinar	Upper Kurram Valley	33° 53'	70° 07'
Peshawar	Capital of NWFP	34° 01'	71° 33'
Punial	North-west of Gilgit town	36° 07'	74° 02'
Rahimyar Khan	Southern Punjab	28° 24'	70° 18'
Raigora or Ragora	Central Balochistan on road to Loralai	30° 18'	68° 14'
Rama Lake	North-west of Astor town	35° 18'	74° 46'
Ratu or Ratoo	Astor, Gilgit	35° 10'	74° 47'
Rawal Lake	Near Islamabad	33° 40'	73° 02'
Rawalpindi	North-east Punjab	33° 36'	73° 03'
Razmak	South Waziristan	32° 42'	69° 52'
Rindi Valley	Tributary of Baltoro Valley	not located	
Rupal Valley	Astor, Gilgit	35° 13'	74° 42'
Saif-ul-Muluk Lake	Upper Kaghan Valley	34° 52'	73° 41'
Saltoro Mountains	East bank of Hushe River	35° 10-29'	76° 32'
Samana Hills	North-western Kohat District	33° 37'	70° 30'-50'
Saranan or Sarawan	Due west of Pishin	30° 34'	66° 53'
Scoro La or Skoro La Pass	Between Askole on Braldo River and Shigar Valley	36° 00'	75° 55'
Shandur Pass	Between Laspur Valley in Chitral and Yasin, Gilgit	36° 03'	72° 35'
Shankagarh	South of Astor on Kamri road	35° 02'	74° 52'
Sheikh Wasil	North-east of Nushki	29° 53'	66° 33'
Shekhupura or Sheikhupura	District north-west of Lahore	31° 43'	73° 59'
Shigar Valley	Valley running north of Skardu	35° 25'	75° 41'
Shimshal Valley	North-eastern Hunza	36° 27'	75° 18'

Name	Approximate location	Co-ordinates Latitude (N)	Longitude (E)
Shingar or Shingai Gah Range	On Zhob-South Waziristan border	31° 20'-40'	69° 45'
Shishi Koh Valley	Leading to Madaglash, north of Drosh, Chitral	35° 37'	71° 50'
Shogran	Lower Kaghan Valley	34° 37'	73° 28'
Sialkot town	North-east Punjab	32° 30'	74° 32'
Sibi	South of Bolan Pass	29° 33'	67° 54'
Skardu	Baltistan main town	35° 17'	75° 38'
Sonoghar	North of Drosh, at junction of Shish Valley	35° 36'	71° 48'
Speraragha	Head of valley north of Kowas, Balochistan	30° 32'	67° 37'
Sukkur	Town in northern Sindh on Indus River	27° 42'	68° 52'
Takht-i-Suleiman	Balochistan-South Waziristan border	31° 36'	69° 59'
Tarben Valley, also called Drosh Gol	East Bank of Kunar River, north of Drosh	35° 34'	71° 47'-52'
Thandiani	Extreme north of Murree Hill range	34° 14'	73° 22'
Thui Pass	Yasin-Chitral border in north	36° 41'	73° 05'
Tochi Valley	North Waziristan	32° 40'-58'	70° 00'-45'
Torghar Range	South-east of Zhob	31° 08'-50'	69° 40'-55'
Tret	Murree foothills	33° 51'	73° 19'
Turikho Valley	North-eastern Chitral	35° 55'	71° 48'
Urak or Uruk Valley	East of Quetta	30° 16'	67° 11'
Urtsun or Utzen Valley	Left bank of Kunar below Drosh	35° 29'	71° 43'
Ushu	East of Kalam, Swat Kohistan	35° 32'	72° 39'
Utrot	West of Kalam, Swat Kohistan	35° 29'	72° 27'
Wam	Northern Sibi District	30° 27'	67° 43'
Yasin	Village on Darkot River, Yasin district	36° 22'	73° 20'
Yarkhun Valley	Northern Chitral	36° 17'	72° 30'
Zarghun Peak	Quetta-Sibi boundary	30° 13'	67° 18
Zhob Town	Northern District of Balochistan (formerly Fort Sandeman)	31° 21'	69° 28'
Ziarat	Balochistan	30° 22'	67° 44'
Ziarat	Southern Chitral	35° 23'	71° 47'
Zizri (Lookout Point)	South of Ziarat, Balochistan	30° 17'	67° 42'

Index

A

acamus - Apharitis 150
acco - Parnassius 39, fig. 9, pl. 12
aconthea - Euthalia 97, pl. 11, pl. 28
ACREINAE 116
Acraea violae 116, pl. 34
actaea - Hipparchia or *Karanasa* 107, pl. 32
actis - Polyommatus 36, 134, pl. 38
actinor - Suada 170, pl. 46
actius - Parnassius 36, pl. 12
adippe - Argynnis 81, pl. 24
Aeromachus inachus 168, pl. 46
Aeromachus stigmata 168, pl. 46
Afghan Fritillary 81, pl. 24
African Babul Blue 125, pl. 35
African Emigrant 63, pl. 20
African Grass Blue 137, pl. 39
African Marbled Skipper 163, pl. 45
agama - Caprona 162, pl. 45
agestis - Aricia 129, pl. 37
Aglais cachmirensis 92, pl. 27
Aglais urticae 92, pl. 27
aglaja - Argynnis 80, pl. 24
aglaja - Mesoacidalia 80
Agriades jaloka 131, pl. 37
Agriades pheretiades 131, pl. 37
ajaka - Pieris 56, pl. 17
ala - Melitaea 84
Albatross - Striped 61, pl. 19
Albulina galathea 132, pl. 38
Albulina metallica 132, pl. 38
Albulina omphisa 132, pl. 37
Albulina pheretes 132, pl. 37
alceae - Carcharodus 167, pl. 45
alcides - Eogenes 175, pl. 47
alexanor - Papilio 46, pl. 14
alexis - Glaucopsyche 133, pl. 38
alexis - Hasora 159, pl. 44
almana - Junonia or *Precis* 85
alpheraky - Colias 68, pl. 21
alpina - Hesperia or *Pyrgus* 166, pl. 45
Alpine Large White 56, pl. 18
alteratus - Tarucus 123, pl. 35
altheae - Carcharodus 166, pl. 45
amata - Colotis 71, pl. 22
ambica - Apatura 75, pl. 4, pl. 23

Amblypodia dodonea 149, pl. 42
Amblypodia ganesa 149, pl. 42
Amblypodia rama 149, pl. 42
Amecera cashmirensis 101, pl. 29
anais - Parenonia 61, pl. 9, pl. 19
Anaphaeis aurota 60, pl. 19
ancila - Telicota 171, pl. 46
Angle - Golden 163, pl. 45
Angle - Spotted 162, pl. 45
annada - Eurebia or *Paralasa* 110, pl. 30
Apatura ambica 73, pl. 4, pl. 23
Apaturinae 75
Apharitis acamas 150, pl. 43
Apharitis epargyros 150, pl. 43
Aphnaeus hypargyrus 150
Apollos 34
Aporia leucodice 58, pl. 18
Aporia nabellica 58, pl. 18
Aporia soracte 58, pl. 18
Appias libythea 61, pl. 19
Araschnia prorsoides 93
arcesia - Melitaea 84, pl. 25
arcturus - Papilio 48, pl. 11, pl. 15
Agriades agestis 129, pl. 37
Agriades jaloka 131, pl. 37
Agriades pheretes 132, pl. 37
argiades - Everes 126, pl. 36
argiolus - Celastrina 128, pl. 36
Argus (Satyrinae) - Common 109, pl. 31
Argus - Mountain 111
Argus - Pallid 110
Argus - Ringed 110, pl. 30
Argus - Scarce Mountain 111, pl. 30
Argus - Yellow 111, pl. 30
Argus (Lycaenidae) - Astor 130, pl. 37
Argus - Brown 129, pl. 37
Argus - Jewel 131, pl. 37
Argus - Streaked 130, pl. 37
Argynnis adippe 81, pl. 24
Argynnis aglaia or *aglaja* 80, pl. 24
Argynnis childreni 79, pl. 4, pl. 24
Argynnis hegemone 81, pl. 25
Argynnis hyperbius 78, pl. 26
Argynnis jerdoni 80, pl. 24
Argynnis kamala 80, pl. 24
Argynnis lathonia 78, pl. 24

Argynnis maia 79, pl. 24
Argynnis niphe 78, pl. 26
Argynnis pales 82, pl. 25
Argynnis pandora 79, pl. 24
Argyreus hyperbius 78, pl. 26
agrospilata - Fabriciana 81, pl. 24
Arhopola dodonea 149, pl. 42
Arhopola ganesa 149, pl. 42
Arhopola rama 149, pl. 42
Ariadne merione 77, pl. 23
Aricia agestis 129, pl. 5, pl. 37
Aricia astorica 130, pl. 37
Aricia chiron 130, pl. 37
aristolochiae - Pachliopta 45, pl. 15
aristolochiae - Tros 45, pl. 3
arletta - Vardhana 127, pl. 36
Artogeia kreuperi 55, pl. 17
asocrates - Varachola 153, pl. 43
aspasia - Gonepteryx 66
assamica - Strymon 146
asterope - Ypthima 112, pl. 33
Astor Argus 130, pl. 37
astorica - Aricia 130, pl. 37
astorica sub-sp. *Lycaena orbitulus* 131
astrarche - Polyommatus 129, pl. 37
Astycus augias 172, pl. 47
Astycus pythias 171, pl. 46
ataxus - Chrysozephyrus or *Thecla* 147, pl. 42
Atella phalantha 77, pl. 26
athamantis - Lycaena 144
athamus - Eriboea 75, pl. 23
athamus - Polyura 75, pl. 23
atkinsoni sub-sp. *Parnassius stoliczkanus* 40, pl. 13
atlanta - Vanessa 87, pl. 26
Atrophaneura polyeuctes 45, pl. 15
Aulocera padma 108, pl. 33
Aulocera saraswati 109, pl. 33
Aulocera swaha 109, pl. 33
aurota - Belenois (syn: *Anaphaeis*) 60, pl. 19
ausonia - Synchloe 53, pl. 16
ausonides - Euchloe 53, pl. 16
avanta - Ypthima 113, pl. 33
Awl - Brown 159, pl. 44
Awl - Common Banded 159, pl. 5, pl. 44

Awl - Orange Tail 159, pl. 44
Awlet - Pale Green 159, pl. 44
Awlking (Indian) 160, pl. 5, pl. 44
Azanus jesous 125, pl. 35
Azanus ubaldus 125, pl. 35
Azanus uranus 126, pl. 35

B

Babul Blue - African 125, pl. 35
Babul Blue - Bright 125, pl. 35
Babul Blue - Dull 126, pl. 35
Badamia exclamationis 159, pl. 44
bakeri - Heliophorus 145, pl. 10, pl. 41
baldiva - Eumenis 105, pl. 32
balkanica - Tarucus 123, pl. 35
balucha sub-sp. *Carcharodus floccifera* 166
balucha sub-sp. *Lycaena athamantis* 144
Balochi Copper 143, pl. 41
Balochi Fritillary 83, pl. 25
Balochi Hairstreak 145, pl. 42
Balochi Heath 104, pl. 31
Balochi Jewel Blue 129, pl. 36
Balochi Meadow Blue 134, pl. 38
Balochi Rockbrown 105, pl. 32
Balochi Skipper 165
Balochi Swallowtail 46, pl. 14
Baltia butleri 51, pl. 16
Baltia shawi 51, pl. 16
bambusae sub-sp. *Astycus pythias* or *Telicota ancila* 171, pl. 46
Banded Awl 159, pl. 44
Banded Tree Brown 99, pl. 29
Baoris bevani 174, pl. 47
Baoris discreta 173, pl. 47
Baoris eltola 173, pl. 47
Baoris guttatus 174, pl. 47
Baoris mathias 173, pl. 47
baroghila - Polyommatus 134, pl. 38
Baron (Common) 97, pl. 4, pl. 11, pl. 28
Bath White 54, pl. 17
Beak - Common 120, pl. 34
Beak - European 121, pl. 5, pl. 34
Beaks (Libytheinae) 120
belemia - Euchloe or *Synchloe* 52, pl. 16
Belenois mesentina 60, pl. 19
belladona - Delias 58, pl. 18
benjaminii - Choaspes 160, pl. 44
bevani - Guttatus or *Baoris* or *Parnara* 174, pl. 47
Bevan's Swift 174, pl. 47
bifasciata - Lobocla 162, pl. 44

Bibasis sena 159, pl. 44
Black Edged Apollo 40, pl. 13
Black Leg Tortoiseshell 91, pl. 26
Black Satyr 107, pl. 32
Black Spotted Pierrot 123, pl. 35
Blackvein - Dusky 58
Blackvein - Himalayan 58
Blackvein Fritillary 84, pl. 25
blanda - Eurema 65, pl. 20
Blue Admiral 89, pl. 10, pl. 26
Blue Apollo 38, pl. 12
Blue Pansy 85, pl. 4, pl. 27
Blue Peacock 48, pl. 11, pl. 15
Blue Spot Arab 71, pl. 22
Blue Tiger 117, pl. 34
bochus - Jamides 140, pl. 40
boedromus - Parnassius 41, pl. 14
boeticus - Lampides 140, pl. 40
bogra - Polyommatus 134, pl. 38
bogra - sub-sp. *Polyommatus actis* 134
bolanica - Ypthima 113, pl. 33
Bolaria pales 82, pl. 25
bolina - Hypolimnas 94, pl. 28
boloricus - Karanasa 108
brassicae - Pieris 57, pl. 17
Branded Meadow Brown 102, pl. 30
Brick Skipper 165, pl. 45
Bright Babul Blue 125, pl. 35
brigitta - Eurema 65, pl. 20
Brilliant Meadow Blue 134, pl. 38
Brimstone - Chitral 67, pl. 22
Brimstone - Common 66, pl. 22
Brimstone - Lesser 66, pl. 22
Broad Bordered Clouded Yellow 69, pl. 21
Brown Argus 129
Brown Awl 159, pl. 5, pl. 44
buddha - Calinaga 77, pl. 23
buddhista - Everes 126, pl. 36
Bush Blue - Tailless 149
Bush Brown (Common) 114, pl. 29
butleri - Baltia 51, pl. 16
Butler's Dwarf 51, pl. 16
butleri sub-sp. *Neptis soma* 96, pl. 28

C

cachmiriensis - Aglais (syn: *Vanessa*) 92, pl. 27
c-album - Polygonia or *Vanessa* 89, pl. 26
cadesia var: *Karanasa huebneri* 107
calais - Colotis 71, pl. 22
CALINAGINAE 76
Calinaga buddha 77, pl. 23

Callerebia nirmala 109, pl. 31
callimachus - Thestor 144, pl. 41
callimachus - Tomares 144, pl. 41
callinara - Tarucus 124
callidice - Parapieris (syn: *Pontia*) 53, pl. 17b
callidice - Pieris 53, pl. 47
Callophrys rubi 146, pl. 5, pl. 42
Calotes versicolor 27, pl. 11
canace - Vanessa or *Kaniska* 89, pl. 10
canidia - Pieris 56, pl. 17
Caprona agama 162, pl. 45
Caprona ransonetti 163, pl. 45
Carcharodus alceae 167, pl. 45
Carcharodus altheae 166, pl. 45
Carcharodus floccifera 166, pl. 45
Cardinal Fritillary 79, pl. 4, pl. 24
cardinalis - Parnassius 39, pl. 12
cardui - Cynthia or *Vanessa* 88, pl. 10, pl. 26
carnea sub-sp. *Hesperia sertorius* 165, pl. 45
cashmiriensis - Amecera 101, pl. 29
cashmiriensis - Vanessa 92, pl. 27
caspius - Lycaena 144, pl. 41
Castalius rosimon 123, pl. 35
Castor (Common) 77, pl. 23
casyapa - Lobocla (syn: *L. liliana*) 162
Catochrysops cnejus 138, pl. 5, pl. 39
Catochrysops dipora 126, pl. 36
Catochrysops strabo 139, pl. 40
Catopsila crocale 62, pl. 3, pl. 19
Catopsila florella 63, pl. 20
Catopsila pomona 62, pl. 19
Catopsila pyranthe 63, pl. 20
Celaenorrhinus leucocera 161, pl. 44
Celaenorrhinus munda 165, pl. 44
Celastrina argiolus 128, pl. 5, pl. 36
Celastrina ladonides 127, pl. 36
Celastrina puspa 127, pl. 36
Cepora nerissa 60, pl. 19
celeno - Jamides 141, pl. 40
celtis - Libythea 121, pl. 34
Cerulean - Common 141
Cerulean - Dark 140
Chaetoprocta odata 148, pl. 42
Chapman's Cupid 126, pl. 36
CHARAXINAE 74
charlonia - Synchloe or *Euchloe* or *Elphinstonia* 52, pl. 16
charltonius - Parnassius 41, pl. 13
Chazara heydenreichii 106, pl. 32
Chazara persephone 106, pl. 32
Chequered Blue 128, pl. 36
Chequered Darter 172, pl. 5, pl. 47

Chilades laius 136, pl. 39
Chilades trochilus 136, pl. 39
Chilasa clytia 49, pl. 3, pl. 15
Childreni childreni 79, pl. 24
childreni - Argynnis 79, pl. 24
chiron - Aricia 130, pl. 37
Chitral Brimstone 67, pl. 22
Chitral Flash 155, pl. 43
Chitral Green-Banded White 55, pl. 17
Chitral Satyr 107, pl. 32
chitralensis sub-sp. *Gonepteryx rhamni* 67
chitralensis - Colias 68
chitralica sub-sp. *Parnassius stoliczkanus* 40, pl. 13
chloridice - Pieris (syn: *Pontia*) 53
chloridice - Parapieris 53
Choaspes benjaminii 160, pl. 44
Choaspes xanthopogon 160, pl. 44
Chocolate Pansy 86, pl. 27
christophi - Plebejus or *Lycaeides* or *Polyommatus* 128, pl. 36
chromus - Hasora 159, pl. 44
chrysippus - Danaus 118, pl. 34
Chrysozephyrus ataxus 147, pl. 42
Chrysozephyrus syla 148, pl. 42
Chrysozephyrus ziha 147, pl. 42
Cigaritis acamus 150, pl. 43
Cigaritis epargyros 150, pl. 43
cippus - Tajuria 152, pl. 42
cloanthus - Graphium 44, pl. 14
Clossiana hegemone 81, pl. 25
Clouded Yellow - Broad Bordered 69
Clouded Yellow - Dark 67
Clouded Yellow - Eastern Pale 68
Clouded Yellow - Fiery 70
Clouded Yellow - Glaucous 70
Clouded Yellow - Greenish 68
Clouded Yellow - Ladakh 70
Clouded Yellow - Marco Polo 68
Clouded Yellow - Orange 69
Clouded Yellow - Pamir 69
clytia - Chilasa 49, pl. 15
cnejus - Euchrysops or *Catochrysops* 138, pl. 39
cocandica - Colias 69, pl. 21
COELIADINAE 158
Coenonympha myops 104, pl. 31
COLIADINAE 62
Colias alpheraky 68, pl. 21
Colias cocandica 69, pl. 21
Colias croceus 69, pl. 21
Colias electo 67, pl. 20
Colias erate 68, pl. 20
Colias eogene 70, pl. 21
Colias fieldi 67, pl. 3, pl. 9, pl. 20

Colias ladakensis 70, pl. 21
Colias leechi 70, pl. 21
Colias marcopolo 68, pl. 21
Colias shipkee 70, pl. 21
Colias stoliczkana 69, pl. 21
Colias wiskotti 69, pl. 21
Colotis amata 71, pl. 3, pl. 22
Colotis calais 71
Colotis danae 73, pl. 22
Colotis etrida 73, pl. 22
Colotis fausta 72, pl. 22
Colotis phisadia 71, pl. 22
Colotis portractus 71, pl. 22
Colotis vestalis 72, pl. 22
Comma - Eastern 89, pl. 26
Comma - European 89, pl. 26
Comma - Tortoiseshell 90, pl. 26
comma - Hesperia or *Pamphila* 172, pl. 47
Common Argus 109, pl. 31
Common Banded Awl 159, pl. 5, pl. 44
Common Baron 97, pl. 11
Common Beak 120, pl. 34
Common Blue 135
Common Blue Apollo 38, pl. 12
Common Brimstone 66, pl. 22
Common Bush Brown 114, pl. 29
Common Castor 77, pl. 23
Common Cerulean 141
Common Copper 142, pl. 5, pl. 41
Common Crow 119, pl. 4, pl. 34
Common Evening Brown 114, pl. 33
Common Flash 155, pl. 43
Common Grass Dart 170, pl. 46
Common Grass Yellow 64, pl. 3, pl. 11, pl. 20
Common Guava Blue 153, pl. 43
Common Gull 60, pl. 19
Common Hedge Blue 127, pl. 36
Common Jezebel 159, pl. 3, pl. 18
Common Lineblue 141, pl. 5, pl. 40
Common Leopard 77, pl. 26
Common Map 94, pl. 23
Common Meadow Blue 135, pl. 5, pl. 39
Common Mime 49, pl. 3, pl. 15
Common Mormon 44, pl. 9, pl. 15
Common Nawab 75, pl. 23
Common Peacock 48, pl. 9, pl. 15
Common Pierrot 123, pl. 35
Common Punch 156, pl. 10, pl. 34
Common Red Apollo 37, pl. 12
Common Rose 45, pl. 3. pl. 15
Common Sailer 95, pl. 4, pl. 28
Common Satyr 109, pl. 33
Common Shot Silverline 151, pl. 43

Common Silverline 151, pl. 43
Common Silverstripe 80, pl. 24
Common Spotted Flat 161, pl. 44
Common Threering 112, pl. 33
Common Tiger 118, pl. 34
Common Tiger Blue 124, pl. 35
Common Tree Brown 99, pl. 10, pl. 29
Common Wall 100, pl. 4, pl. 29
Common Wanderer 61, pl. 9. pl. 19
Common Windmill 45, pl. 15
Common Yellow Swallowtail 46
confusa - Lethe 99
connae - Neolycaena 145, pl. 42
contracta - Euchrysops 139
Coster (Tawny) 116, pl. 34
Copper - Balochi 143, pl. 41
Copper - Common 142, pl. 41
Copper - Golden 143, pl. 41
Copper - Green 143, pl. 41
Copper - Purple 144, pl. 41
Copper - Red 144, pl. 41
Copper - Small 142, pl. 41
Copper - White Bordered 142, pl. 41
core - Euploea 119, pl. 34
Cornelian 152, pl. 5, pl. 43
Crimson Tip 73, pl. 22
crocale - Catopsila 62, pl. 19
croceus - Colias 67, pl. 20
Crow - Common 119, pl. 34
Cupid - Plains 139, pl. 40
Cupid - Small 139, pl. 40
Cupid - Tailed 126, pl. 5, pl. 36
Cupido sebrus 126
cydippe - Fabriciana 81
cyllarus - Polyommatus 133
Cynthia cardui 88, pl. 10, pl. 26
Cyrestis thyodamas 94, pl. 23
cytis - Turanana 130, pl. 37

D

DAININAE 117
daksha var: *Erebia nirmala* 109
danae - Colotis 73, pl. 22
Danaid Eggfly 93, pl. 4, pl. 28
Danais or *Danaus - chrysippus* 118, pl. 4, pl. 34
Danais or *Danaus - genutia* 118
Danais or *Danaus - limniace* 117, pl. 34
Danais or *Danaus - plexippus* 118, pl. 34
danna - Taractrocera 170, pl. 46
daphalis - Synchloe 53
daplidice - Pieris or *Pontia* 54, pl. 17
dara - Padraona 171

Dark Blue Jewel 130, pl. 37
Dark Cerulean 140, pl. 40
Dark Clouded Yellow 67, pl. 3, pl. 9, pl. 20
Dark Grass Blue 137, pl. 39
Dark Green Silverspot or Fritillary 80, pl. 24
Dark Himalayan Oakblue 149, pl. 42
Dark Palm Dart 171, pl. 46
Dark Rockbrown 106, pl. 32
Dark Satyr 107, pl. 32
Dark Wall 101, pl. 29
Dart - Common Grass 170, pl. 46
Dart - Dark Palm 171, pl. 46
Dart - Himalayan 171, pl. 46
Dart - Himalayan Grass 170, pl. 46
Dart - Pale Palm 172, pl. 47
Dart - Veined 170, pl. 46
davendra - Maniola 103, pl. 31
Delias belladona 58, pl. 18
Delias eucharis 59, pl. 3, pl. 18
Delias sanaca 59, pl. 18
delphius - Parnassius 39, pl. 12
demoleus - Papilio 47, pl. 14
Demon - Spotted 169
deota - Pieris 56, pl. 18
Desert Apollo 36, pl. 3, pl. 12
Desert Bath White 54, pl. 17
Desert Fourring 113, pl. 33
Desert Fritillary 82, pl. 25
Deudoryx epijarbus 152, pl. 5, pl. 43
devanica - Polyommatus 134
devta - Pieris 55, pl. 17
dichroa - Sephisa 76, pl. 23
didyma - Melitaea 83, pl. 25
digna - Karanasa or *Hipparchia* or *Kanesita* 107
Dingy Swift 175, pl. 47
dipoea - Dodona 156, pl. 34
dipora - Catochrysops 126, pl. 36
diporides sub-sp. *Everes argiades* 126
discobolus - Parnassius 37, pl. 12
discreta - Baoris 173, pl. 47
dissimilis (form), *Chilasa clytia* 49, pl. 15
Dodona dipoea 156, pl. 34
Dodona durga 156, pl. 10, pl. 34
Dodona eugenes 157, pl. 5, pl. 34
dodonea - Amblypodia 149, pl. 42
dohertyi - Sesera 162
doris - Spialia 165
dorripus - Danaus 119
dravira sub-sp. *Carcharodus floccifera* 166
dubiosa - Nacaduba or *Prosotas* 142
Dull Babul Blue 126, pl. 35

durga - Dodona 156, pl. 10, pl. 34
Dusky Blackvein 58, pl. 18
Dusky Green Underwing 132
Dusky Hedge Blue 127, pl. 36
Dusky Meadow Brown 101, pl. 30
Dusky Meadow Blue 134

E

Eastern Comma 89, pl. 26
Eastern Pale clouded Yellow 68
egea - Polygonia or *Vanessa* 89, pl. 26
egena - Halpe 169
electo - Colias 67
elima - Spindasis 151
elma - Gomalia 163, pl. 45
Elphinstonia charlonia 52
eltola - Baoris 173
Emigrant - African 63, pl. 20
Emigrant - Common 62, pl. 3, pl. 19
Emigrant - Lemon 62, pl. 19
Emigrant - Mottled 63, pl. 20
eogene - Colias 70, pl. 21
Eogenes alcides 175, pl. 47
Eogenes leslei 175, pl. 47
epaphus - Parnassius 37, pl. 12
epargyros - Apharitis 150, pl. 43
Epicopia polydorus 28, 45
epijarbus - Deudoryx 152
Epinephele interposita 102
Epinephele mandane 103
Epinephele wagneri 103
erate - Colias 68
Erebia annada 110, pl. 30
Erebia kalinda 111, pl. 30
Erebia mani 111, pl. 30
Erebia nirmala 109, pl. 31
Erebia scanda 110
Erebia shallada 111
Ergolis merione 77
Eriboea athamus 75
Eros Blue 135, pl. 5, pl. 39
eros - Polyommatus 135, pl. 5, pl. 39
ERYCNIDAE 156
Erynnis marloyi 167, pl. 46
etrida - Colotis 73, pl. 22
Euaspa miliona 147, pl. 42
Euaspa ziha 147, pl. 42
eucharis - Delias 59, pl. 18
Euchloe belemia 52, pl. 16
Euchloe charlonia 52, pl. 16
Euchloe lucilla 52, pl. 16
Euchrysops cnejus 138, pl. 39
Euchrysops pandava 139, pl. 40
Euchrysops parrhasius 139, pl. 40
eugenes - Dodona 157, pl. 34

Eumenis baldiva 105, pl. 32
Eumenis heydenreichi 106, pl. 32
Eumenis mniszechii 105, pl. 4, pl. 32
Eumenis parisatis 104, pl. 31
Eumenis persephone 106, pl. 32
Eumenis thelephassa 105, pl. 32
Euploea core 119, pl. 4, pl. 34
Eurema blanda 65, pl. 20
Eurema brigitta 65, pl. 20
Eurema hecabe 64, pl. 3, pl. 11
Eurema laeta 65, pl. 20
European Comma 89, pl. 26
European Beak 121, pl. 5, pl. 34
European Red Admiral 87, pl. 26
European Tortoiseshell 92, pl. 27
Euthalia garuda or *aconthea* 97, pl. 4, pl. 11, pl. 28
evanidus - Syrichtus 165, pl. 45
Evening Brown (Common) 114, pl. 4, pl. 33
Everes argiades 126, pl. 5, pl. 36
Everes buddhista 126, pl. 36
Everes hugelii 126, pl. 36
Everes shandura 126, pl. 36
eversmanni - Pararge or *Kirinia* 101
exclamationis - Badamia 159, pl. 44
extensa - Rapala 155, pl. 43
extricatus - Tarucus 123

F

Fabriciana argyrospilata 81, pl. 24
Fabriciana (syn: *Argynnis*) *kamala* 80, pl. 24
False Comma 90, pl. 26
farinosa - Gonepteryx 67, pl. 22
fausta - Colotis 72, pl. 22
fieldi - Colias 67, pl. 9
Fiery Clouded Yellow 70, pl. 21
Fiery Fritillary 83
fiesthamelii - Notocrypta 169, pl. 46
Flash - Common 155, pl. 43
Flash - Chitral 155, pl. 43
Flash - Indian Red 154, pl. 43
Flash - Red Himalayan 155, pl. 43
Flash - Slate 154, pl. 43
Flat - Common Spotted 161, pl. 44
Flat - Himalayan Spotted 161, pl. 44
Flat - Himalayan White 162, pl. 44
Flat - Marbled 162, pl. 44
Flat - Spotted Small 161, pl. 44
floccifera - Carcharodus 166, pl. 45
florella - Catopsila 63
florenciae - Polyommatus 135, pl. 39
Forget-Me-Not 139, pl. 40
Freak 77, pl. 23
Freyeria galba 137, pl. 39

INDEX

Freyeria trochilus 136, pl. 39
Fritillary - Afghan 81, pl. 24
Fritillary - Balochi 83, pl. 25
Fritillary - Blackvein 84, pl. 25
Fritillary - Dark Green 80, pl. 24
Fritillary - Desert 82, pl. 25
Fritillary - Fiery (Redband) 83, pl. 25
Fritillary - High Brown 81, pl. 24
Fritillary - Indian 78, pl. 26
Fritillary - Lesser Spotted 83, pl. 25
Fritillary - Pamir 84, pl. 25
Fritillary - Queen of Spain 78, pl. 24
Fritillary - Red Band 83, pl. 25
Fritillary - Shandur 82, pl. 25
Fritillary - Shepherd's 82, pl. 25
Fritillary - Whitespot 81, pl. 25

G

galathea - *Albulina* or *Lycaena* or *Polyommatus* 132
galba - *Spialia* or *Hesperia* or *Syrichtus* 164
galba - *Zizeeria* or *Freyeria* 137, pl. 39
ganesa - *Arhopala* or *Narathura* or *Panchala* 149, pl. 42
garuda - *Euthalia* 97, pl. 11, pl. 28
Gegenes nostrodamus 175
genutia - *Danaus* 118
geron - *Hesperia* 165, pl. 45
geron - *Spialia* 165
gigantea - *Iolana* 133, pl. 38
Gilgit Meadow Blue 133, pl. 38
gilgitica sub-sp. *Eumenis mniszechii* 105
gilgitica sub-sp. *Limenitis trivena* 46, pl. 23
gisca sub-sp. *Celastrina puspa* 127
Glassy Bluebottle 44, pl. 14
glauconome - *Pontia* (syn: *Pieris*) 54, pl. 17
Glaucopsyche alexis 133, pl. 38
Glaucous Clouded Yellow 70, pl. 21
Golden Angle 163, pl. 5, pl. 45
Golden Copper 143, pl. 41
Gomalia elma 163, pl. 45
Gonepteryx aspasia 66, pl. 22
Gonepteryx farinosa 67, pl. 22
Gonepteryx mahaguru 66, pl. 9, pl. 22
Gonepteryx rhamni 66, pl. 22
Gonepteryx zaneka 66, pl. 9, pl. 22
gracilis - *Polyommatus* 134, pl. 38
Gram Blue 138, pl. 5, pl. 39
Graphium cloanthus 44, pl. 14

Grass Blue - African 137, pl. 39
Grass Blue - Dark Grass Blue 137, pl. 39
Grass Blue - Lesser 138, pl. 39
Grass Blue - Pale 137, pl. 39
Grass Blue - Persian 134, pl. 39
Grass Blue - Tiny 138, pl. 39
Grass Dart - Common 170
Grass Dart - Himalayan 171
Grass Jewel 136, pl. 39
Grass Yellow - Common 64, pl. 3, pl. 20
Grass Yellow - Small 65, pl. 20
Grass Yellow - Spotless 65, pl. 20
Grass Yellow - Three-Spot 65, pl. 20
Great Eggfly 94, pl. 28
Great Satyr 108, pl. 33
Green Banded White 55, pl. 17
Green Copper 143, pl. 41
Green Hairstreak 146, pl. 5, pl. 42
Green Striped White 52, pl. 16
Green Underside Blue 133
Green Veined White 55, pl. 17
Green Underwing - Dusky 132, pl. 37
Green Underwing - Large 132, pl. 38
Green Underwing - Small 132, pl. 38
Green Underwing - Western 133, pl. 38
Greenish Black Tip 52, pl. 16
Greenish Clouded Yellow 68, pl. 21
Greenish Mountain Blue 131, pl. 37
gremius - *Saustus* 168
Gull (Common) 60, pl. 19
Guava Blue 153, pl. 43
Guttatus bevani 174, pl. 47

H

Hairstreak - Balochi 145, pl. 42
Hairstreak - Green 146, pl. 42
Hairstreak - Silver 148, pl. 42
Hairstreak - Water 147, pl. 42
Hairstreak - White-Line 146, pl. 42
Hairstreak - White-Spotted 147, pl. 42
Hairstreak - Wonderful 147, pl. 42
Halpe egena or *homolea* 169, pl. 46
hardwickei - *Parnassius* 38, pl. 12
Hasora alexis 159, pl. 5, pl. 44
Hasora chromus 159, pl. 44
hecabe - *Eurema* or *Terias* 64, pl. 11, pl. 20
Hedge Blue - Common 127, pl. 36
Hedge Blue - Dusky 127, pl. 36
Hedge Blue - Hill 128, pl. 36
Hedge Blue - Silvery 127, pl. 36

hegemone - *Argynnis* or *Clossiana* 81, pl. 25
Heliophorus bakeri 145, pl. 10, pl. 41
Heliophorus sena 144, pl. 41
Heliophorus tamu 145, pl. 41
helios - *Hypermnestra* 36, pl. 12
Hesperia alpina 166, pl. 45
Hesperia comma 172, pl. 47
Hesperia galba 164, pl. 5, pl. 45
Hesperia geron 165, pl. 45
Hesperia poggei 166, pl. 45
Hesperia sertorius 165, pl. 45
Hesperia zebra 164, pl. 45
HESPERINAE 163
HESPERIDAE 158
heydenreichi - *Eumenis* 106, pl. 32
hierta - *Junonia* or *Precis* 85, pl. 27
High Brown Silverspot or Fritillary 81, pl. 24
hilaris - *Maniola* 104, pl. 30
Hill Hedge Blue 128, pl. 5, pl. 36
Hill Jezebel 58, pl. 18
himalaya sub-sp. *Baoris discreta* 173
Himalayan Blackvein 58, pl. 18
Himalayan Dark Oakblue 149, pl. 42
Himalayan Dart 171, pl. 46
Himalayan Fivering 113, pl. 33
Himalayan Grass Dart 170, pl. 46
Himalayan Pale Oakblue 149, pl. 42
Himalayan Pierrot 124, pl. 35
Himalayan Sailer 96, pl. 28
Himalayan Sergeant 95, pl. 28
Himalayan Spotted Flat 161, pl. 44
Himalayan Swift 173, pl. 47
Himalayan White Flat 162, pl. 44
Hipparchia actaea 107, pl. 32
Hipparchia digna 107, pl. 32
Hipparchia mniszechii 105, pl. 32
Hipparchia moorei 108, pl. 32
Hipparchia parisatis 104, pl. 31
Hipparchia persephone 106, pl. 32
Hipparchia shandura 106, pl. 32
Hipparchia thelephassa 105, pl. 32
Holly Blue 128, pl. 5, pl. 36
homolea - *Halpe* 169, pl. 46
Hopper - Leslie's 175
Hopper - Torpedo 175
huebneri - *Karanasa* 107
hugeli - *Everes* 126
Huphina nerissa 60
hydaspes sub-sp. *Limenitis trivena* 76, pl. 23
hylas - *Neptis* 95, pl. 28
hylax - *Zizula* 138, pl. 39
hyperbius - *Argynnis* 78, pl. 26
hypergyrus - *Aphnaeus* 150

hypergyrus - Spindasis 150
Hypermnestra helios 36, pl. 3, pl. 12
Hypolimnus bolina 94, pl. 28
Hypolimnus missipus 93, pl. 4, pl. 28
Hyponephele hilaris 104
Hyponephele lupinus 102
Hyponephele narica 102
Hyponephele pulchra 101
hyppia - Parenonia 61
hyrcana - Vaccinia 130
hyrcanana - Lycaena 144, pl. 41

I

iarbus - Rapala 154
icarus - Plyommatus 135, pl. 38
ictis - Spindasis 151
Ilerda viridipunctata 145, pl. 41
inachus - Aeromachus 168
inaria var: *Hypolimnas missipus* 93
Indian Ace 169, pl. 46
Indian Awlking 160, pl. 5, pl. 44
Indian Cabbage White 56, pl. 17
Indian Fritillary 78, pl. 26
Indian Palm Bob 168, pl. 46
Indian Purple Emperor 75, pl. 4, pl. 23
Indian Red Admiral 87, pl. 10, pl. 26
Indian Red Flash 154, pl. 5, pl. 43
Indian Skipper 164, pl. 5, pl. 45
Indian Tortoiseshell 92, pl. 4, pl. 27
Indian White Admiral 76, pl. 23
indica sub-sp. *Everes argiades* 126, pl. 36
indica - Tarucus 124
indica - Vanessa 87, pl. 10, pl. 26
indica - Zizera 138
inica - Ypthima 112, pl. 33
Inky Skipper 167, pl. 46
inopinatus - Parnassius 42, pl. 14
interposita - Epinephele 102
Iolana gigantea 133, pl. 38
iphita - Junonia or *Precis* 86, pl. 27
iris - Vaccinia or *Polyommatus* 131, pl. 37
isocrates - Virachola 153
Issoria lathonia 78, pl. 24
Ixias pyrene 63, pl. 19
Ixias marianne 64

J

jacquemontii - Parnassius 37, pl. 12
jaloka - Agriades 131, pl. 37
jaloka sub-sp. *Polyommatus orbitulus* 131, pl. 37
Jamides bochus 140, pl. 40

Jamides celeno 141, pl. 40
jerdoni - Argynnis 80, pl. 24
Jerdon's Silverspot 80, pl. 24
jermyni sub-sp. *Aricia astorica* 130
jesous - Azanus or *Syntarucus* 125, pl. 35
Jewel Blue - Balochi 129, pl. 36
Jewel Blue - Large 134, pl. 38
Jewel Blue - Pale 133, pl. 38
Jewel Blue - Small 128, pl. 36
Jewel Argus 131, pl. 37
Jewel Fourring 113, pl. 33
Jezebel - Common 59, pl. 3, pl. 18
Jezebel - Hill 58, pl. 18
Jezebel - Pale 59, pl. 18
Judies 156
Junonia almana 85, pl. 27
Junonia iphita 86, pl. 27
Junonia hierta 85, pl. 27
Junonia lemonias 86, pl. 27
Junonia orithya 85, pl. 4, pl. 27

K

Kafir Banded Apollo 39, pl. 12
kalinda - Erebia 111
kamala - Fabriciana or *Argynnis* 80, pl. 24
Kanetsia digna 107, pl. 37
Kaniska canace 89
Karanasa actaea 107, pl. 32
Karanasa boloricus 108, pl. 32
Karanasa digna 107, pl. 32
Karanasa huebneri 107, pl. 32
Karanasa regeli 108, pl. 32
Kashmir White 56, pl. 18
kasyapa - Lycaena 143, pl. 41
Keeled Apollo 37, pl. 12
Kirinia eversmanni 101, pl. 29
Kite Swallowtails 43
klugi - Limnas 119
knysa - Zizeeria 137, pl. 39
kreuperi - Pieris or *Artogeia* 55, pl. 17
Kreuper's Small White 55, pl. 17
khwaja (kwaja) - Vaccinia 130, pl. 37

L

Ladak Clouded Yellow 70, pl. 21
ladakensis - Colias 70, pl. 21
ladakensis - Papilio 47, pl. 14
Ladaki Swallowtail 47, pl. 14
ladonides - Celastrina 127, pl. 36
laeta - Eurema 65, pl. 20
laius - Chilades 136, pl. 39

l-album - Vanessa 90
Lampides boeticus 140, pl. 40
Large Cabbage White 57, pl. 17
Large Green Underwing 132, pl. 38
Large Jewel Blue 34
Large Salmon Arab 72, pl. 22
Large Silverstripe 79, pl. 24
Large Threering 112, pl. 33
Large Tortoiseshell 91, pl. 10, pl. 27
Larger Keeled Apollo 37, pl. 12
Lassiomata eversmanni 101, pl. 29
Lassiomata maerula 100, pl. 29
Lassiomata menava 101, pl. 29
Lassiomata schakra 100, pl. 4, pl. 29
lathonia - Argynnis or *Issoria* 78, pl. 24
latistigma sub-sp. *Maniola davendra* 103, pl. 31
lativita sub-sp. *Colias erate* 68, pl. 20
leda - Melanitis 114, pl. 33
leechi - Colias 70, pl. 21
lehana - Pseudochazara 105
leisliei - Eogenes 175, pl. 47
Lemon Emigrant 62, pl. 19
Lemon Pansy 86, pl. 27
Lemon White 52, pl. 16
lemonias - Junonia or *Precis* 86, pl. 27
lepita - Libythea 120, pl. 34
Leopard (Common) 77, pl. 26
Leptosia nina or *xiphia* 51, pl. 16
Leslie's Hopper 175, pl. 47
Lesser Bath White 53, pl. 16
Lesser Brimstone 66, pl. 9, pl. 22
Lesser Grass Blue 138, pl. 39
Lesser Punch 156, pl. 34
Lesser Spotted Fritillary 83
Lesser Threering 112, pl. 33
Lesser Whitering Meadow Brown 103, pl. 31
Lethe confusa 99, pl. 29
Lethe rohria 99, pl. 10, pl. 29
Lethe verma 98, pl. 29
leuca var: *Colias wiskotti* 69, pl. 21
Leucocera leucocera 161, pl. 44
Leucocera munda 161, pl. 44
leucocera - Celaenorrhinus 161, pl. 44
leucodice - Aporia 58, pl. 18
libythea - Appias 61, pl. 19
Libythea celtis 121, pl. 5, pl. 34
Libythea lepita 120, pl. 34
libythea - Eurema 65
LIBYTHEINAE 120
liliana - Lobocla 162, pl. 44
Lime Blue 136, pl. 39

Lime Butterfly (Common) 47, pl. 14
Limenitis trivena 76, pl. 23
limniace - Tirumala or *Danais* 117, pl. 34
Limnis klugi 119
Lineblue - Common 141, pl. 5, pl. 40
Lineblue - Tailess 142, pl. 40
lisandra - Ypthima 113
Little Orange Tip 73, pl. 22
Little Tiger Blue 123, pl. 35
Lobocla liliana 162, pl. 44
Lobeless Hairstreak 145
loewii - Polyommatus 134
Lofty Bath White 53, pl. 17
Long Tailed Blue 140
loxias - Parnassius 42, pl. 13
lucilla - Euchloe 52, pl. 16
lunulata sub-sp. *Melitaea saxatilis* 83
lupinus - Maniola or *Hyponephele* 102, pl. 30
lutko - Melitaea 83, pl. 25
Lycaeides christophi 128, pl. 36
Lycaena athamantis 144, pl. 41
Lycaena caspius 144, pl. 41
Lycaena callimachus 144, pl. 41
Lycaena galathea 132
Lycaena hyrcana 144, pl. 41
Lycaena kasyapa 143, pl. 41
Lycaena pavana 143, pl. 41
Lycaena phlaeas 142, pl. 5, pl. 41
Lycaena phoenicurus 143, pl. 41
Lycaena solskyi 143, pl. 41
Lycaena thetis 143, pl. 41
Lycaene orbitulus astorica 131, pl. 37
Lycaenopsis vardhana 127
LYCAENIDAE 122
lyela - Myops 104
lysimon - Zizeeria 137

M

machaon - Papilio 46, pl. 9, pl. 14
maerula - Pararge or *Lasiommata* 100, pl. 29
maevius - Taractrocera 170
mahaguru - Gonepteryx 66, pl. 9, pl. 22
mahendra - Neptis 96, pl. 28
maha - Zizeeria 137
maia - Argynnis 79, pl. 24
Mallow Skipper 67
mandane - Epinephele 103
manea - Rapala 154, pl. 43
mani - Erebia 111
Maniola davendra 103, pl. 31

Maniola hilaris 104, pl. 30
Maniola lupinus 102, pl. 30
Maniola narica 102, pl. 31
Maniola pulchella 102, pl. 4, pl. 30
Maniola pulchra 101, pl. 30
Maniola tenuistigma 103, pl. 31
Maniola wagneri 103, pl. 30
Map (Common) 94, pl. 23
Marbled Flat 162, pl. 44
Marco Polo's Clouded Yellow 68, pl. 21
marcopolo - Colias 68, pl. 21
marianne - Ixias 64
marloyi - Nisionades 167
materta var: *Erebia nirmala* 109
mathias - Pelopidas or *Baoris* 173
Meadow Brown - Branded 102, pl. 30
Meadow Brown - Dusky 101, pl. 30
Meadow Brown - Lesser Whitering 103, pl. 31
Meadow Brown - Tawny 102, pl. 30
Meadow Brown - Tawny Branded 102, pl. 31
Meadow Brown - Oval Spot 103, pl. 30
Meadow Brown - Pamir 104, pl. 30
Meadow Brown - White-Ringed 103, pl. 31
Meadow Blue - Balochi 134, pl. 38
Meadow Blue - Brilliant 134, pl. 38
Meadow Blue - Dusky 134, pl. 38
Meadow Blue - Common 135, pl. 39
Meadow Blue - Silvery 135, pl. 39
Meadow Blue - Violet 135
Mediterranean Pierrot 124, pl. 35
mediterraneae - Tarucus 124, pl. 35
melampus - Rapala 154, pl. 43
Melanitis leda 114, pl. 4, pl. 33
Melitaea ala 84, pl. 25
Melitaea arcesia 84, pl. 25
Melitaea didyma 83, pl. 4, pl. 25
Melitaea lutko 83, pl. 25
Melitaea minerva 84, pl. 25
Melitaea persea 82, pl. 25
Melitaea pseudoala (syn:) *ala* 84, pl. 25
Melitaea robertsi 83, pl. 25
Melitaea saxatilis (syn: *persea*) 82, pl. 25
Melitaea schoenis 83, pl. 25
Melitaea shandura 82, pl. 25
Melitaea trivia 83, pl. 25
menava - Pararge or *Lasiommata* 101, pl. 29

merione - Ariadne or *Ergolis* 77, pl. 23
mesentina - Belenois 60, pl. 19
Mesoacidalia aglaja 80
metallica - Albulina 132, pl. 38
mettasuta sub-sp. *Caprona agama* 162
micans - Rapala 155, pl. 43
miliona - Euaspa 147, pl. 42
Milkweed Butterflies 117
Mime (Common) 43, 49, pl. 3, pl. 15
minerva - Melitaea 84, pl. 25
minuta sub-sp. *Euchrysops parrhasius* 139
missipus - Hypolimnas 93, pl. 28
mniszechii - Eumenis or *Hipparchia* 105, pl. 32
Mongol 93
montana sub-sp. *Pieris napi* 55, pl. 17
moorei - Hipparchia 108, pl. 32
Mormon (Common) 44, pl. 9, pl. 15
Mottled Emigrant 63, pl. 20
Mountain Argus 111
Mountain Blue 132, pl. 37
Mountain Dappled White 53, pl. 16
Mountain Skipper 166, pl. 45
Mountain Tortoiseshell 92
munda - Celaenorrhinus or *Leucocera* 161, pl. 44
Murree Green-Veined White 56, pl. 17
Muschampia staudingeri 166, pl. 45
Muschampia poggei 166, pl. 45
Mycalesis perseus 114, pl. 29
myops - Coenonympha or *Lyela* 104, pl. 31

N

nabellica - Aporia 58, pl. 18
Nacaduba dubiosa 142, pl. 40
Nacaduba nora 142, pl. 40
napi - Pieris 55, pl. 17
nara - Tarucus 123, pl. 35
Narathura ganesa 149, pl. 42
Narathura rama 149, pl. 42
Narathura sanesa 149, pl. 42
nareda - Ypthima 112, pl. 33
narica - Maniola 102, pl. 31
Nawab (Common) 75, pl. 23
Neolycaena connae 145, pl. 42
neoza sub-sp. *Maniola pulchra* 101
Neptis hylas 95, pl. 4, pl. 28
Neptis mahendra 96, pl. 28

Neptis soma 96, pl. 28
Neptis yerbury 97
Neptis zaida 96, pl. 28
nerissa - *Cepora* or *Huphina* 60, pl. 19
Nettle Tree Butterfly 121, pl. 34
nigra - *Tarucus* 123, pl. 35
nina - *Leptosia* 51, pl. 16
niphe - *Argynnis* 78
nirmala - *Erebia* or *Callerebia* 109, pl. 31
nisionades - *Marloyi* 167
nissa - *Rapala* 155
nora - *Nacaduba* 141, pl. 40
nostrodamus - *Gegenes* 175, pl. 47
Notocrypta fiesthamelii 169, pl. 46
NYMPHALIDAE 74
NYMPHALINAE 77
Nymphalis xanthomelas 91, pl. 27
Nymphalis vau-album 90

O

Oak Blue - Dark Himalayan 149, pl. 42
Oak Blue - Pale Himalayan 149, pl. 42
Oak Blue - Tailless 149, pl. 42
odata - *Chaetoprocta* 148, pl. 42
omphisa - *Albulina* 132, pl. 32
omphisa - *Polyommatus* 132, pl. 37
opalina - *Parathyma* or *Pantoporia* 95, pl. 28
Orange Clouded Yellow 69, pl. 21
Orange Bordered Argus 129, pl. 5, pl. 37
Orange Tail Awl 159, pl. 44
orbitulus - *Polyommatus* 131, pl. 37
orbitulus - *Lycaene* 131
ordinata - *Ypthima* 113
orithya - *Junonia* or *Precis* 85, pl. 27
otis - *Zizina* 138
Oval Spot Meadow Brown 103, pl. 30

P

Pachliopta aristolochiae 45, pl. 3, pl. 15
padma - *Aulocera* 108, pl. 33
Padraona dara 171, pl. 46
Padraona rectifasciata (syn: *danna*) 170, pl. 46
Painted Lady 88, pl. 10, pl. 26
Pale Clouded Yellow 68, pl. 20
Pale Grass Blue 137, pl. 39

Pale Green Awlet 159
Pale Green Sailer 96, pl. 28
Pale Himalayan Oakblue 149, pl. 42
Pale Lemon White 52, pl. 16
Pale Jewel Blue 133, pl. 38
Pale Jezebel 59, pl. 18
Pale Palm Dart 172, pl. 47
pales - *Bolaria* or *Argynnis* 82, pl. 25
Pallid Argus 110
pallida sub-sp. *Colias erate* 68, pl. 20
Pamir Clouded Yellow 69, pl. 21
Pamir Fritillary 84, pl. 25
Pamir Meadow Brown 104, pl. 30
Pamphila comma 172, pl. 47
panchala - *Ganesa* 149
pandava - *Euchrysops* 139
pandora - *Argynnis* 79, pl. 24
Pansy - Blue 85, pl. 4, pl. 27
Pansy - Chocolate 86, pl. 27
Pansy - Lemon 86, pl. 27
Pansy - Peacock 85, pl. 27
Pansy - Yellow 85, pl. 27
Pantoporia opalina 95, pl. 28
Papilio alexanor 46, pl. 14
Papilio arcturus 48, pl. 11, pl. 15
Papilio demoleus 47, pl. 14
Papilio ladakensis 47, pl. 14
Papilio machaon 46, pl. 3, pl. 9, pl. 14
Papilio polyctor 48, pl. 3, pl. 9, pl. 15
Papilio polytes 44, pl. 9, pl. 15
PAPILIONIDAE 34
PAPILIONINAE 34, 43
Paralasa annada 110, pl. 30
Paralasa scanda 110
Paralasa shakti 111, pl. 30
Pararge eversmanni 101, pl. 29
Pararge maerula 100, pl. 29
Pararge menava 101, pl. 29
Pararge schakra 100, pl. 4, pl. 29
Parathyma opalina 95
Parenonia anais 61, pl. 9, pl. 19
Parenonia valeria 61
Parenonia hyppia 61
parisatis - *Hipparchia* or *Eumenis* 104, pl. 31
Parnara guttatus 174, pl. 47
Parnara bevani 174, pl. 47
PARNASSIINAE 34
Parnassius acco 39, pl. 12, fig. 9
Parnassius actius 36, pl. 12
Parnassius boedromius 41, pl. 14

Parnassius cardinalina or *cardinalis* 39, pl. 13
Parnassius charltonius 41, pl. 9, pl. 13
Parnassius delphius 39, pl. 12, fig. 9
Parnassius discobolus 37, pl. 12
Parnassius epaphus 37, pl. 12, fig. 9
Parnassius hardwickei 38, pl. 12, fig. 9
Parnassius inopinatus 42, pl. 14, fig. 9
Parnassius jacquemontii 37, pl. 12, fig. 9
Parnassius loxias 42, pl. 13, fig. 9
Parnassius simo 40, pl. 13
Parnassius staudingeri 40, pl. 13
Parnassius stoliczkanus 40, pl. 13, fig. 9
Parnassius tianschanikus 37, pl. 12,
parrhasius - *Euchrysops* 139
pasargades sub-sp. *Argynnis maia* 79
pavana - *Lycaena* 142, pl. 41
Pea Blue 140, pl. 40
Peacock - Common 48, pl. 3, pl. 9, pl. 15
Peacock - Blue 48, pl. 15
Peacock Royal 152, pl. 42
Peacock Pansy 85, pl. 27
Pearl White 53, pl. 16
Peak White 53, pl. 17
Pelopidas mathias 173, pl. 47
persea - *Melitaea* 82, pl. 25
persephone - *Chazara* or *Hipparchia* or *Eumenis* 106, pl. 32
perseus - *Mycalesis* 114, pl. 29
Persian Grass Blue 137, pl. 39
Persian Skipper 165, pl. 45
Philarcta shandura 106, pl. 32
Phalantha phalantha 77, pl. 26
pheretes - *Albulina* 132
pheretiades - *Agriades* 131
Philotes vicrama 128, pl. 36
philoxenus - *Tros* 45
phisadia - *Colotis* 71
phlaeas - *Lycaena* 142, pl. 41
phlomidis - *Spialia* 165, pl. 45
phoenicurus - *Lycaena* 143, pl. 41
Phragis 34, fig. 9
PIERIDAE 50
PIERINAE 50
Pieris ajaka 56, pl. 17
Pieris brassicae 57, pl. 17
Pieris callidice 53, pl. 17
Pieris canidia 56, pl. 17
Pieris chloridice 53, pl. 16
Pieris daplidice 54, pl. 17

Pieris deota 56, pl. 18
Pieris devta 55, pl. 17
Pieris glauconome 54, pl. 17
Pieris kreuperi 55, pl. 17
Pieris montana 55, pl. 17
Pieris napi 55, pl. 17
Pieris rapae 57, pl. 18
Pierrot - Black Spotted 123, pl. 35
Pierrot - Common 123, pl. 35
Pierrot - Himalayan 124, pl. 35
Pierrot - Pointed 124, pl. 35
Pierrot - Rounded 123, pl. 35
Pierrot - Spotted 124, pl. 35
pimpla - Satyrus 107, pl. 32
Pioneer 60, pl. 19
Pir Panjal Banded Apollo 40, pl. 13
Plain Marbled Skipper 167, pl. 45
Plain Tiger 118, pl. 34
Plains Cupid 139
Plebejus christophi 128, pl. 36
Plebejus pylaon 129, pl. 36
Plebejus sephyrus 129, pl. 36
plexippus - Danaus 118, pl. 34
plinius - Syntarucus 125, pl. 35
plurimacula sub-sp. *Muschampia staudingeri* 166
plurimacula - Syrichtus 166
poggei - Syrichtus or *Muschampa* or *Hesperia* 166
Pointed Pierrot 124, pl. 35
polychloros - Vanessa 91, pl. 27
polyctor - Papilio 48, pl. 15
polydorus - Epicopia 28, 45
Polygonia c-album 89
Polygonia canace 89
Polyommatus actis 134, pl. 38
Polyommatus astrarche 129, pl. 37
Polyommatus baroghila 134, pl. 38
Polyommatus bogra 134, pl. 38
Polyommatus christophi 128, pl. 38
Polyommatus cyllarus 133, pl. 38
Polyommatus devanica 134, pl. 38
Polyommatus eros 135, pl. 5, pl. 39
Polyommatus florenciae 134
Polyommatus galathea 132, pl. 38
Polyommatus gracilis 134, pl. 38
Polyommatus icarus 135, pl. 38
Polyommatus iris 131, pl. 37
Polyommatus loewii 134, pl. 38
Polyommatus omphisa 132, pl. 37
Polyommatus orbitulus 131, pl. 37
Polyommatus orbitulus jaloka 131, pl. 37
Polyommatus pylaon 129, pl. 36
Polyommatus sarta 134, pl. 38
Polyommatus sieversi 133, pl. 38
Polyommatus vicrama 128, pl. 36

polytes - Princeps 44, pl. 9, pl. 15
Polyura athamus 75, pl. 23
polyeuctes - Atrophaneura 45, pl. 15
pomona - Catopsila 62, pl. 19
Pontia chloridice 53, pl. 16
Pontia daplidice 54, pl. 17
Pontia glauconome 54, pl. 17
Potanthus dara 171, pl. 46
Powdery Green Sapphire 145, pl. 41
Precis almana 85, pl. 27
Precis hierta 85, pl. 27
Precis lemonias 86, pl. 27
Precis iphita 86, pl. 27
Precis orithya 85, pl. 27
Princeps demoleus 47, pl. 14
Princeps polytes 44, pl. 15
prorsoides - Araschnia 92
Prosotas dubiosa 142
Prosotas nora 141
protractus - Colotis 71
pseudoala - Melitaea 84, pl. 25
Pseudochazara lehana 105
Psyche 51, pl. 16
pulchella - Maniola (syn: *Hyponephele*) 102, pl. 30
pulchra - Maniola (syn: *Hyponephele*) 101, pl. 30
Punch - Common 156, pl. 34
Punch - Lesser 156, pl. 34
Punch - Tailed 157, pl. 34
pupillata var: *Karanasa huebneri* 107
purendra - Sarangesa 161
Purple Copper 144, pl. 41
Purple Emperor 73, pl. 23
puspa - Celastrina 127, pl. 36
pylaon - Plebejus 129, pl. 36
pyranthe - Catopsila 63
pyrene - Ixias 63, pl. 19
PYRGINAE 160
Pyrgus alpina 166, pl. 45
pythias - Astycus 171, pl. 46

Q

Queen of Spain Fritillary 78, pl. 24

R

radians sub-sp. *Actinor suada* 170, pl. 46
rahmni - Gonepteryx 66
rama - Narathura or *Amblypodia* or *Arhopola* 149, pl. 42
ransonetti - Caprona 163
rapae - Pieris 57, pl. 18
Rapala extensa 155, pl. 43
Rapala iarbus 154, pl. 5, pl. 43

Rapala melampus 154, pl. 43
Rapala manea 154, pl. 43
Rapala micans 155, pl. 43
Rapala nissa 155, pl. 43
Rapala selira 155, pl. 43
Rapala schistacea 154, pl. 43
rectifasciata - Padraona 170, pl. 46
Redband Fritillary 83, pl. 25
Red Copper 144, pl. 41
Red Himalayan Flash 155, pl. 43
Red Underwing Skipper 165, pl. 45
Regal Apollo 41
regeli - Karanasa 108, pl. 32
Rhopalocampta (syn: *Choaspes*) *xanthopogon* 160, pl. 44
Ringed Argus 110, pl. 30
RIODININAE 156
robertsi - Melitaea 83, pl. 25
Rockbrown - Balochi 105, pl. 32
Rockbrown - Dark 106, pl. 32
Rockbrown - Shandur 106, pl. 32
Rockbrown - Tawny 105, pl. 32
Rockbrown - White-Edged 104, pl. 31
rohria - Lethe 99, pl. 10, pl. 29
romulus var: *Princeps polytes* 44
roschana - Colias 68
rosaea - Tarucus 124, pl. 35
Rose (Common) 45, pl. 15
rosimon - Castalius 123, pl. 35
Rounded Pierrot 123, pl. 35
Royal Peacock 152
rubi - Callophrys 146, pl. 42

S

Sailer - Common 95, pl. 28
Sailer - Himalayan 96, pl. 28
Sailer - Pale Green 96, pl. 28
Sailer - Sullied 96, pl. 28
Sailer - Yerbury's 97
Salmon Arab - Large 72
Salmon Arab - Small 71, pl. 3
sakra - Ypthima 113, pl. 33
sanaca - Delias 59, pl. 18
sanesa - Narathura 149
Sapphire - Powdery Green 145, pl. 41
Sapphire - Sorrel 44, pl. 41
Sapphire - Western Blue 145, pl. 10, pl. 41
Sarangesa purendra 161
saraswati - Aulocera 109, pl. 33
sarta - Polyommatus 134, pl. 38
sassanides - Strymon 146, pl. 42
Satyr - Black 107, pl. 32
Satyr - Chitrali 107, pl. 32

Satyr - Common 109, pl. 33
Satyr - Dark 107, pl. 32
Satyr - Great 108, pl. 33
Satyr - Striated 109, pl. 33
Satyr - Tawny 107, pl. 32
Satyr - Turkestan 108, pl. 32
SATYRINAE 98
Satyrus padma 108, pl. 33
Satyrus pimpla 107. pl. 32
Saustus gremius 168, pl. 46
saxatilis (syn: *perseus*) - *Melitaea* 82, pl. 25
Scarce Mountain Argus 111, pl. 30
Scarce Shot Silverline 151, pl. 43
Scarce Wall 100, pl. 29
scanda - *Paralasa* or *Erebia* 110
schakra - *Pararge* or *Lasiommata* 100, pl. 29
schistacaea - *Rapala* 154, pl. 43
schoenis - *Melitaea* 83, pl. 25
Scrub Hopper - Veined 168
sebrus - *Cupido* 126
selira - *Rapala* 155, pl. 43
selira - *Hysudra* 155, pl. 43
sena - *Bibasis* 159, pl. 44
sena - *Heliophorus* 144, pl. 41
Sephisa dichroa 76, pl. 23
sephyrus - *Plebejus* 129, pl. 36
Sergeant - Himalayan or Hill 95, pl. 28
sertorius - *Hesperia* or *Spialia* 165
Seseria dohertyi 162, pl. 44
shakti - *Paralasa* 111, pl. 30
shallada - *Erebia* 111
Shandur Cupid 126
Shandur Fritillary 82, pl. 25
Shandur Rockbrown 106, pl. 32
Shandur Satyr 108, pl. 32
shandura - *Everes* 126
shandura - *Melitaea* 82, pl. 25
shandura - *Philarcta* or *Hipparchia* 106, pl. 32
shawi - *Baltia* 51, pl. 16
Shaw's Dwarf 51, pl. 16
Shepherd's Fritillary 82, pl. 25
shipkee - *Colias* 70
Short - Tailed Blue 126
Short - Tailed Cupid 126, pl. 5, pl. 36
seiversi - *Polyommatus* 133, pl. 38
silhetana - *Terias* 65
Silver Hairstreak 148, pl. 42
Silver Spotted Skipper 172
Silverline - Common 151, pl. 43
Silverline - Common Shot 151, pl. 43
Silverline - Scarce Shot 151, pl. 43
Silverline - Tawny 150, pl. 43
Silverline - Yellow 150, pl. 43
Silverstripe - Common 80, pl. 24
Silverstripe - Large 79, pl. 24
Silverstripe - Western 79, pl. 24
Silvery Hedge Blue 127, pl. 36
Silvery Meadow Blue 135, pl. 39
simo - *Parnassius* 40, pl. 13
Sindh Skipper 165, pl. 45
Skipper - African Marbled 163, pl. 45
Skipper - Brick 165, pl. 45
Skipper - Inky 167, pl. 46
Skipper - Indian 164, pl. 5, pl. 45
Skipper - Mallow 167, pl. 45
Skipper - Mountain 166, pl. 45
Skipper - Persian 167, pl. 45
Skipper - Plain Marbled 167, pl. 45
Skipper - Red Underwing 165, pl. 45
Skipper - Sindh 165, pl. 45
Skipper - Syrian 166, pl. 45
Skipper - Tufted Marbled 166, pl. 45
Skipper - Zebra 164, pl. 45
Slate Flash 154, pl. 43
Small Bath White 53, pl. 17
Small Branded Swift 173, pl. 47
Small Cabbage White 57, pl. 18
Small Copper 142, pl. 5, pl. 41
Small Cupid 139, pl. 40
Small Grass Yellow 65, pl. 20
Small Green Underwing 132, pl. 38
Small Jewel Blue 128, pl. 36
Small Salmon Arab 71, pl. 3, pl. 22
solskyi - *Lycaena* 143, pl. 41
soma - *Neptis* 96, pl. 28
soracte - *Aporia* 58
Sorrel Sapphire 144, pl. 41
Southern Swallowtail 46
Spialia doris 165, pl. 45
Spialia galba 164, pl. 45
Spialia geron 165, pl. 45
Spialia phlomidis 165, pl. 45
Spialia sertorius 165, pl. 45
Spialia zebra 164, pl. 45
Spindasis acama 150, pl. 5, pl. 43
Spindasis elima 151, pl. 43
Spindasis hypergyrus 150
Spindasis ictis 151, pl. 43
Spindasis vulcanus 151, pl. 43
Spotless Grass Yellow 65, pl. 20
Spotted Angle 62, pl. 45
Spotted Argus Blue 130, pl. 37
Spotted Demon 169, pl. 5, pl. 46
Spotted Fritillary 83, pl. 4, pl. 25
Spotted Himalayan Spotted Flat 161, pl. 44
Spotted Pierrot 124, pl. 35
Spotted Small Flat 161, pl. 5, pl. 44
staudingeri - *Muschampia* 166
staudingeri - *Parnassius* 40, pl. 13
stichius var: *Princeps polytes* 44, pl. 15
stigmata - *Aeromachus* 168, pl. 46
stoliczkana - *Colias* 69, pl. 21
stoliczkana sub-sp. *Polyommatus eros* 135, pl. 39
stoliczkanus - *Parnassius* 40, pl. 13
strabo - *Catachrysops* 139, pl. 40
Straight Banded Tree Brown 98, pl. 29
Straight Swift 174, pl. 47
Straightwing Silverspot 82
Streaked Argus 130, pl. 37
Streaked Skipper 166, pl. 5, pl. 45
Striated Satyr 109, pl. 33
stichius - *Papilio polytes* 44
Striped Albatross 61, pl. 19
Striped White 52, pl. 16
Strymon assamica or *sassanides* 146, pl. 42
suada - *Actinor* 170
Sullied Sailer 96, pl. 28
swaha - *Aulocera* 109, pl. 33
Swallowtail - Balochi 46, pl. 14
Swallowtail - Common Yellow 46, pl. 14
Swallowtail - Ladaki 47, pl. 14
Swallowtail - Southern 46, pl. 14
Swift - Bevan's 174
Swift - Dingy 175
Swift - Himalayan 173
Swift - Small Branded 173
Swift - Straight 174
Swift - Yellow Spot 173
swinhoe sub-sp. *Carcharodus alceae* 167
syla - *Chrysozephyrus* or *Thecla* 148, pl. 42
Synchloe ausonia 53, pl. 16
Synchloe belemia 52, pl. 16
Synchloe charlonia 52, pl. 16
Synchloe daphalis 53, pl. 16
Syntarucus jesous 125, pl. 35
Syntarucus plinius 125, pl. 35
Syrian Skipper 166, pl. 5, pl. 45
Syrichtus galba 164
Syrichtus geron 165
Syrichtus evanidus 165
Syrichtus plurimaculus 166, pl. 5

INDEX

Syrichtus poggei 166
Syrichtus zebra 164

T

Tailed Cupid 126, pl. 36
Tailed Hairstreaks 146
Tailed Punch 157, pl. 5, pl. 34
Tailless Lineblue 142, pl. 40
Tailless Oakblue or Bushblue 149, pl. 42
Tajuria cippus 152, pl. 42
tamu - Heliophorus 145, pl. 41
Taractrocera danna 170, pl. 46
Taractrocera maevius 170, pl. 46
Tarucus alteratus (syn: *nara*) 123, pl. 35
Tarucus balkanica 123, pl. 35
Tarucus callinara 124, pl. 35
Tarucus extricatus 123, pl. 35
Tarucus indica 124, pl. 35
Tarucus mediterraneae 124, pl. 35
Tarucus nara 123, pl. 35
Tarucus nigra 123, pl. 35
Tarucus rosimon 123, pl. 35
Tarucus theophrastus 124, pl. 35
Tarucus venosus 124, pl. 35
Tawny Branded Meadow Brown 102, pl. 31
Tawny Coster 116, pl. 34
Tawny Meadow Brown 102, pl. 3, pl. 30
Tawny Rockbrown 105, pl. 4, pl. 32
Tawny Satyr 107, pl. 32
Tawny Silverline 105, pl. 5, pl. 43
Telchinia violae 116, pl. 34
Telicota ancila 171
Telicota augias 172
tenuistigma - Maniola (syn: *Coenonympha*) 103, pl. 31
Terias hecabe 64, pl. 20
Terias laeta 65, pl. 20
Terias libythea 65, pl. 20
Terias sylhetana 65, pl. 20
Terias venata 65, pl. 20
Thecla ataxus 147
Thecla syla 148
Thecla ziha 147
thelephassa - Eumenis or *Hipparchia* 105
theophrastus - Tarucus 124
Thestor callimachus 144, pl. 41
Thestor pavana 142, pl. 41
thetis - Lycaena 143, pl. 41
Threering - Common 112, pl. 32
Threering - Large 112, pl. 33
Threering - Lesser 112, pl. 33

Three Spot Grass Yellow 65, pl. 20
thyodamus - Cyrestis 94, pl. 22
tianschanikus - Parnassius 37, pl. 12
Tiger - Blue 117, pl. 35
Tiger - Common 118, pl. 34
Tiger - Plain 118, pl. 34
Tiny Grass Blue 138, pl. 39
Tirumala limniace 117
Tomares callimachus 144, pl. 41
Torpedo 175, pl. 47
Tortoiseshell - Black Leg 91
Tortoiseshell - European 92, pl. 27
Tortoiseshell - Indian 92, pl. 4, pl. 27
Tortoiseshell - Large 92, pl. 27
Tortoiseshell - Yellow Legged 91, pl. 27
Tree Brown - Banded 99, pl. 29
Tree Brown - Common 99, pl. 29
Tree Brown - Straight-Banded 98, pl. 29
trivena - Limenitis 76, pl. 23
trivia - Melitaea 83, pl. 25
trochilus - Zizeeria or *Chilades* or *Freyeria* 136
Tros aristolochiae 45, pl. 15
Tros philoxenus 45
Tufted Marbled Skipper 166, pl. 45
Turanana cytis 130, pl. 37
Turkestan Satyr 108
tytlerianus sub-sp. *Parnassius stoliczkanus* 40, pl. 13

U

ubaldus - Azanus 125, pl. 35
uranus - Azanus 126, pl. 35
urticae - Aglais or *Vanessa* 92, pl. 27

V

Vaccinia kwaja 130, pl. 37
Vaccinia hyrcana 130, pl. 37
Vaccinia iris 131, pl. 37
valeria - Parenonia or *Valeria* 61, pl. 19
Vanessa atlanta 87, pl. 26
Vanessa c-album 89, pl. 26
Vanessa canace 89, pl. 10, pl. 26
Vanessa cardui 88, pl. 10
Vanessa cashmirensis 92, pl. 4, pl. 27
Vanessa egea 89, pl. 26
Vanessa indica 87, pl. 10, pl. 26
Vanessa l-album 90, pl. 27
Vanessa polychloros 91, pl. 27

Vanessa urticae 92, pl. 27
Vanessa vau-album 90, pl. 26
Vanessa xanthomelas 91, pl. 10, pl. 27
vardhana - Arletta or *Lycaenopsis* 127, pl. 36
Varnished Apollo 39, pl. 12
vau-album or *v-album - Polygonia* or *Vanessa* 90
Veined Dart 170, pl. 46
Veined Scrub Hopper 168, pl. 46
venata - Terias 65
venosus - Tarucus 124, pl. 35
verma - Lethe 98, pl. 29
versicolor - Calotis 27, pl. 11
vestalis - Colotis 72, pl. 22
vicrama - Philotes or *Polyommatus* 128, pl. 36
violae - Telichinia 116, pl. 34
Violet Meadow Blue 135, pl. 38
Virachola isocrates 153, pl. 43
viridipunctata - Ilerda 145, pl. 41
vulcanus - Spindasis 151, pl. 43

W

wagneri - Maniola (syn: *Hyponephele*) 103, pl. 30
walli sub-sp. *Polyommatus orbitulus* 131
Wanderer - Common 61, pl. 19
Wall - Common 100, pl. 4, pl. 29
Wall - Dark 101, pl. 29
Wall - Scarce 100, pl. 29
Wall - Yellow 101, pl. 29
Walnut Blue 148, pl. 42
Water Hairstreak 147, pl. 42
Western Blue Sapphire 145, pl. 10, pl. 41
Western Courtier 76, pl. 32
Western Green Underwing 133, pl. 38
Western Silverstripe 79, pl. 24
White Admiral 76, pl. 23
White Arab 72, pl. 22
White Bordered Copper 142, pl. 41
White Edged Rockbrown 104, pl. 31
White Line Hairstreak 146, pl. 42
White Orange Tip 64
White Ringed Meadow Brown 103, pl. 31
White Spot Fritillary 81, pl. 25
White Spotted Hairstreak 147, pl. 42
Windmill (Common) 45, pl. 15
wiskotti - Colias 69, pl. 21
Wonderful Hairstreak 147, pl. 42

X

xanthomelas - *Vanessa* or *Nymphalis* 91, pl. 9, pl. 27
xanthopogon - *Choaspes* 160, pl. 44
xiphia - *Leptosia* 51, pl. 16

Y

Yellow Argus 111, pl. 30
Yellow Orange Tip 63, pl. 19
Yellow Pansy 85, pl. 27
Yellow Silverline 150, pl. 43
Yellow Spot Swift 173, pl. 47
Yellow Swallowtail (Common) 46, pl. 3, pl. 9, pl. 14
Yellow Wall 101, pl. 29
'Yellows' (Coliadinae) 62
Yellowleg Tortoiseshell 91, pl. 10, pl. 27
yerbury - *Neptis* 97
yerbury sub-sp. *Caprona ransonetti* 163, pl. 45
Yerbury's Sailer 97
Ypthima asterope 112, pl. 33
Ypthima avanta 113, pl. 33
Ypthima bolanica 113, pl. 33
Ypthima inica 112, pl. 33
Ypthima nareda 112, pl. 33
Ypthima sakra 113, pl. 33

Z

zaida - *Neptis* 96, pl. 28
zaneka - *Gonepteryx* 66
Zebra Blue 125, pl. 35
Zebra Skipper 164, pl. 45
zebra - *Hesperia* 164, pl. 45
zebra - *Spialia* 164, pl. 45
zebra - *Syrichtus* 164, pl. 45
Zephyr Blue 129, pl. 36
zetides - *Cloanthus* 44, pl. 14
ziha - *Chrysozephyrus* or *Euaspa* or *Thecla* 147, pl. 42
Zizeeria galba 137, pl. 39
Zizeeria knysa 137, pl. 39
Zizeeria lysimon 137, pl. 39
Zizeeria maha 137, pl. 39
Zizeeria trochilus 136, pl. 39
Zizera indica 138, pl. 39
Zizina otis 138, pl. 39
Zizula hylax 138, pl. 39